Replacement copy
August '88

XLV. 2.

/1

ÆDES CHRISTI
in Academia Oxoniensi

The Theory of

THE PROPERTIES OF
METALS AND ALLOYS

by

N. F. MOTT, M.A., F.R.S.

and

H. JONES, Ph.D.

DOVER PUBLICATIONS, INC., NEW YORK

First published 1936 by Clarendon Press, Oxford, England.
Reprinted from corrected sheets by Dover Publications, Inc.
New York, U.S.A. 1958.

Oxford University Press, Amen House, London, E.C.4
GLASGOW   NEW YORK   TORONTO   MELBOURNE   WELLINGTON
BOMBAY   CALCUTTA   MADRAS   KARACHI   KUALA LUMPUR
CAPE TOWN   IBADAN   NAIROBI   ACCRA

Standard Book Number: 486-60456-X
Library of Congress Catalog Card Number: 58-11279

Manufactured in the United States of America
Dover Publications, Inc.
180 Varick Street
New York, N.Y. 10014

PREFACE

SINCE the introduction of quantum mechanics the theory of metals has developed rapidly. The first stage of the development led to an understanding of many of the electric and magnetic properties common to all metals. In more recent investigations attempts have been made to explain the differences between individual metals and alloys in terms of the properties of their constituent atoms. It is our principal aim in this book to give an account of the properties of individual metals and alloys, and in particular to show how their crystal structure, magnetic susceptibility, and electrical and optical properties are related to one another and to their more chemical properties. For this reason we have given no description of the phenomenon of supraconductivity, since it has not yet proved possible to relate its occurrence to any of the other properties of the supraconducting materials. We have also omitted any discussion of the properties of surfaces (thermionic emission, adsorption of gas atoms, work function, etc.), since these phenomena bear only a small relation to the subject-matter of the rest of the book.

We should like to express our thanks to Dr. W. Heitler and to Mr. K. Fuchs, who have read the book in manuscript and have helped us in the correction of the proofs, and to Dr. H. H. Potter, Dr. R. Peierls, and Dr. E. J. Williams, who have advised us on various points.

N. F. M.
H. J.

BRISTOL. *May* 1936

CONTENTS

VI. HEAT CAPACITY AND MAGNETIC PROPERTIES OF THE METALLIC ELECTRONS

VII. THE ELECTRICAL RESISTANCE OF METALS AND ALLOYS

INTRODUCTION

METALS are distinguished from other solids primarily by their high electrical and thermal conductivities. Since the discovery of the electron in 1897 by J. J. Thomson[†] and by Wiechert,[‡] it has been recognized that the electric current in metals is carried by electrons. Before the introduction of quantum mechanics several attempts were made to give a quantitative theory of conduction, some of the most important being due to Riecke,[||] Drude,[††] J. J. Thomson,[‡‡] and Lorentz.[||||] In these theories a certain number of electrons were supposed to become detached from their atoms and to be free to move through the solid metal. Perhaps the greatest success of the theory was the explanation of the fact, discovered empirically by Wiedemann and Franz, that the ratio of the thermal to the electrical conductivity is approximately the same for all metals. The theory, however, led to inconsistent estimates of the number of electrons which are actually free in a metal. On the one hand, the optical properties of metals could be accounted for qualitatively[†††] on the assumption that the number of electrons is comparable with the number of atoms; on the other hand, according to the equipartition theorem, each electron ought to have thermal energy $\frac{3}{2}kT$, and should thus make a contribution $\frac{3}{2}k$ to the heat capacity. Except at high temperatures, however, the heat capacity of most metals does not rise appreciably above the value given by the law of Dulong and Petit, which is reached also for insulators and which can be accounted for by the vibration of the atoms. It follows, therefore, either that the number of electrons is very much less than the optical measurements suggest, or that the heat capacity per electron is much less than $\frac{3}{2}k$.

No satisfactory explanation of these divergent results could be obtained until the introduction of quantum mechanics. Then,

---

[†] *Phil. Mag.* **44** (1897), 298.
[‡] *Verh. d. Phys. Ges. zu Königsberg i. Pr.* (1897).
[||] *Ann. d. Phys. u. Chem.* **66** (1898), 1199; *Ann. d. Physik,* **2** (1900), 835.
[††] *Ann. d. Physik,* **1** (1900), 566.
[‡‡] *Int. Phys. Congress, Paris, Report,* **3** (1900), 138.
[||||] *Proc. Amsterdam Acad.* **7** (1905), 684.
[†††] Lorentz, *Proc. Amsterdam Acad.* **5** (1903), 666; Bohr, *Studier over Metallernes Elektrontheori,* Copenhagen, 1911; Lindemann, *Phil. Mag.* **24** (1915), 127; Livens, ibid. **30** (1915), 434; H. A. Wilson, ibid. **20** (1910), 835; Schuster, ibid. **7** (1904), 151. Cf. also Chapter III of this book.

however, Sommerfeld[†] applied the newly formulated Fermi-Dirac statistics[‡] to the electrons in metals and was able to show that the heat capacity per electron ought to be very much less than its value according to the equipartition theorem. Following Sommerfeld, other research workers built up a quantum theory of electrical conductivity, of which an account is given in Chapter VII of this book. This theory has been able to account qualitatively for the dependence of resistance on temperature, pressure, and constitution of many metals and alloys, and in fact leaves only the major phenomenon of supraconductivity unexplained in principle.

Quantum mechanics has also enabled great progress to be made in the understanding of the magnetic properties of metals. Pauli's[||] treatment of the weak paramagnetism of metals was the first application of the new theories in this field, and Heisenberg's work on ferromagnetism[††] gave a quantum-mechanical background to the phenomenological theory previously developed by Weiss. These theories are discussed in Chapter VI of this book.

Possibly even more fundamental than the problems of electrical conductivity and magnetism is that of metallic cohesion. For polar crystals of the alkali-halide type, a quantitative theory based on electrostatics has been given by Born[‡‡] and his co-workers; for many insulating crystals, such as diamond, the forces holding the molecules together are similar to the homopolar bonds familiar in chemistry; solid rare gases and certain molecular lattices seem to be held together by the van der Waals forces between the constituent atoms or molecules. None of these modes of description has proved suitable for metals; many attempts have been made to explain the structures of metallic alloys in terms of the homopolar bond, but without very much success.

The first successful application of quantum mechanics to the problem of metallic cohesion was made by Wigner and Seitz,[||||] and is discussed in Chapter IV of this book. Their calculations show that a typical metal may be regarded as an array of positive ions embedded in a cloud of negative electrons, and that the cohesive force is mainly the electrostatic attraction between the electrons and the ions.

† *Zeits. f. Phys.* **47** (1928), 1.
‡ Fermi, *Zeits. f. Phys.* **36** (1926), 902; Dirac, *Proc. Roy. Soc.* A, **112** (1926), 661.
|| *Zeits. f. Phys.* **41** (1926), 81.　　　　　　　　†† Ibid. **49** (1928), 619.
‡‡ Cf., for example, Born, *Atomtheorie des festen Zustandes*, Leipzig (1923).
|||| *Phys. Rev.* **43** (1933), 804.

According to quantum mechanics these electrons are moving rapidly through the metal, even at the absolute zero of temperature, and therefore any purely electrostatic theory of metallic cohesion is impossible. Calculations based on quantum mechanics, however, enable the heat of sublimation, compressibility, and elastic constants to be estimated for certain metals, and also (cf. Chap. V) the crystal structure of some metals and alloys to be accounted for.

Any theoretical account of the properties of metals must be based on a theory of the behaviour of electrons in a crystal lattice. Chapters II and III are devoted to the development of such a theory. The optical properties of metals, discussed also in Chapter III, give perhaps the most direct experimental evidence in favour of the preceding theoretical conclusions. Chapter I deals with those thermal properties of metals which can be discussed without knowing the nature of the interatomic forces.

In this book we make no mention of that property of metals which is of the greatest technical importance, namely that of strength. It has not yet been possible to apply the methods of atomic physics to this problem, though recent work by G. I. Taylor† inspires the hope that it may soon be possible to do so. No apology is needed, however, for the appearance of a book at this stage, because a theory of metallic cohesion is certainly a necessary preliminary to an attack on the problem of strength.

† *Proc. Roy. Soc.* A, **145** (1934), 362.

THE THEORY

OF THE PROPERTIES

OF METALS

AND ALLOYS

THERMAL PROPERTIES OF THE CRYSTAL LATTICE

1. The specific heat at low temperatures

1.1. *Theories of Einstein and Debye.* In a solid at any temperature the mass centres or nuclei of the atoms are not at rest but vibrate about mean positions. In insulators the specific heat of the solid is, in general,† due entirely to these vibrations; in metals a part of the specific heat is due to the motion of free electrons, but this part is usually‡ very small and will be neglected in this section. In order to calculate the specific heat and many other properties of solids, we must know the frequencies with which the atoms of the solid vibrate.

A solid built of N similar atoms bound together by elastic forces has $3N$ degrees of freedom and can therefore vibrate in $3N$ independent normal modes,‖ each with its characteristic frequency. According to classical statistics, the mean energy of each normal mode will be†† kT, so that the total thermal energy of the solid is $3NkT$. Thus, if N is the number of atoms in a gramme atom $(6 \cdot 062 \times 10^{23})$, the heat capacity per gramme atom (atomic heat) is

$$C_v = 3Nk = 5 \cdot 96... \text{ calories per degree,}$$

which is the same for all substances (law of Dulong and Petit). When experimental determinations of the specific heat at sufficiently high temperatures are corrected to apply to constant volume, the corrected values for most substances have almost this value. We shall refer to this quantity as the 'classical' value of the atomic heat.

According to the quantum theory, the mean energy of a normal mode of the crystal with characteristic frequency ν is

$$E(\nu) = \tfrac{1}{2}h\nu + \frac{h\nu}{e^{h\nu/kT} - 1}. \tag{1}$$

The term $\tfrac{1}{2}h\nu$ represents the 'zero-point energy', or the energy which a vibrator will have at the absolute zero of temperature. It is easily

---

† Except when an excited electronic state of the atom lies near to the ground state, as in salts of the rare earths.
‡ Cf. Chap VI, § 2. A large part of the specific heat of a ferromagnetic near the Curie point is due to electrons in incomplete shells.
‖ More accurately in $3N - 5$, since the solid as a whole has 5 degrees of freedom.
†† Cf. Jeans, *Dynamical Theory of Gases*, 4th ed., p. 394, Cambridge (1925).

seen that (1) tends to the classical value kT at high temperatures:

$$E(\nu) \sim kT\left[1 + \frac{1}{12}\left(\frac{h\nu}{kT}\right)^2 + O\left(\frac{h\nu}{kT}\right)^4\right].$$

The entropy per normal mode is

$$S(\nu) = \int_0^T \frac{d}{dT} E(\nu) \frac{dT}{T}$$

$$= \frac{E(\nu)}{T} - k\log(1 - e^{-h\nu/kT}).$$

The free energy is

$$F(\nu) = E - TS$$

$$= kT\log(1 - e^{-h\nu/kT}), \tag{2}$$

which, for $kT \gg h\nu$, tends to

$$F(\nu) = kT\log(h\nu/kT). \tag{3}$$

In order, therefore, to calculate the internal energy and the specific heat, we must know the number of normal modes of the crystal with frequency between the values ν and $\nu + d\nu$. Let this number be $f(\nu)d\nu$; since the total number of normal modes is $3N$, the function f must satisfy

$$\int f(\nu)\,d\nu = 3N, \tag{4}$$

the integration being over all frequencies of the crystal. The internal energy U is then given by

$$U = \int f(\nu)E(\nu)\,d\nu. \tag{5}$$

The calculation of the internal energy and hence of the specific heat thus involves the problem of finding the vibration spectrum $f(\nu)$ of the crystal.

The simplest possible assumption about the spectrum is that of Einstein,[†] that all vibrations have the same frequency ν_0. This is equivalent to the assumption that all atoms vibrate independently. The internal energy U is then $3NE(\nu_0)$, and the heat capacity

$$c_v = \frac{dU}{dT} = 3NkH\left(\frac{\Theta_E}{T}\right), \tag{6}$$

where

$$H(x) = x^2 e^x (e^x - 1)^{-2} \tag{7}$$

† *Ann. d. Physik*, **22** (1907), 180 and 800; **34** (1911), 170.

and $h\nu_0 = k\Theta_E$. The temperature Θ_E will be termed the Einstein characteristic temperature. Einstein's formula for the specific heat is in fair agreement with experiment for $T \sim \Theta_E$, but the agreement is poor for low temperatures. On account of its simplicity, however, it is often useful for the discussion of other phenomena, e.g. thermal expansion (cf. § 4).

The calculation of $f(\nu)$ for any real crystal is at present an unsolved problem. In order to obtain the specific heat at very low temperatures, however, it is only necessary to know $f(\nu)$ for small values of ν, since the mean energy $E(\nu)$ of the vibrations of frequency ν becomes small when $h\nu/kT$ is large. If ν is so small that the corresponding wave-length λ is large compared with the interatomic distance, then it may be shown† quite generally that for a solid of volume V

$$f(\nu)\, d\nu = BV\nu^2\, d\nu, \tag{8}$$

where B is a constant. B may be calculated in terms of the elastic constants of the solid; for an isotropic solid,

$$B = 4\pi\left(\frac{1}{c_l^3} + \frac{2}{c_t^3}\right), \tag{9}$$

where c_l, c_t are the velocities of longitudinal and transverse sound waves in the solid; these are given in terms of the compressibility χ and Poisson's ratio σ by

$$c_l^2 = \frac{3(1-\sigma)}{(1+\sigma)\chi\rho}, \qquad c_t^2 = \frac{3(1-2\sigma)}{2(1+\sigma)\chi\rho},$$

where ρ is the density. For a non-isotropic solid the velocities of sound will be different in different directions; B is then given by‡

$$B = \sum_{i=1}^{3} \iint \frac{d\omega}{c_i^3}, \tag{10}$$

where $d\omega$ denotes an element of solid angle and c_1, c_2, c_3 are the velocities of the two transverse waves and of the longitudinal wave in the corresponding direction. A method of evaluating this integral numerically has been given by Hopf and Lechner,‖ and an approximate analytical formula by Blackman.††

† Born and von Kármán, *Phys. Zeits.* **13** (1912), 297; **14** (1913), 15.
‡ Ibid. Cf. also the report by Schrödinger, *Handb. d. Phys.* **10** (1926), 308.
‖ *Verh. d. deuts. phys. Ges.* **16** (1914), 643.
†† *Proc. Roy. Soc.* A, **149** (1935), 126.

Using formula (8) for $f(\nu)$, we have for the internal energy

$$BV \int\limits_0^\infty \nu^2 E(\nu)\, d\nu = 6{\cdot}495\ldots\, BVh\left(\frac{kT}{h}\right)^4,$$

and hence for the heat capacity of a volume V

$$c_v = 25{\cdot}980\ldots\, BVk(kT/h)^3. \tag{11}$$

The result expressed in formula (11), that at low temperatures c_v is proportional to T^3, was first obtained by Debye[†] and almost simultaneously by Born and von Kármán.[‡]

The formula (8) is a good approximation only at very low frequencies; various attempts have been made to find a form valid over the whole frequency range. Of these the first was that of Debye,[||] who set

$$
\begin{aligned}
f(\nu) &= BV\nu^2 \quad (\nu < \nu_D) \\
&= 0 \qquad (\nu > \nu_D)
\end{aligned}
\tag{12}
$$

(cf. Fig. 1), where ν_D is defined by equation (4), which gives

$$\tfrac{1}{3}BV\nu_D^3 = 3N. \tag{13}$$

The internal energy is then

$$BV \int\limits_0^{\nu_D} \nu^2 E(\nu)\, d\nu.$$

Introducing the Debye characteristic temperature Θ_D, defined by

$$k\Theta_D = h\nu_D, \tag{14}$$

a short calculation gives for the heat capacity of a solid containing N atoms

$$c_v = 3Nk\left[D\!\left(\frac{\Theta_D}{T}\right) - \frac{\Theta_D}{T} D'\!\left(\frac{\Theta_D}{T}\right)\right], \tag{15}$$

where

$$D(x) = \frac{3}{x^3} \int\limits_0^x \frac{\xi^3\, d\xi}{e^\xi - 1}.$$

c_v is shown as a function of T/Θ_D in Fig. 99.

For low temperatures ($T/\Theta_D \ll 1$), (15) reduces to the form (11); on substituting for V from (13) this gives

$$c_v = 233{\cdot}82\ldots\, Nk(T/\Theta_D)^3. \tag{16}$$

It is to be emphasized that this approximate formula is *not* dependent on the particular distribution function used by Debye, because at low temperatures only sound waves of long wave-length are excited, and

---

† *Ann. d. Physik*, **39** (1912), 789.   ‡ loc. cit. || loc. cit.

for these the effect of the atomic structure of the material is unimportant and formulae (8) and (16) are accurate.

From formulae (9), (13), and (14) we have for Θ_D

$$\Theta_D = \frac{h}{k}\left\{\frac{4\pi\Omega_0}{9}\left(\frac{1}{c_l^3}+\frac{2}{c_t^3}\right)\right\}^{-\frac{1}{3}}, \tag{17}$$

where $\Omega_0 = V/N$ = atomic volume. Formula (17) may be written

$$\Theta_D = \frac{3\cdot6\times10^{-3}}{A^{\frac{1}{3}}\rho^{\frac{1}{6}}\chi^{\frac{1}{2}}\{f(\sigma)\}^{\frac{1}{3}}}, \tag{17.1}$$

where $\qquad f(\sigma) = 2\left\{\frac{2(1+\sigma)}{3(1-2\sigma)}\right\}^{\frac{3}{2}} + \left\{\frac{1+\sigma}{3(1-\sigma)}\right\}^{\frac{3}{2}},$

and where A is the atomic weight, ρ the density, χ the compressibility, and σ Poisson's ratio. For a non-isotropic substance we have from (10), (13), and (14)

$$\Theta_D = \frac{h}{k}\left\{\frac{\Omega_0}{9}\sum_1^3\iint\frac{d\omega}{c_i^3}\right\}^{-\frac{1}{3}}. \tag{17.2}$$

Comparison with experiment. Nearly all specific heat measurements at low temperatures have been compared with the Debye formula (15).[†]

It is found that the T^3 law for the specific heat at low temperatures is usually in good agreement with experiment in the low-temperature region.

If one fits a Debye curve to the experimental specific heat curve in the region $T \sim \Theta_D$, the value of Θ_D obtained is in quite good agreement with that obtained from the low-temperature measurements, as the following table[‡] shows:

TABLE I

Debye Θ_D (degrees)

Element	In the region of the T^3 law	In the region where $T \sim \Theta_D$
Au	162	180
Sn	127	160
Mo	379	379
Cu	321	315
Al	385	398
Diamond	2,230	1,840

† For the results see the full account given by Eucken, *Handb. d. exp. Phys.* **8/1** (1929), p. 239 et seq.; also Fowler, *Statistical Mechanics*, 1st ed., p. 82; Schrödinger, *Handb. d. Phys.* **10** (1926), p. 304; and Born, ibid., **24/2** (1933), p. 623.

‡ Part of the table given by Eucken, loc. cit. 245.

This suggests that the Debye distribution function (12) is a fair approximation to the true form.

These values of Θ_D are usually in fair agreement with the results obtained by substituting in formula (17) the elastic constants χ, σ of the polycrystalline metal in bulk.† Since, however, even cubic metal crystals are anisotropic,‡ this fact must be mainly accidental.

The only calculations of Θ_D from the elastic constants of single metal cubic crystals are those of Röhl‖ for Ag and Au, Fuchs†† for Cu, and Honnefelder‡‡ for W. The values prove to be in fair agreement with those obtained from the elastic constants of the metal in bulk, as the following table shows:

	Cu	Ag	Au	W
from elastic constants of crystal	342	212	158	384
Θ_D from elastic constants of polycrystalline material	325	215	161	372
from measurement of specific heat at low temperatures	316–23	215	170	310

For non-cubic metals Grüneisen and Goens‖‖ have calculated Θ_D for Zn and Cd and Grüneisen and Hoyer††† for Hg. Grüneisen and Goens have compared their results with the observed specific heats, the agreement being poor. The explanation, according to Blackman, is that the observed values of the specific heat were not in the true T^3 region.‡‡‡

It is usual to express experimental specific heats at low temperatures by Θ_D, T curves; Θ_D is calculated by equation (15), from the observed specific heat at each temperature. Θ_D is then independent of T if c_v is given exactly by the Debye formula. Some Θ_D, T curves are shown in Figs. 2, 3, and 4.

1.2. *Extensions of the theory.* The Debye distribution function (12) represents, of course, a very crude approximation to the true form, and various attempts have been made to find a better approximation. Born and von Kármán‖‖‖‖ found the true distribution function for a linear lattice; they proved that, for a chain of N atoms distant a from each other, the frequencies are given by

$$\nu = \frac{v}{\pi a} \sin \frac{\pi k}{N} \quad (k = 1, 2, ..., N), \tag{18}$$

where v is the velocity of sound for wave-lengths large compared with a.

---

† Cf. Eucken, loc. cit. 242.   ‡ Cf. Chap. IV, § 4.
‖ *Ann. d. Physik*, **16** (1933), 887.
†† Unpublished; the experimental values were those of Goens, *Zeits. f. Instrumentenkunde*, **52** (1932); 167.   ‡‡ *Zeits. f. phys. Chem.* B, **21** (1933), 53.
‖‖ *Zeits. f. Phys.* **26** (1924), 250.   ††† *Ann. d. Physik*, **22** (1935), 663.
‡‡‡ See p. 40.   ‖‖‖‖ Loc. cit.

The maximum frequency is

$$v/\pi a,$$

which is smaller by $\qquad 2/\pi = 0\cdot637...$

than the maximum frequency given by Debye's theory.

Recently the frequency spectrum for real crystals has been investigated by Blackman.[†] According to Blackman, for simple crystals containing one type of atom only, the distribution function differs from that assumed by Debye in the following ways:

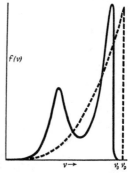

(1) Debye's maximum frequency v_D is in general greater than the true maximum frequency v_0 of the lattice.

(2) The true distribution function $f(v)$ increases more rapidly with v than that assumed by Debye.

(3) The distribution function $f(v)$ shows two or more maxima.

The two functions are illustrated in Fig. 1. These results are derived from a study of the two-dimensional lattice, but are probably true

FIG. 1. Frequency spectrum of a solid according to Debye (dotted line) and Blackman (full line).

also in three dimensions, though it is possible that other weak maxima occur.

Blackman gives the following formulae for the relative frequencies at which these two strong maxima occur:

$$v_0 = \frac{1}{2\pi}\left(\frac{4\alpha}{M}\right)^{\frac{1}{2}}, \qquad v_1 = \frac{1}{2\pi}\left(\frac{8\gamma}{M}\right)^{\frac{1}{2}},$$

where M is the mass of the atom of which the lattice is built up, and α and γ are determined by the elastic constants of the crystal. If the elastic constants of a cubic lattice in Voigt's notation[‡] are c_{11}, c_{12}, c_{44}, then

$$\frac{\gamma}{\alpha} = \frac{c_{12}+c_{44}}{4(c_{11}-c_{12}-c_{44})}.$$

These formulae depend, however, on the Cauchy relations,[||] which are not applicable to metals.

We shall now consider the difference between the specific heat calculated with the function $f(v)$ illustrated in Fig. 1 and with the

† *Proc. Roy. Soc.* A, **159** (1937), 416; *Proc. Camb. Phil. Soc.* **23** (1937), 93.
‡ Cf. Geckeler, *Handb. d. Phys.* **6** (1928), 407. || Cf. Chap. IV, § 4.1.

Debye function. We shall use the Θ_D, T diagram (p. 6) to express both the calculated and experimental specific heats. Thus, for example, if the observed specific heats exactly fitted the Debye formula, the corresponding Θ_D, T diagram would consist of a straight line, denoting a value of Θ_D constant for all temperatures. We shall consider a few special types of vibration spectrum and show the corresponding Θ_D, T diagrams. First we consider the simplest of all spectra, namely that consisting of a single line at frequency ν. This gives the Einstein formula (6) for the specific heat. If we take an example in which $\Theta_E = h\nu/k = 200°$, the corresponding curve is shown in Fig. 2.

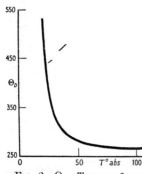

FIG. 2. Θ_D, T curve for an Einstein function with $\Theta_E = 200°$.

From this we can conclude that if there is a tendency for the vibrations to heap up about one particular frequency the Θ_D, T curve will fall off with increasing temperature. For moderate or high temperatures the Einstein function approximates fairly well to the Debye form and gives, therefore, a fair representation of the specific heats in this region. The reason for this is clearly the strong maximum in $f(\nu)$ assumed by Debye and shown in Fig. 1. In the moderate temperature region we may take Θ_E to be proportional to the mean frequency, so that

$$\Theta_E = \frac{h}{k} \int \nu^3 \, d\nu \bigg/ \int \nu^2 \, d\nu = \tfrac{3}{4}\Theta_D.$$

Secondly, we consider a spectrum composed of two lines of frequencies ν_1 and ν_2 and of equal intensities. The specific heat will then be given by the formula

$$c_v = 3Nk\left[H\!\left(\frac{\Theta_1}{T}\right) + H\!\left(\frac{\Theta_2}{T}\right)\right],$$

where $H(\Theta/T)$ denotes the Einstein specific heat function (7) with characteristic frequency ν equal to $k\Theta/h$. This form, which was suggested some years ago by Nernst and Lindemann,[†] who assumed that $\nu_1 = \tfrac{1}{2}\nu_2$, has now an added interest in view of Blackman's discovery that a spectrum consisting of two lines is a fair approximation

† *Preuss. Akad. Wiss. Berlin, Sitz. Ber.* **22** (1911), 494.

to the true vibration spectrum (Fig. 1). Fig. 3 shows how Θ_D varies with T, (a) when $\nu_1 < \frac{1}{2}\nu_2$, (b) when $\nu_1 > \frac{1}{2}\nu_2$, and (c) when $\nu_1 = \frac{1}{2}\nu_2$.

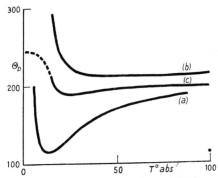

Fig. 3. Θ_D, T curves. (a) $\Theta_1 = 60°$, $\Theta_2 = 200°$, (b) $\Theta_1 = 120°$, $\Theta_2 = 200°$, (c) $\Theta_1 = 100°$, $\Theta_2 = 200°$. The dotted line shows a possible form for a real crystal.

In the case (c) the value of Θ_D is remarkably constant over a wide range of temperature, as the following table shows:

$$\Theta_1 = \tfrac{1}{2}\Theta_2 = 100°$$

Temp. °K.	Atomic heat C_v	Θ_D
15	0·1708	209
20	0·5218	190
30	1·441	190
40	2·321	193
50	3·062	196
60	3·651	196
200	5·658	204

These results show that either a rising or a falling Θ_D, T curve, both of which types are met with experimentally (cf. Fig. 4), may be interpreted quite well in terms of a particular form of the vibration spectrum, and that an almost constant Θ_D value need not necessarily imply that the true T^3 part of the specific-heat curve is reached.

We consider next the specific heat at very low temperatures. We have already seen that, at sufficiently low temperatures, the Debye T^3 form should be exact. Thus, for a few degrees above the absolute zero of temperature, Θ_D should be constant. Moreover, the value should be exactly equal to that calculated from the elastic constants at low temperatures from formulae (17), (17.1), and (17.2). According to Blackman, as T increases from zero, Θ_D will initially *decrease*. The reason for this is that, for small ν, $f(\nu)$ is always greater than formula (12) implies; i.e. if we write

$$f(\nu) = a\nu^2 + b\nu^3 + ...,$$

b will be positive. Thus we see that the form of Θ_D, T curve shown by the dotted line in Fig. 3 is possible. This result has interesting consequences, since it implies that if experiment shows a Θ_D curve which falls as the temperature decreases down to low temperatures, say 20° abs., and then appears to become constant, this constant

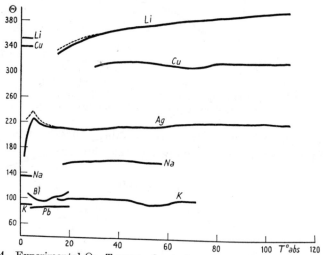

Fig. 4. Experimental Θ_D, T curves for various metals.† The dotted lines show the values with the theoretical electronic specific heat (Chap. VI, formula (17)) subtracted from the observed specific heat. For Cu the value calculated from the observed elastic constants of single crystals by formula (17.2) is shown, and for Li, Na, K the values calculated from the theoretical elastic constants (Chap. IV, § 3).

value is not the ultimate constant value predicted by the Debye theory, but represents only a minimum in the Θ_D, T curve. Before the ultimate constant value is reached, therefore, Θ_D must rise as the temperature falls.

In Fig. 4 we show the observed Θ_D, T curves for Li, Na, K, Cu, Ag, Pb, and Bi. We show also for certain metals the limiting Θ_D values as $T \to 0$, calculated from the elastic constants at low temperatures; those for Cu are taken from the measurements of Goens (cf. Chap. IV),

† The experimental results are due to: Simon and Swain, *Zeits. f. phys. Chem.* B, **28** (1935), 189 (Li); Simon and Zeidler, ibid. A, **123** (1926) 383 (Na, K); Nernst, *Ann. d. Physik*, **36** (1911), 395 (Ag, high temperatures); Keesom and Kok, *Physica*, **1** (1934), 770 (Ag, low temperatures); Keesom and van den Ende, *Proc. Amsterdam Acad.* **33** (1930), 243, and **34** (1931), 210 (Pb and Bi).

and those for the alkalis are obtained by theoretical methods.† The dotted lines show the values deduced from the observed specific heats with the theoretical contribution from the electrons subtracted from them. They therefore represent the heat capacity of the lattice alone.

As Simon‡ has emphasized, the Θ_D values for Li and Na are less constant than those for Cu, Ag, and most other regular metals. We must deduce that for the alkalis the divergence between the true vibration spectrum and the Debye form is particularly great.‖ In Chapter IV we shall see that the alkalis are more anisotropic in their elastic properties than is copper, and to this fact their peculiar Θ_D, T curves must doubtless be ascribed (cf. Fuchs, *Proc. Roy. Soc.* A, **157** (1936), 444).

2. Specific heats at high temperatures

The atomic heats (heat capacity per gramme atom) at constant volume and constant pressure are connected by the equation

$$C_v = C_p - TV\alpha^2/\chi, \tag{19}$$

where T is the absolute temperature, V the atomic volume,†† α the thermal expansion coefficient, and χ the compressibility. The following table gives values of C_p and C_v at various temperatures centigrade (in cals. per degree):

TABLE II

		C_p	C_v		C_p	C_v		C_p	C_v
Pd‡‡	0°	5·74	5·60	500°	6·97	6·59	1,500°	8·17	7·23
Pt‖‖	20°	6·1	5·95	500°	6·8	6·4	1,600°	7·45	6·65
Ag‖‖	20°	6·0	5·75	500°	6·7	6·0	900°	7·3	6·1
Au‖‖	20°	6·15	5·8	500°	6·7	6·0	1,000°	7·4	6·1

We give also a table††† of $(C_p - C_v)/C_v$, according to equation (19), at 20° C.:

	W	Cu	Pb	Pt	Ag	Fe(α)	Au	Na	K	Ni	Zn	Cd	Sn
$\dfrac{C_p - C_v}{C_v} \times 100$	0·6	2·8	6·7	2·0	4·0	1·6	3·8	7·9	10·0	2·1	5·25	6·0	4·0

The increase in C_v above the classical value

$$3R = 5·955 \text{ cal./degree} \tag{20}$$

† Fuchs, *Proc. Roy. Soc.* A, **153** (1936), 622.
‡ *Ergebnisse d. exakt. Naturwiss.* 9 (1930), 256.
‖ A recent X-ray investigation of the amplitudes of the atomic vibrations in lithium supports this hypothesis (Pankow, *Helv. Phys. Act.* 9 (1936), 87).
†† i.e. volume per gramme atom.   ‡‡ *Landolt-Börnstein's Tabellen*, II b (1931).
‖‖ *Handb. d. Metallphysik*, 1 (1935), 246.
††† Eucken, *Handb. d. exp. Phys.* 8/1 (1929), 211.

at high temperatures may be due to one of two causes. According to Born and Brody[†] the anharmonic terms in the lattice vibrations will cause a deviation from the value (20) at high temperatures; they consider a system of oscillators with potential energy of the form

$$\phi(x) = \tfrac{1}{2}m\omega^2 x^2 + gx^3 + fx^4, \tag{21}$$

and find the heat capacity per oscillator to be[‡]

$$c_v = k\left[1 + \left(\frac{15g^2}{m^3\omega^6} - \frac{6f}{m^3\omega^4}\right)kT\right]. \tag{22}$$

The second term may be positive or negative.

The authors quoted considered the excess specific heat shown in Table II for Pt to be due to this cause. On the other hand, more modern research has shown (cf. Chap. VI) that in metals at high temperatures the free electrons will make to the atomic heat a small contribution proportional to T.

By extrapolating the results obtained by Keesom at low temperatures (cf. Chap. VI, § 2), one finds that for silver at 900° C. the contribution from the electrons will be 0·12 cal./gm. atom. For the transition metals it may be larger, as we shall see in Chap. VI, § 5.2. We believe that in most metals the rise of C_v above the classical value is due to the electrons, rather than to the effect investigated by Born and Brody.

3. Numerical values of the characteristic temperature

We have seen that the Debye Θ_D may be obtained from the specific-heat curve at low temperatures, from the specific-heat curve in the neighbourhood of Θ_D, and from the elastic constants from formulae (17), (17.1), and (17.2). The considerations of § 1.2 show that exact agreement between the values obtained in different temperature ranges is not to be expected. On the other hand, approximate values of the atomic frequencies are of importance in the theory of conductivity and elsewhere. We give, therefore, methods of estimating the mean atomic frequency v and hence $\Theta\ (= hv/k)$ when thermal data do not exist.

According to Einstein,[‖] there must be a simple relation between

---

† *Zeits. f. Phys.* **6** (1921), 132. Cf. also the article by Born and Göppert-Mayer, *Handb. d. Phys.* **24**/2 (1933), 676.

‡ A recent theoretical paper by Damköhler (*Ann. d. Physik*, **24** (1935), 1) reaches the conclusion that c_v should first increase and then decrease.

‖ *Ann. d. Physik*, **34** (1911), 170. See also Müller-Pouillet, *Lehrbuch d. Phys.*, 11th ed., **3**/1 (1925), 382.

the compressibility of a solid and the characteristic frequencies of the atoms, since the same forces are responsible for both. Einstein in this way obtained the formula

$$\Theta_E = \frac{13 \cdot 25 \times 10^{-4}}{A^{\frac{1}{2}} \rho^{\frac{1}{6}} \chi^{\frac{1}{2}}}. \tag{23}$$

This formula is clearly analogous with (17). The numerical factor is of course only approximate.

According to Lindemann[†] there exists a relation between the melting-point T_M (in degrees absolute) and the characteristic temperature. The relation may be written

$$\Theta_D = C \sqrt{\left(\frac{T_M}{A V^{\frac{2}{3}}} \right)}, \tag{24}$$

where V is the atomic volume, A the atomic weight, and C a constant which is roughly the same for all metals. To show the agreement with experiment, we deduce C from formula (24) for a number of metals:

Metal	Li	Na	K	Cu	Ag	Au	Ca
T_M (degrees K.)	459	370	335	1,356	1,233	1,336	1,083
Θ_D (observed)	400	160	100	315	215	170	230
C	115	115	116	134	140	142	131

In using the formula (24) to determine Θ_D for any metal, we shall obtain C from the element in the periodic table chemically most similar to it, for which Θ_D is known from thermal data.

Lindemann's formula has not at present received a theoretical explanation.

According to Grüneisen[‡] the electrical resistance of a metal divided by the temperature is very nearly proportional to c_v, except at very low temperatures. Measurements of the resistance may therefore be used to determine Θ_D in the temperature range $T \sim \Theta_D$. The connexion between resistance and specific heat is discussed further in Chapter VII.

Values may also be obtained from the observed elastic constants of single crystals, as explained in § 1.2.

We give below a table of values of Θ_D derived by the methods enumerated above. The Einstein temperature in each case is $\frac{3}{4}\Theta_D$, as explained on p. 8.

† *Phys. Zeits.* **11** (1910), 609.   ‡ *Handb. d. Phys.* **13** (1928), 1.

TABLE III

Characteristic temperatures of the metallic elements

Element	Θ_D	Method	Reference	Element	Θ_D	Method	Reference
Li	328–430†	S. H.	(3)	Zr	288	E. C.	
	363	E. C.		Mo	380	S. H.	
Be	1,000	S. H.		Ru	426	E. C.	
C (dia-	2,340	S. H.		Rh	370	E. C.	
mond)					315	M. P. F.	
Ne (solid)	63	S. H.		Pd	275	S. H.	(5)
					270	E. C.	
Na	140–160†	S. H.	(1)				
	202	E. C.		Ag	215	S. H.	
Mg	290	S. H.			212	E.	
Al	390	S. H.		Cd	172	S. H.	
	394	E. C.		In	106	M. P. F.	
A (solid)	85	S. H.		Sn	260	S. H.	
K	100†	S. H.		Sb	140	S. H.	
	163	E. C.		Te	120	M. P. F.	
Ca	230	S. H.					
Ti	342	E. C.		Cs	42	M. P. F.‖	
	396	M. P. F.		Ba	113	M. P. F.††	
V	300	E. C.		La	152	M. P. F.	
	413	M. P. F.		Hf	213‡‡	S. H.	(2)
				Ta	245	S. H.	
Cr	485	S. H.		W	310	S. H.	
Mn	368	M. P. F.			384	E.	
Fe	420	S. H.		Re	310	E. C.	
Co	385	S. H.		Os	256	M. P. F.	
Ni	375	S. H.		Ir	285	S. H.	
Cu	315	S. H.		Pt	225	S. H.	
	333	E. C.			240	E. C.	
	342	E.					
Zn	200–300	S. H.	(4)	Au	170	S. H.	
	213	M. P. F.			175	E. C.	
	305	E.		Hg	96	S. H.	
Ga	125	M. P. F.			37	E. C.	
Ge	290‡	S. H.	(3)		68·$_6$	E.	
As	224	M. P. F.		Tl	100	S. H.	
Se	135	M. P. F.		Pb	88	S. H.	
				Bi	100	S. H.	(4)
Rb	58	M. P. F.‖			62	E. C.	
Sr	171	E. C.			110	C.	
Nb	301	M. P. F.		Th	168	E. C.	

† Cf. Fig. 4.

‡ From measurements above 90° K. At lower temperatures the specific heat is anomalous.

‖ $C = 115$, as for Na and K.

†† $C = 131$, as for Ca.

‡‡ The specific heat of Hf is anomalous, cf. ref. (2).

and the second term in the series vanishes (cf. Fig. 5). If N is the number of atoms in the volume V_0, we have therefore for the total work required to compress the solid

$$N(\epsilon - \epsilon_0) = \tfrac{1}{2}N(V-V_0)^2\left(\frac{d^2\epsilon}{dV^2}\right)_{V=V_0}.$$

Equating the right-hand side of this equation to (25), we have

$$\frac{1}{\chi_0} = NV_0\left(\frac{d^2\epsilon}{dV^2}\right)_{V=V_0}. \tag{27}$$

If we denote by v the volume per atom, we have

$$\frac{1}{\chi_0} = v_0\left(\frac{d^2\epsilon}{dv^2}\right)_{v=v_0}.$$

4.1. Thermal expansion. The curve in Fig. 5 shows the potential energy of an atom in a solid, plotted against the total volume V of the solid. In other words, the ordinate is equal to the total energy of an atom at rest. At the absolute zero of temperature, therefore, when the atoms are at rest,† the actual volume of the solid will be that for which ϵ is a minimum, as we have seen. At any other temperature T, when the atoms are not at rest, the volume of the solid is greater, because of its thermal expansion. To calculate the thermal expansion, the most direct course is to use the thermo-dynamical theorem that, for a substance in equilibrium at zero pressure and constant temperature, the free energy

$$F = U - TS$$

is a minimum, where U is the internal energy and S the entropy. For our solid, consisting of N oscillators with frequency ν, the free energy is‡

$$F = N\left(\epsilon + 3kT\log\frac{h\nu}{kT}\right), \tag{28}$$

and, since this is a minimum, we have

$$0 = \left(\frac{\partial F}{\partial V}\right)_T = N\left(\frac{d\epsilon}{dV} + 3kT\frac{d(\log\nu)}{dV}\right). \tag{29}$$

From equation (29) we may at once obtain the thermal-expansion coefficient, as will be shown below. In view of its importance, however, we shall first obtain equation (29) without quoting formula

---

† Neglecting the quantum-mechanical zero-point energy.
‡ Cf. equation (3). The formula is only valid if $h\nu < kT$.

(28) for the free energy. We hope that this course will give a greater insight into the mechanism of the phenomenon. The argument is simplified if we take account of the fact that the vibrational motion is quantized.

Let us then consider any atom of the solid, and let us suppose that it has n quanta of vibrational energy, so that its energy is $nh\nu$. If $\bar{n}$ is the mean value of n for all atoms, then, for $kT \gg h\nu$,

$$\bar{n}h\nu = 3kT. \tag{30}$$

For any particular atom, however, n may have any integral value.

Now the internal energy per atom of the solid at zero pressure is

$$U = \epsilon + \bar{n}h\nu,$$

and the change of U due to any small change dV in the volume is

$$dU = d\epsilon + \bar{n}h\,d\nu + h\nu\,d\bar{n}.$$

The last term, $h\nu\,d\bar{n}$, representing the change in the numbers of atoms in the respective quantum states, is equal to the heat dQ flowing into the solid; hence

$$d\epsilon + \bar{n}h\,d\nu = dU - dQ.$$

Now $dU - dQ$ represents the external work done, which vanishes for a solid in equilibrium. Thus the solid is in equilibrium when the energy $\epsilon + \bar{n}h\nu$ is a minimum for displacements in which n, the number of atoms in any quantum state, is kept constant. We have therefore

$$\frac{d\epsilon}{dV} + \bar{n}h\frac{d\nu}{dV} = 0, \tag{31}$$

and, substituting for $\bar{n}$ from (30), we obtain (29).

It will be realized that the assumption that the vibrations are quantized need not be used in deriving formula (30), and was only introduced in order that the process of thermal expansion might be more easily visualized.

To obtain the thermal-expansion coefficient, we expand $d\epsilon/dV$ by Taylor's theorem; if the total expansion $V - V_0$ is small compared with V_0, we have, with sufficient accuracy,

$$\frac{d\epsilon}{dV} = \left(\frac{d\epsilon}{dV}\right)_{V=V_0} + (V - V_0)\left(\frac{d^2\epsilon}{dV^2}\right)_{V=V_0}.$$

The first term vanishes, as is shown in Fig. 5. The second term, by (27), is equal to $(V - V_0)/NV_0\chi_0$, where N is equal to the number of atoms in the solid, and χ_0 is the compressibility at zero temperature.

We have therefore from (29)

$$\frac{V-V_0}{V_0\chi_0} = -3NkT\frac{d(\log\nu)}{dV}. \tag{32}$$

But $3NkT$ is the thermal energy of the solid; hence we obtain for the thermal-expansion coefficient α, differentiating with respect to T and multiplying by χ_0,

$$\alpha = \frac{1}{V_0}\frac{dV}{dT} = -\chi_0 c_v\frac{d(\log\nu)}{dV}, \tag{33}$$

where (if V refers to one gramme) c_v is the specific heat. Rearranging, we obtain

$$\frac{\alpha V_0}{\chi_0 c_v} = -\frac{d(\log\nu)}{d(\log V)}, \tag{34}$$

where V_0 is equal to the volume of one gramme of the solid, i.e. to the reciprocal of the density. α denotes the thermal-expansion coefficient and χ_0 the compressibility at zero temperature. The quantity (34) will be denoted by γ.

Theoretical predictions of the value of the quantity $d(\log\nu)/d(\log V)$, and hence of α, can only be made on the basis of a detailed calculation of the vibration spectrum, which has not yet been carried out.[†] We shall therefore calculate $d(\log\nu)/d(\log V)$ from the experimental values of α, χ_0, etc. The following are a few typical values; a more complete table is given in Appendix II.

	Al	Ni	Ag	Au
$-\dfrac{d(\log\nu)}{d(\log V)} = \gamma$	2·06	1·70	2·60	2·93.

Thus, in gold, a change in the volume of 3 per cent., which would produce a change in the interatomic distance of only 1 per cent., would nevertheless change the atomic frequency by nearly 9 per cent.

Low temperatures.[‡] Formulae (33) and (34) have been obtained subject to the assumption that T is large compared with the characteristic temperature. We shall now show that, with certain assumptions, these formulae are valid at all temperatures.

We can no longer assume that the vibrations of the solid can be represented approximately by a single frequency ν. We suppose that the $3N$ normal modes of the solid may be divided up into N_1 of

† Cf. § 1.1.
‡ Ratnowsky, *Ann. d. Physik*, **38** (1912), 637; *Verh. d. deuts. Phys. Ges.* **15** (1913), 75; Ornstein, *Proc. Amsterdam Acad.* **14** (1912), 983.

frequency v_1, N_2 of frequency v_2, and so on. The mean energy per normal mode of frequency v_i is

$$E_i(T) = \frac{hv_i}{e^{hv_i/kT}-1} \tag{35}$$

and the free energy

$$F_i = kT \log(1-e^{-hv_i/kT}),$$

which clearly tends to $kT \log(hv_i/kT)$ at high temperatures (cf. equations (2) and (3)). To this we must add the energy† $N\epsilon$ of the crystal at the absolute zero of temperature; the free energy for the whole crystal is thus

$$F = N\epsilon + kT \sum N_i \log(1-e^{-hv_i/kT}). \tag{36}$$

Since in equilibrium at zero pressure F is a minimum, we have, differentiating (36) with respect to the volume and substituting from (35),

$$0 = N\frac{d\epsilon}{dV} + \frac{1}{V} \sum_i N_i E_i \frac{d(\log v_i)}{d(\log V)}. \tag{37}$$

Grüneisen‡ has assumed that $d(\log v_i)/d(\log V)$ is the same for all frequencies of the solid. In that case, since $\sum N_i E_i$ is the total internal energy U of the solid, we have, using (37),

$$0 = \frac{V-V_0}{V\chi_0} + \frac{U}{V}\frac{d(\log v)}{d(\log V)},$$

which is the same expression as (32). Differentiating with respect to T, we obtain as before equation (33) or (34). It follows that the quantity γ $(= \alpha V_0/\chi_0 c_v)$ *is independent of temperature* even at low temperatures; and hence that the *thermal-expansion coefficient is proportional to the specific heat*. The agreement between this prediction and experiment has been shown by Grüneisen‖ for various solids over a wide range of temperature.

4.2. *Variation of compressibility with temperature.* We have already obtained formulae for the compressibility at $T = 0$ and for the volume at temperature T and zero pressure. We require still the general equation of state, i.e. the volume of the solid at any temperature and pressure, from which the compressibility at any temperature may be obtained. The equation of state may be obtained at once from

---

† Including the zero-point energy.
‡ *Handb. d. Phys.* **10** (1926), 43; *Ann. d. Physik*, **26** (1908), 211.
‖ Loc. cit. Cf. also Adenstedt, *Ann. d. Physik* **26** (1936), 69.

the thermodynamic equation

$$p = -\left(\frac{\partial F}{\partial V}\right)_T,$$

where p is the pressure and F the free energy at constant volume $(U-TS)$. Assuming as before that $d(\log\nu)/dV$ is the same for all normal modes of the solid, we obtain from (36) the Mie-Grüneisen† equation of state

$$p = -N\frac{d\epsilon}{dV} - U\frac{d(\log\nu)}{dV}, \tag{38}$$

valid at all temperatures, where U is the internal energy. Hence for the compressibility χ_T at temperature T we have, if $T > \Theta$,

$$\frac{1}{\chi_T} = -V\left(\frac{\partial p}{\partial V}\right)_T = \frac{1}{\chi_0} + UV\frac{d^2(\log\nu)}{dV^2}. \tag{39}$$

It follows that
$$\frac{d}{dT}\left(\frac{1}{\chi}\right) = Vc_v\frac{d^2(\log\nu)}{dV^2}, \tag{40}$$

or, making use of formula (33) for the thermal-expansion coefficient,

$$-\frac{1}{\chi}\frac{d\chi}{dT} = \frac{\alpha}{\gamma}V^2\frac{d^2(\log\nu)}{dV^2}. \tag{41}$$

We see, since the right-hand side is positive (cf. Fig. 5), that the compressibility *increases* with increasing temperatures.

Grüneisen‡ has measured by an indirect method the compressibilities of Al, Ag, Cu, Fe, and Pt between $-190°$ and $165°$; he finds that χ increases faster at high than at low temperatures, as formulae (40) and (41) suggest. The following table gives the measured compressibilities.‖

$$\chi \times 10^{12} \ (c.g.s. \ units)$$

Temp. °C.	−273	−190	17	131	165
Cu	(0·710)	0·718	0·773	0·815	0·828
Pt	(0·371)	0·374	0·392	0·401	0·404
Fe	(0·600)	0·606	0·633	0·664	0·675

Bridgman has measured the compressibilities of most metals at $30°$ and $75°$, but for the harder metals the differences are too small to be reliable. For certain of the softer metals the results are given on p. 22.

† Mie, *Ann. d. Physik*, **11** (1903), 657; Grüneisen, *Ann. d. Physik*, **26** (1908), 393.
‡ *Ann. d. Physik*, **33** (1910), 1239, **39** (1912), 284; *Verh. d. deuts. phys. Ges.* **13** (1911), 491; *Handb. d. Phys.* **10** (1926), 37.
‖ Bridgman's values at room temperature differ by about 10 per cent.

Element	Li	Na	Ca	Al	Pb
$10^3 \dfrac{d(\log \chi)}{dT}$	0·71	1·2	0·60	0·55	0·56
$\chi . 10^{12}$ (c.g.s. units) at 20° C.	8·8	15·8	5·6	1·37	2·30

We may deduce from (41) that for sodium, for instance, $-V^2 d^2(\log \nu)/dV^2$ has the value 7, for aluminium 18.

4.3. *Relation between the thermal-expansion coefficient and the change of compressibility with pressure.* According to Einstein's formula (23), the characteristic frequency ν of a solid varies in the following way:
$$\nu \propto \rho^{-\frac{1}{3}} \chi^{-\frac{1}{2}},$$
where ρ is the density and χ the compressibility. If we take logarithms and differentiate both sides with respect to the volume, we obtain
$$-\frac{d(\log \nu)}{d(\log V)} = \frac{1}{2}\frac{d(\log \chi)}{d(\log V)} - \frac{1}{6}.$$
The left-hand side is the quantity γ, depending on the thermal-expansion coefficient, tabulated in Appendix II. For the quantity on the right, if we write the relation between V and p in the form
$$-\frac{\Delta V}{V_0} = Ap - Bp^2, \qquad (42)$$
we obtain
$$B/A^2 - \tfrac{1}{6}.$$
We thus have
$$\gamma = B/A^2 - \tfrac{1}{6}. \qquad (43)$$

The quantities B, A have been tabulated for a number of metals by Bridgman.† For the more compressible metals equation (43) is in fair agreement with his values, as the following table shows:

TABLE IV

Element	γ	$B/A^2 - \tfrac{1}{6}$
Li	1·17	1·1
Na	1·25	1·0
K	1·34	0·8₄
Cs	1·29	0·8₄
Pb	2·73	2·9
Al	2·17	2·6
Ca	..	1·3
Sr	..	0·9₃
Ba	..	1·1

† *The Physics of High Pressures*, p. 160 et seq. (1931). The measurements are for room temperature.

On the other hand, for most of the hard metals (Cu, Ni, Pt), the values of B/A^2 measured by Bridgman are 10 to 100 times larger than γ. We conclude that the change of compressibility with volume for these metals at the pressures used is determined by some factor such as the microcrystalline structure of the metal, rather than by the interatomic forces.[†]

5. Binding energy of the metallic elements

By the binding energy is meant the work required to separate one gramme atom of the crystalline substance into its constituent *atoms*; in other words, for monatomic substances, the sublimation energy at the absolute zero of temperature. Table V shows the binding energy of certain elements. The table is taken from an article by Grimm and Wolf.[‡] The values are obtained from latent heats of melting and vaporization, corrected to apply to the absolute zero of temperature.

TABLE V

Binding energies of the metallic elements, in kilo-calories per gramme atom. (One kilo-calorie per gramme atom is equal to $0\cdot0434$ electron volts per atom.)

Li	46	Cu	76	Pb	51
B	80	Zn	32·5	Bi	48·5
Na	30	Rb	25	C	150
Mg	41	Sr	39	Si	81
Al	60	Ag	64·5	Sn	76
K	26·5	Cd	28		
Ca	39	Cs	24	Sb	49
Cr	83	Ba	39	Ne	0·5
Mn	63	Pt	122	Ar	1·5
Fe	108	Au	83	Kr	2·4
Co	105	Hg	18·5	X	3·4
Ni	101	Tl	44	Rn	4·0

From the experimental data discussed in this chapter and its theoretical interpretation, we may draw curves showing the total energy of a metal for various atomic volumes. The curves have been obtained in the following way: Denoting by W the total energy of the metal, we write

$$W = W_0 + W_2\left(\frac{\Delta V}{V_0}\right)^2 - W_3\left(\frac{\Delta V}{V_0}\right)^3 \dots .$$

[†] Grüneisen, *Handb. d. Phys.* **10** (1926), 38, has pointed out that Bridgman's pressure coefficients of compressibility are surprisingly high, in agreement with the conclusions of this section.

[‡] *Handb. d. Phys.* **24**/2 (1934), 1073.

W_0 is the total binding energy (Table V). V_0 is the volume of the crystal for zero pressure and temperature. Writing the compressibility in the form (42) we have, in ergs,

$$W_2 = \tfrac{1}{2}V_0^2\left(\frac{d^2W}{dV^2}\right)_0 = \frac{V_0}{2A},$$

$$W_3 = -\tfrac{1}{6}V_0^3\left(\frac{d^3W}{dV^3}\right)_0 = \frac{BV_0}{3A^3},$$

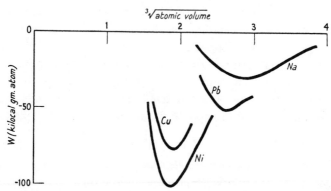

FIG. 6. Binding energy W in kilo-cal./gm. atom as a function of atomic volume (volume of one gramme atom).

or, for hard metals where B is not known directly, from (43),

$$W_3 = \frac{V_0}{6A}(2\gamma + \tfrac{1}{3}),$$

where γ is given in Appendix II.

Fig. 6 shows the binding energy of various metals plotted in this way.

6. Alloys: the phase diagram

In Chapter V of this book we shall give a discussion of the factors which determine the crystal structure of an alloy of given composition. In this section we shall derive the thermodynamical equations which determine the boundaries of a given phase.

Fig. 7 shows part of a typical phase diagram; for certain compositions and temperatures the alloy consists of a phase α, characterized by definite crystal structure and lattice parameters. The phase β will have, in general, different structure and parameters. For compositions and temperatures marked $\alpha+\beta$, the alloy consists of a *mixture* of small crystals of α and β.

It is typical of most metallic alloys that the atoms are distributed at random over the lattice points of the structure. This is the case, for instance, for Ag-Au for all compositions. For other alloys, e.g. Cu-Au for certain compositions, a superstructure forms at low enough temperatures, in some cases only after annealing. In other cases, e.g. Mg_2Sn, the alloy forms only for a fairly definite composition and a superstructure always exists. Such alloys are more properly called compounds.

The question of the superstructure is considered in the next section. In this section we assume that the temperature is too high for any superstructure to form.

In the following general considerations we need not, in the first place, specify the nature of the two phases, which are to be regarded as in thermal equilibrium. In the first phase, α, let there be N_A atoms of type A and N_B atoms of type B; in the second phase, β, let the numbers be N'_A and N'_B respectively. The free energy of the alloy in the first phase is a function of N_A and N_B, of the absolute temperature T and of the volume V_1. We may write it $F_1(N_A, N_B, T, V_1)$ and the free energy in the second phase $F_2(N'_A, N'_B, T, V_2)$.

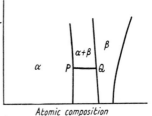

Fig. 7. Phase-equilibrium diagram of two solid phases.

Let us consider the two phases at their respective boundaries at a given temperature, say at P and Q in Fig. 7, at which compositions they are in equilibrium with one another. According to the second law of thermodynamics, we know that in equilibrium the total free energy F_1+F_2 must be a minimum with respect to any internal change of the whole system. For example, if a certain number of atoms of type A are transferred from the first to the second phase, the total free energy must increase. The conditions of equilibrium for constant temperature and volume may therefore be expressed:

$$\frac{\partial F_1}{\partial N_A} - \frac{\partial F_2}{\partial N'_A} = 0, \qquad \frac{\partial F_1}{\partial N_B} - \frac{\partial F_2}{\partial N'_B} = 0. \qquad (44)$$

In practice we are concerned with equilibrium at constant temperature and pressure. The appropriate conditions for equilibrium would then be given by equations similar to (44) but in which the thermodynamic potential ζ replaces the free energy F. The relation between

F and ζ is the following:
$$\zeta = F + pV,$$
where p denotes the pressure and V the volume. For alloys we are principally interested in the liquid and solid phases, so that the change in volume in a phase change is relatively small; thus, when p is the atmospheric pressure, the change in pV is completely negligible compared with other changes in F.

Let c_1 be the concentration of atoms B in the first phase, and c_2 the concentration of atoms B in the second phase, defined by:

$$c_1 = \frac{N_B}{N_A + N_B}, \qquad c_2 = \frac{N'_B}{N'_A + N'_B}. \tag{45}$$

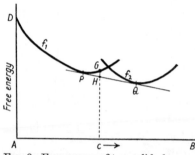

FIG. 8. Free energy of two solid phases as a function of the atomic composition c.

Since in a particular phase at a given concentration both the energy and the entropy are proportional to the total number of atoms in the phase, we may define for each phase a free energy per atom, f, which is a function of the concentration and of T and V only. Thus for example:

$$F_1 = (N_A + N_B) f_1(c_1, T, V).$$

The conditions of equilibrium (44) may therefore be written in the form:

$$\frac{\partial f_1}{\partial c_1} = \frac{\partial f_2}{\partial c_2} = \frac{f_1 - f_2}{c_1 - c_2}. \tag{46}$$

This form of the equations is particularly convenient as the starting-point for the interpretation of a phase-equilibrium diagram. Let us suppose that the free energies of two phases α and β are known as functions of the concentrations of atoms of type B. Fig. 8 shows two curves intended to represent f_1 and f_2 as functions of c at a particular temperature T. Equations (46) show that the α phase at the concentration P in the figure is in thermal equilibrium at temperature T with the β phase at the concentration Q. To the left of P the pure α phase is stable, to the right of Q the pure β phase. At any point between P and Q the stable state consists of a mixture of the two phases, α at concentration P and β at concentration Q. For suppose the phase α to exist at concentration G; then, by breaking it up into α at P and β at Q, the free energy is diminished to the value corresponding to H. If the construction is repeated for different temperatures, we find the concentrations of the phase boundaries P and Q as functions of T. This gives us part of the usual phase-equilibrium diagram, as for example in Fig. 7.

If c_v denotes the heat capacity per atom at constant volume and $E(c)$ the energy at the absolute zero,† the free energy per atom of a

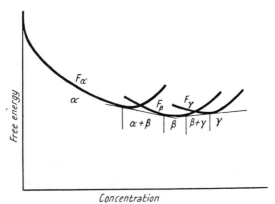

Fig. 9. Free energy of a number of phases, showing the ranges in which the pure phase can exist.

phase at concentration c and temperature T may be written:

$$f(c, T) = kT\{c\log c + (1-c)\log(1-c)\} + E(c) + \int_0^T c_v\,dT - T\int_0^T c_v\,dT/T.$$

The first bracketed term in this expression, which may be called the disorder term, arises in the following way. The weight of a macroscopic state is increased in a phase containing two kinds of atoms by the possibility of permuting the atoms of different sorts. In this way the entropy per atom is increased by a term $k\log\dfrac{(N_A+N_B)!}{N_A!\,N_B!}$, which by Stirling's formula reduces to $-k\{c\log c + (1-c)\log(1-c)\}$. This formula assumes a completely random distribution of the atoms, and will not be valid if this does not exist at all temperatures.

Since the disorder term is independent of the structure, the characteristics of various phase diagrams must depend on $E(c)$, the energy at $T = 0$. We give in Chap. V, § 3.1, reasons for believing that, at certain values of c, $E(c)$ will always rise rapidly. Such a rapid rise of $E(c)$ would account for the existence of alloys with a narrow range of composition, as Fig. 9 shows.

† The zero taken for the energy is arbitrary; we may take, for instance, the energy of free atoms, or the energy of the pure metals.

7. Alloys: thermal equilibrium of the superstructure

We have seen that a pure phase of an alloy is characterized by a definite lattice structure. In general the different atoms of which the alloy is composed are distributed at random amongst the lattice points of the structure. When the composition can be expressed by some simple formula, for example AB or A_3B, or when the atomic percentages are in the neighbourhood of these values, it is possible in certain cases for the atoms to take an *ordered* arrangement within the lattice, somewhat in the same way as the Na and Cl ions are arranged in order with respect to each other in the crystalline salt.

The hypothesis that under certain circumstances the atoms of an alloy segregate into regular positions was first put forward by Tammann[†] in 1919. By X-ray analysis the process was proved to occur for the solid solution Au-Cu by Johansson and Linde[‡] in 1925.

When order is established, for example in AuCu, new lines occur in the X-ray reflection spectrum which are characteristic of the lattice formed by the Au atoms alone or by the Cu atoms alone. In the state of complete order the phase is said to have developed a *superlattice*, and the new lines in the X-ray reflection spectrum are referred to as superlattice lines.

The development of order within a phase at fixed composition affects the physical properties of the alloy. A very striking change occurs in the electrical resistance as the order in the lattice changes. The resistance of a metal is due to the scattering of the electron waves by the irregularities of the lattice. These may be irregularities produced by heat motion or by the disorder of the atomic arrangement in the lattice.[||] Thus the resistance of an ordered alloy is less than that of a disordered alloy.

Fig. 10 shows a graph of the resistance of $AuCu_3$ due to Borelius, Johansson, and Linde.[††]

These results suggest that above a critical temperature T_c there is no order, whilst, as the temperature falls below T_c, order is gradually developed. The critical temperature is to be regarded as analogous to the Curie point of a ferromagnetic.[‡‡] At the critical temperature order disappears, just as at the Curie point ferromagnetism disappears.

Attempts to calculate the dependence of order on the temperature

† *Zeits. f. an. Chem.* **107** (1919), 1.

‡ *Ann. d. Physik*, **78** (1925), 439. For a recent discussion of this alloy by these authors, cf. *Ann. d. Physik*, **5** (1936), 1.

|| Cf. Chap. VII, § 13.   †† *Ann. d. Physik*, **86** (1928), 291.   ‡‡ Cf. Chap. VI, § 7.

have been made by Borelius,[†] Johansson and Linde,[‡] Gorsky,[||] and Dehlinger and Graf.[††]   The subject has been considered afresh from the theoretical standpoint by Bragg and Williams,[‡‡] and by Williams[||||]; recently a more rigorous theory has been developed by Bethe.[†††] We shall show in this section, following Bragg and Williams, and Bethe, how to calculate the degree of order appropriate to thermal equilibrium at temperature T. It is important to remember that, in an alloy in thermal equilibrium, the atoms are not in general

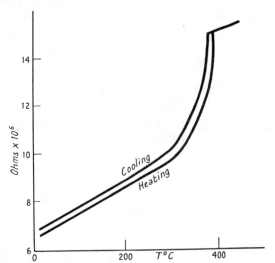

Fig. 10.   Resistance of the copper-gold alloy with the composition Cu_3Au.

rigidly fixed to given lattice points but are continually changing places. Of course at low temperatures the rate of exchange becomes very slow; at room temperature, for instance, for many alloys the rate of exchange is so slow that they cannot be regarded as being in thermal equilibrium.

7.1. *Long-distance order*. We shall in the first place restrict our considerations to the case of an alloy whose composition may be represented by a formula AB. An example is provided by β-brass (CuZn). We suppose that in this alloy there are altogether N atoms.

† *Ann. d. Physik*, **20** (1934), 57.   ‡ Loc. cit.
|| *Zeits. f. Phys.* **50** (1928), 64.
†† Ibid. **64** (1930), 359; **68** (1931), 535; **74** (1932), 267; **79** (1932), 550; **83** (1933), 832.
‡‡ *Proc. Roy. Soc.* A, **145** (1934), 699; **151** (1935), 540.
|||| Ibid. **152** (1935), 231.   ††† Ibid. **150** (1935), 552.

When the alloy is perfectly ordered, the $\frac{1}{2}N$ atoms of type A will lie on a regular lattice of their own and the $\frac{1}{2}N$ atoms of type B will also lie on a regular lattice as shown, for example, in Fig. 11. The lattice points occupied by the A atoms when perfect order exists we shall call a positions, and the lattice points occupied by the B atoms b positions.

We shall use the following notation: when an A atom is in an a position or a B atom in a b position, we shall call such an atom an R atom (right atom); when an A atom is in a b position or a B atom in an a position, we shall call the atom a W atom (wrong atom). At any moment of time in an alloy at temperature T, there will be a certain number of atoms which are R, and if we divide this number by the total number of atoms, N, we obtain the probability that a particular atom is R. We shall denote this probability by r. The probability that an atom is W is defined in a similar way and denoted by w, so that

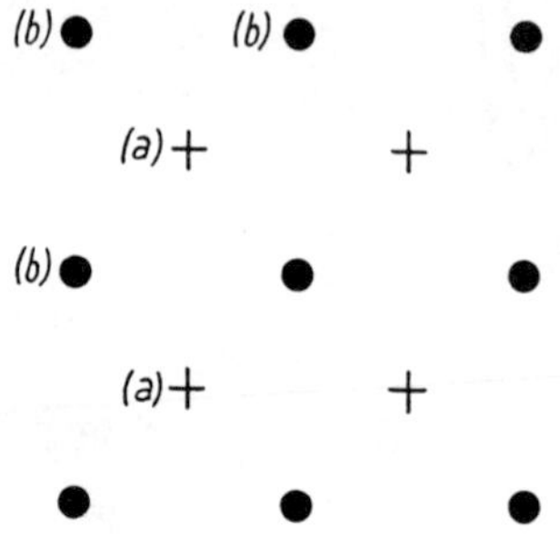

$$r+w = 1.$$

The quantity $$S = \frac{r-w}{r+w} \qquad (47)$$

Fig. 11.

is clearly a measure of the order of the lattice, though other measures are possible. When all the atoms are R, we have perfect order, so that $S = 1$; when all the atoms are W, $S = -1$, which also represents perfect order. Complete disorder occurs when as many atoms are R as W, so that $S = 0$. Thus a change in the sign of S does not represent a change in the physical state of the system.

We shall call S the 'long-distance order' of the lattice, since S can only be defined by considering the state of the lattice over a large volume.

7.2. *The order of neighbours*. The quantity S which we have just defined is not by itself sufficient to determine fully the order existing in a lattice. For instance, S vanishes when there are as many R as W atoms, but if all the R atoms occur together and all the W atoms occur together the state is really one of very high order, although according to our definition the long-distance order is zero. Also, as we shall show later, there exists a temperature T_c above which S vanishes, but it is clear on physical grounds that, if a tendency to order exists, then, even at very high temperatures greater than T_c, it will still be more probable that an A atom will have a B than another A as nearest neighbour. There exists therefore an 'order of neighbours', which remains finite at all temperatures. It may be defined in the following way: the order of neighbours σ is the difference of the probabilities of finding an unequal and an equal neighbour beside a given atom. Hence, if z is the number of nearest neighbours to any atom, there will be, among N atoms of both kinds,

$$\tfrac{1}{4}Nz(1+\sigma) \text{ pairs of neighbours } AB,$$
$$\tfrac{1}{8}Nz(1-\sigma) \text{ pairs } AA, \text{ and an equal number of pairs } BB. \tag{48}$$

7.3. *The method of Bragg and Williams.* In order to obtain S as a function of T, Bragg and Williams define an energy V which is the energy required to interchange *two* atoms, viz.: an A atom from an a to a b position and a B atom from a b to an a position, so that the number of W atoms increases by 2 and the number of R atoms decreases by 2. We may express the ratio w/r and hence S in terms of the Boltzmann factor. Assuming that the R atoms and the W atoms are distributed at random among the a positions, and also among the b positions, the increase in the free energy when a pair of atoms is interchanged is:

$$\delta F = V - NkT(\log w \, \delta w + \log r \, \delta r),$$

where $\delta w = 2/N$, $\delta r = -2/N$ are the changes in w and r. Since for equilibrium δF must vanish, this gives for the ratio of the probabilities, w/r,

$$\frac{w}{r} = \exp\left(-\frac{V}{2kT}\right), \tag{49}$$

and hence directly from the definition of S (equation (47)) we obtain

$$S = \tanh\left(\frac{V}{4kT}\right). \tag{50}$$

Bragg and Williams make the assumption that V is directly proportional to the degree of long-distance order already existing;[†] i.e.

$$V = V_0 S. \tag{51}$$

With this assumption (50) and (51) together determine S as a function of T. It will be observed that there is a striking formal similarity to the Weiss theory of ferromagnetism,[‡] S corresponding to the

---

[†] It is of course obvious that V vanishes with S, but the assumption that V is a linear function of S is merely introduced in the interests of simplicity.

[‡] Cf. Chap. VI, § 7.

intensity of magnetization and V to the intrinsic field. In particular equations (50) and (51) show that there is a critical temperature beyond which S is zero, which is quite analogous to the Curie temperature. This critical temperature is given by

$$T_c = V_0/4k. \tag{52}$$

In Fig. 12, curve (a), we show the degree of long-distance order S plotted against† $12kT/V_0$, i.e. against $3T/T_c$.

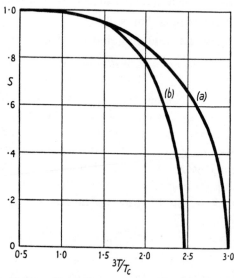

FIG. 12. Degree of long-distance order S as a function of $3T/T_c$.
(a) According to Bragg and Williams.
(b) According to Bethe.

If an alloy has a composition represented by a formula A_3B, for instance, or more generally whenever the composition differs from AB, Bragg and Williams find that in place of equation (50) the dependence of S upon $V/4kT$ is of the form shown in Fig. 13. In a certain limited temperature range there are therefore three solutions, i.e. three values of S for a particular temperature.

It has been shown that, of the three solutions, that corresponding to the intersection P has the lowest free energy and gives therefore the stable state.‡ At the critical temperature the free energies of the states O and P will be equal; this may be shown to be the case for the

---

† Bethe has shown that a comparison between his theory and that of Bragg and Williams leads to the conclusion that $V_0 = 2zV_1$ (cf. § 7.4); Fig. 12 refers to a simple cubic lattice, for which $z = 6$.   ‡ Williams, loc. cit. 243.

value of T such that the areas of the two loops cut off by the line OP are equal. Solutions such as that represented by Q have higher free energy than O and are unstable.

In contradistinction to an alloy of the type AB, therefore, an alloy of the type A_3B can exist at the critical temperature in *two* states in equilibrium with each other, one having zero order and the other a finite degree of order. The condition of the alloy at this temperature is therefore analogous to that of, say, ice and water in thermal equilibrium at $0°$ C.; two distinct phases are in equilibrium

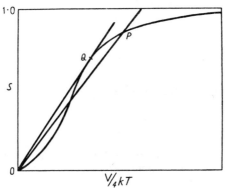

FIG. 13. S as a function of $V/4kT$.

with each other. We shall show later that an alloy AB has a finite discontinuity in the specific heat at the transition point. For an alloy A_3B, on the other hand, there must be a latent heat at the transition temperature.

These conclusions for alloys of the type A_3B only apply when all positions in the lattice may be regarded as equivalent, as, for example, in Cu_3Au. In the alloy Fe_3Al, which has a body-centred cubic lattice, two aluminium atoms can never be nearest neighbours, and therefore half the lattice points, namely those which lie on one simple cubic lattice, are permanently occupied by Fe atoms. This alloy therefore behaves, with respect to the development of order, like an alloy of composition AB.

7.4. *Method of Bethe.* Bethe† has given a more detailed theory which avoids the somewhat arbitrary assumption expressed by equation (51). He bases his theory on the assumption that the order is controlled by the interaction between atoms which are

† Loc. cit.

nearest neighbours.† If we denote the interaction energy between an A and a B atom by V_{AB}, and that between two A atoms and two B atoms by V_{AA} and V_{BB} respectively, the total energy of the crystal becomes, according to (48),

$$E = \tfrac{1}{4}Nz(1+\sigma)V_{AB}+\tfrac{1}{8}Nz(V_{AA}+V_{BB})(1-\sigma);$$

this may be written

$$E = \text{const.}+\tfrac{1}{4}NzV_1(1-\sigma), \tag{53}$$

where
$$V_1 = \tfrac{1}{2}(V_{AA}+V_{BB})-V_{AB}. \tag{54}$$

We shall assume the quantity V_1 to be positive. Bethe confines himself to the case where the alloy has a formula of the type AB. We fix our attention, then, on a given A atom which we call the 'central' atom. Let p be the probability that a given nearest neighbour to the central A atom is itself an A atom. Then the probability that this particular neighbour is a B atom is equal to $1-p$. From the definition of the 'order of neighbours', σ, we have

$$\sigma = 1-2p. \tag{55}$$

If we were to assume that p did not depend upon the nature of the other atoms in the neighbourhood of the central A atom, then we should have

$$\frac{p}{1-p} = e^{-V_1/kT},$$

or, according to (55),

$$\sigma = \tanh(V_1/2kT).$$

It is to be observed that there is then no critical temperature, but that σ tends slowly to zero at high temperatures.

We now calculate the long-distance order S as a function of T and of the parameter V_1 defined by equation (54). In this calculation no *ad hoc* assumption is made, such as that expressed by (51) in the treatment of Bragg and Williams. We consider again a particular lattice point, surrounded by z nearest neighbours, and calculate the probability that the atom at this point is R or W. When two R atoms or two W atoms are nearest neighbours, we may suppose the interaction energy to be zero (putting the constant in (53) zero); on the same scale the interaction energy between an R and a W atom is V_1. Since we consider only interaction between nearest neighbours, it follows that the probability that the central atom is

---

† A discussion of the validity of this assumption for metals has not yet been given.

R or W depends *directly* only on the nature of the z neighbours. For example, if there were as many R as W atoms amongst the z neighbours, the probability of the central atom being R or W would be just $\frac{1}{2}$. It is the effect of the outside atoms on the z neighbours, causing more R than W atoms among them, which in turn makes the probability of an R atom being the centre greater than $\frac{1}{2}$. Bethe makes use of the following device. He assumes that the effect of the outer atoms alone on the z neighbours of the central atom is to multiply the probability that one of these z atoms is W by a factor ϵ. It is then possible to express the probability that the central atom is W in terms of ϵ and of the Boltzmann factor

$$x = e^{-V_l/kT}.$$

One can also find the total probability that one of the z neighbours is W. Equating these two probabilities, since the central atom is not distinguished in the lattice, we obtain an equation to determine ϵ in terms of x, and hence ultimately S as a function of x.

The relative probability that the number of W atoms in the shell of z neighbours is n and that the central atom is R will be denoted by r_n; then, by the definition of ϵ,

$$r_n = {}^zC_n x^n \epsilon^n,$$

the factor ${}^zC_n$ being the number of ways of putting n atoms into z equivalent places. The relative probability that the number of W atoms in the shell of z neighbours is n and that the central atom is W is similarly:
$$w_n = {}^zC_n x^{z-n} \epsilon^n.$$

Allowing the shell of z neighbours to take every possible arrangement of R or W atoms, we obtain for the total probabilities that the central atom is R or W:

$$r = \sum_{n=0}^{z} r_n = (1+\epsilon x)^z,$$

$$w = \sum_{n=0}^{z} w_n = (\epsilon+x)^z,$$

(56)

respectively.† The relative probability that the number of W atoms in the shell of z neighbours is n is equal to $r_n + w_n$; hence the relative

---

† The relative probabilities of equation (56) are not identical with the r and w previously defined and used in (47). The difference, however, is only the trivial one of some constant factor which does not affect any subsequent calculation. It is for this reason that we use the term 'relative probability' instead of simply 'probability'.

probability that a given one of these is W is equal to $\dfrac{n}{z}(r_n+w_n)$. Thus the total relative probability that one of these atoms is W is given by w', where

$$w' = \sum_{n=0}^{z} \frac{n}{z}(r_n+w_n) = \frac{\epsilon xr}{1+\epsilon x} + \frac{\epsilon w}{\epsilon+x}. \tag{57}$$

Since a 'central' atom and one in the shell are, of course, entirely equivalent, we may equate w and w'; hence, substituting from (56) for r and w, we obtain the following equation for ϵ:

$$\epsilon = \left(\frac{\epsilon+x}{1+\epsilon x}\right)^{z-1}. \tag{58}$$

If we introduce a new variable δ in place of ϵ, defined by the equation

$$\epsilon = e^{-2\delta(z-1)},$$

(58) reduces to

$$x = \frac{\sinh(z-2)\delta}{\sinh z\delta}. \tag{59}$$

Equation (58) determines ϵ in terms of x and therefore r and w, according to (56), in terms of x. Hence by (47) we obtain S as a function of x, and thus of kT/V_1. The result of the calculation made in this way is shown in Fig. 12, curve (b). We can easily determine the value of the critical temperature T_c. We observe that $\epsilon = 1$ corresponds to complete disorder, because in that case $r = w$. We need therefore the value of x as $\epsilon \to 1$, i.e. as $\delta \to 0$. According to (59) this value is

$$x_c = 1-\frac{2}{z},$$

and therefore the critical temperature is given by

$$\frac{kT_c}{V_1} = 1\Big/\log\Big(\frac{z}{z-2}\Big). \tag{60}$$

For β-brass, $T_c = 743°$ K., $z = 8$, and therefore $V_1 = 0\cdot018$ electron volts.†

7.5. *The contribution to the specific heat arising from disorder.* We shall make the calculation according to the method of Bragg and Williams, and merely quote the results obtained by that of Bethe.

The degree of long-distance order is defined by (47), which may be written alternatively

$$S = (\text{number of } R \text{ atoms}-\text{number of } W \text{ atoms})/N.$$

When the number of W atoms increases by 2 and the number of

† See p. 40.

R atoms decreases by 2 (i.e. by interchange of two atoms from wrong to right places) the change in the energy of the crystal is V. Hence, if S changes by dS, the energy E changes by dE in such a way that

$$dE = -\tfrac{1}{4}NV\,dS = -\tfrac{1}{4}NV_0 S\,dS.$$

The additional specific heat per atom, Δc_v, due to the growth of disorder, is given, substituting from (52), by

$$\Delta c_v = \frac{1}{N}\frac{dE}{dT} = -kT_c S\frac{dS}{dT}. \tag{61}$$

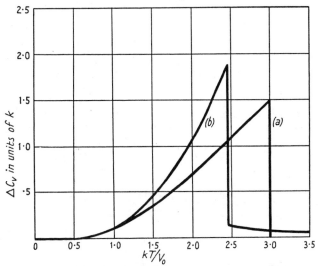

FIG. 14. The contribution Δc_v to the specific heat arising from disorder:
  (a) According to the approximation of Bragg and Williams.
  (b) According to the approximation of Bethe.

This result, together with the previous calculation of S, gives us Δc_v as a function of T. The result of such a calculation is shown in Fig. 14, curve (a). We can obtain the discontinuity in Δc_v at T_c as follows: We may expand $\tanh(V/4kT)$ in powers of (T_c-T), and thus obtain from (50) for S in the neighbourhood of T_c

$$S^2 = 3(T_c-T)/T_c. \tag{62}$$

It follows from (61) and (62) that the jump in the specific heat is $\tfrac{3}{2}k$ per atom.

Fig. 14, curve (b), shows Δc_v calculated by Bethe. The tail of the specific heat curve (b) is due to the ordering of neighbours and does not vanish abruptly at any temperature.

Fig. 15 shows the specific heat of β-brass, recently measured by Sykes,[†] for comparison with the corresponding theoretical curves.

A simple check on the basic idea of the theory is obtained by calculating the total entropy change due to the transition from order to complete disorder. According to the previous section this is $k \log_e 2$ per atom. Hence we should have:

$$\int \frac{\Delta c_v}{T} \, dT = k \log_e 2.$$

Sykes's observations are in fair agreement with this condition.

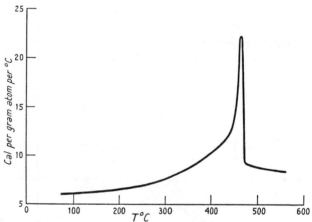

FIG. 15. Observed specific heat of β-brass (CuZn).

It is interesting to observe that the theory of Bragg and Williams shows, when the composition of the alloy is $A_3 B$, or indeed anything other than AB, that there is not merely a finite discontinuity in the specific heat at the critical temperature, but that at this temperature a latent heat is evolved. The recent measurements of the specific heat of Cu_3Au by Sykes, quoted by Bragg and Williams,[‡] suggest that this deduction from theory is, in fact, correct.

7.6. *The rate of approach to the equilibrium state.* The properties of an alloy, in general, depend upon its previous heat treatment, for instance on whether it has been quenched or annealed. This implies that the alloy during a heating or cooling process is not in thermal equilibrium at each moment of the process. This can readily be understood, since, if an alloy is suddenly cooled from a given temperature at which it is in thermal equilibrium to a lower temperature T, a

[†] *Proc. Roy. Soc.* A, **148** (1935), 422.   [‡] Ibid., **151** (1935), 540.

definite time must elapse before the equilibrium appropriate to T can be established. Bragg and Williams suppose that the alloy goes through a succession of states each of which may be regarded as an equilibrium state appropriate to a certain temperature Θ. Θ therefore may be supposed to approach T steadily according to the equation

$$\frac{d\Theta}{dt} = -\frac{\Theta - T}{\tau}. \tag{63}$$

τ is the 'time of relaxation' of the alloy, or the time taken for the departure from equilibrium to be reduced to $1/e$ of its initial value.

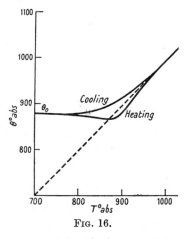

Fig. 16.

τ will clearly depend upon the energy which an atom must have in order to exchange places with a neighbouring atom. This 'activation energy' will of course be different from the V_0 already introduced; we shall denote it by W. It is reasonable to suppose that τ depends upon the temperature in the following way:

$$\tau = Ae^{W/kT}. \tag{64}$$

We may regard $e^{-W/kT}$ as a measure of the probability that an atom will exchange places with a neighbour in a single oscillation, and $1/\tau$ roughly as a measure of the frequency of the exchange process. Hence $1/A$ must be of the order of the frequency of oscillation, so that we may set $A \sim 10^{-12}$ sec. Further details of a calculation of τ are given in the paper of Bragg and Williams.

Let us now consider an alloy which is being cooled (or heated) at the rate of σ degrees centigrade per second. We write

$$\frac{dT}{dt} = \pm\sigma, \tag{65}$$

the positive sign denoting heating and the negative cooling.

By equations (63) and (65) we obtain:

$$\frac{d\Theta}{dT} = \pm\frac{\Theta - T}{\sigma\tau}.$$

Fig. 16 shows a solution of this differential equation, when τ is given by equation (64). A has been assumed to be equal to 10^{-12} sec. Θ_0 is the temperature which determines the extent of the disorder which is 'frozen in' at the absolute zero. If A is regarded as known, then Θ_0 is a function only of the parameter W and the rate of cooling σ. Although the value of W has not been calculated, the theory can yield useful information in the following way. If a particular alloy is cooled from above the critical temperature at a known rate and the residual disorder is estimated, it is possible to obtain Θ_0 and therefore the value of W for that alloy. The theory will now tell us the extent to which Θ_0 can be affected by varying the rate of cooling σ. As an example we may suppose that, for a particular alloy, W has been estimated to be $k \times 12,000°$, where k is Boltzmann's constant. The following are some values of Θ_0 for different rates of cooling for this alloy:

σ (degrees/sec.)	Θ_0 (degrees centigrade)
10^3	852
10	685
10^{-1}	560
10^{-3}	464

The critical temperature of this alloy might be between 600° and 700° C.; for convenience let it be 685° C. The theory shows that by quenching ($\sigma > 10$ degrees per sec.) it would be possible, in this case, to preserve complete disorder. On the other hand, careful annealing ($\sigma = 10^{-3}$ degrees per sec.) would give for the maximum degree of order attainable $S = 0.77$.

According to the above theory the development of order within a lattice shows hysteresis. At a given temperature the degree of order is less during a steady cooling than during a steady heating process. According to Bragg and Williams, this is the way in which the electrical-resistance temperature curves of Borelius, Johannsson, and Linde shown in Fig. 10 are to be interpreted. A different explanation of the hysteresis has been proposed by Borelius (loc. cit.), but this explanation does not appear to be altogether free from objection.[†]

[†] Cf. Bragg and Williams, *Proc. Roy. Soc.* A, **151** (1935), 540.

Notes of recent developments.

p. 6. The anisotropy of the atomic vibrations in zinc crystals has recently been demonstrated by X-ray methods by Brindley, *Phil. Mag.* **21** (1936), 790; the amplitude is greatest parallel to the principal axis.

p. 36. A discussion on similar lines of the formation of a superlattice when the concentrations of the two components are unequal has been given by Peierls, *Proc. Roy. Soc.* A, **154** (1936), 207.

II

ELECTRONS IN EQUILIBRIUM IN THE CRYSTAL LATTICE

1. Wave-mechanical principles

1.1. *The hydrogen atom and the alkali atoms.* A hydrogen atom consists of a single electron moving in the field of a proton. The potential energy $V(r)$ of the electron at a distance r from the proton is given by the equation

$$V(r) = -e^2/r.$$

An alkali atom consists of a single valence electron moving in the field of a 'core', i.e. of a nucleus surrounded by a number of electrons forming a closed shell. In the mathematical discussion both of the spectrum of the atom and of its chemical properties, we may to a certain approximation treat the valence electron as moving in the field of the nucleus and in the *average* field of all the electrons forming the core. If Ze is the charge on the nucleus, the number of electrons constituting the core is $Z-1$; therefore, if the valence electron is at any moment right outside the core, it moves in the field of a positively charged sphere carrying a charge $+e$. By a well-known theorem of electrostatics, the field of such a sphere is the same as it would be if the charge were concentrated at the centre; therefore, at large distances from the nucleus, the field in which the valence electron moves is the same as in a hydrogen atom. At smaller distances, on the other hand, when the valence electron comes within the core of the atom, it is no longer so completely screened from the nucleus, and therefore the force pulling the electron to the nucleus is greater than it would be in a hydrogen atom at the same distance. The potential energy function $V(r)$, which gives the work that must be done to bring up the valence electron from infinity to a distance r from the nucleus, is therefore equal to $-e^2/r$ at large distances, but is numerically greater (algebraically less) than $-e^2/r$ for small r. This is illustrated in Fig. 17, where the potential energy $V(r)$ is plotted against r.

The energies that the valence electron may have are determined by Schrödinger's equation,

$$\nabla^2\psi + \frac{2m}{\hbar^2}[E - V(r)]\psi = 0. \tag{1}$$

Only for certain values of the energy E is it possible to obtain a bounded solution ψ tending to zero at infinity. For every such solution ψ_n of the equation, a stationary state of the atom exists with energy E_n. The wave function ψ has the following physical significance: If the atom is known to be in the stationary state n, then $|\psi_n(\mathbf{r})|^2 d\tau$ is equal to the *probability* that the valence electron is in the volume element $d\tau$ at the point $\mathbf{r}$. An equivalent statement is that $-e|\psi_n(\mathbf{r})|^2$ is the *average* charge density due to the valence electron in the atom at the point $\mathbf{r}$. By average is meant either a time average for a single atom, or an average taken at a moment of time for a large number of atoms.

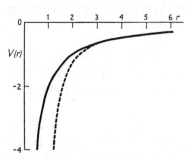

FIG. 17. Potential energy of electron in hydrogen and alkali atoms.

Full line = hydrogen.
Dotted line = potassium (from Hartree, *Proc. Roy. Soc.* A, **143** (1934), 506.)
r in atomic units, $V(r)$ in Rydberg units (13·53 e.v.).

Solutions of Schrödinger's equation may be obtained in spherical polar coordinates, and are of the form

$$\psi = f(r)S_l(\theta, \phi),$$

where $S_l(\theta, \phi)$ is a rational integral harmonic of order† l; $f(r)$ then satisfies the equation

$$\frac{d^2f}{dr^2} + \frac{2}{r}\frac{df}{dr} + \frac{2m}{\hbar^2}\left[E - V(r) - \frac{l(l+1)}{r^2}\right]f = 0.$$

If $l = 0$, S_l is equal to a constant, and the solution is *spherically symmetrical*. The solution then corresponds to an S state of the atom. The function f may have 0, 1, 2,... zeros, so that the wave function ψ will have a number of nodal spheres surrounding the origin. Fig. 18 shows a cross-section through the centre of the atom, the nodes being marked by full lines; below is shown the radial part $f(r)$ of the wave function plotted outwards as a function of r.

If $l = 1$, the wave function corresponds to a P state; S_l may then have either of the three values

$$\cos\theta = z/r,$$
$$\sin\theta\cos\phi = x/r,$$
$$\sin\theta\sin\phi = y/r,$$

† Cf. Jeans, *Electricity and Magnetism*, 4th ed., p. 209, or Whittaker and Watson, *Modern Analysis*, Chap. XV.

and the wave function ψ will therefore have, in addition to the spherical nodes caused by the zeros of $f(r)$, a nodal plane passing through the origin. This again is shown in Fig. 18, the nodes being shown above and the function $f(r)$ below. Since a nodal surface passes through the nucleus, the wave function must vanish at the nucleus $(r = 0)$.

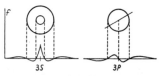

FIG. 18. Nodal surfaces and wave functions f of a hydrogen atom.

Similarly, for D states, *two* nodal planes pass through the nucleus, the function S_l taking any of the forms†

$$\frac{x^2-y^2}{r^2}, \quad \frac{z^2-x^2}{r^2}, \quad \frac{yz}{r^2}, \quad \frac{zx}{r^2}, \quad \frac{xy}{r^2}.$$

The total number of nodal surfaces plus unity is called the principal quantum number n. Thus the two wave functions illustrated in Fig. 18 have each two nodal surfaces, and therefore principal quantum number three.

For the hydrogen atom, where $V(r) = -e^2/r$, the mathematical solution of Schrödinger's equation shows that the energy of a state depends only on the principal quantum number n, and is given by Bohr's formula

$$E = -\frac{C}{n^2}, \qquad C = \frac{me^4}{2\hbar^2}.$$

Thus, for hydrogen, each of the states illustrated in Fig. 18 will have the same energy. This is not true of any other type of field; for the field illustrated by the dotted line in Fig. 17 the S state will always have lower energy than the P state, and the P state will have lower energy than the D state.

1.2. *The helium atom and the self-consistent field.* The energy levels W of an atom with two electrons, helium, are determined by a Schrödinger equation containing *six* independent variables. This equation is

$$(H-W)\Psi = 0, \tag{2}$$

where H denotes the operator

$$H = -\frac{\hbar^2}{2m}(\nabla_1^2+\nabla_2^2)-\frac{e^2}{r_1}-\frac{e^2}{r_2}+\frac{e^2}{|\mathbf{r}_1-\mathbf{r}_2|}, \tag{2.1}$$

and where $\qquad \mathbf{r}_1 = (x_1,y_1,z_1), \qquad \mathbf{r}_2 = (x_2,y_2,z_2)$

are the coordinates of the electrons. This equation has not been

† The expressions $P_2^u(\cos\theta)e^{iu\phi}$ may be formed by linear combination of these functions.

solved exactly, and approximate methods must be used to find the energy levels W, the wave functions, and hence the charge density in the atom. Of these, the method of the 'self-consistent field', due originally to Hartree,† is particularly important, and will form the basis of many of the considerations of this book.

In Bohr's orbital theory it was considered that, to a certain approximation, each electron in an atom moved in an orbit which was independent of all the others; the interaction between the orbits could be considered as a second approximation. Guided by this idea, Hartree assumed that each electron in a helium atom moved in a *static* field, so that each electron could be considered to have its own wave function. This static field, in which a given electron is supposed to move, is not of course the field of the nucleus alone, but the field of the nucleus together with the *average* field of the other electron. Thus, if the two electrons have wave functions $\psi_1(\mathbf{r})$ and $\psi_2(\mathbf{r})$, the first electron, with wave function ψ_1, would move in a field made up of two parts:

(1) The field of the nucleus, giving a potential $-2e^2/r$.

(2) The average field of the other electron; this is obtained by treating this electron as though it were smeared out into a uniform charge distribution, of density $-e|\psi_2(\mathbf{r})|^2$ at the point $\mathbf{r}$. The potential energy of an electron (the first electron) in the field of this uniform charge distribution is, by direct integration,

$$e^2 \iiint \frac{|\psi_2(\mathbf{r}')|^2\, dx'dy'dz'}{|\mathbf{r}-\mathbf{r}'|}, \tag{3}$$

where $\mathbf{r}$ denotes the point where the first electron is supposed to be.

It follows that the first electron moves in a field in which its potential energy is

$$V_1(r) = -\frac{2e^2}{r} + e^2 \iiint \frac{|\psi_2(\mathbf{r}')|^2\, dx'dy'dz'}{|\mathbf{r}-\mathbf{r}'|}.$$

Its wave function, therefore, satisfies the Schrödinger equation

$$\nabla^2\psi_1 + \frac{2m}{\hbar^2}[E - V_1(r)]\psi_1 = 0. \tag{4}$$

Similarly, the other electron moves in the somewhat different field

$$V_2(r) = -\frac{2e^2}{r} + e^2 \iiint \frac{|\psi_1(\mathbf{r}')|^2\, dx'dy'dz'}{|\mathbf{r}-\mathbf{r}'|}, \tag{5}$$

† *Proc. Camb. Phil. Soc.* **24** (1928), 89. See also, for instance, *Handb. d. Phys.* **24**/1 (1933), 368.

which gives a Schrödinger equation similar to (4). These two equations determine the wave function of each electron.

The fields V_1 and V_2 are called 'self-consistent' for the following reason. In the practical application of the method, one estimates, say, V_1; Schrödinger's equation (4) may then be solved for ψ_1, and hence V_2 may be calculated from equation (5). This enables the Schrödinger equation for ψ_2 to be written down and solved, and from the solution ψ_2 of this equation we may calculate V_1. If the function V_1 obtained is not the one originally estimated, then the original estimate was incorrect, and must be modified, and the whole process repeated.

If the helium atom is in the ground state, ψ_1 and ψ_2 are the same function, and the potential $V(r)$ is the same for either electron.

It is important to realize that Hartree's method does *not* neglect entirely the influence of one electron on the other. For instance, if we take as an example the ground state of helium, the potential $V(r)$ of the field in which either electron is supposed to move will approximate to $-2e^2/r$ at small distances and $-e^2/r$ at large distances, owing to the screening effect of the other electron (i.e. to a term of the type (3)). On the other hand, if one electron is on the right-hand side of the atom, the other electron is more likely to be on the left than on the right because of the repulsion between the two electrons. This fact is entirely neglected in the Hartree approximation.

The parameter E which occurs in Hartree's equation (4) *is not even approximately equal to the ionization energy of the atom.* This may be seen as follows: we confine ourselves to the case where both electrons are in the same state, so that the wave function of the atom may be written

$$\Psi(\mathbf{r}_1, \mathbf{r}_2) = \psi(\mathbf{r}_1)\psi(\mathbf{r}_2).$$

The energy of the atom is then

$$W = \int\int \Psi^*H\Psi \, d\tau_1 \, d\tau_2, \tag{6}$$

where H is the operator (2.1). Now H may be written $H_1 + H_2 + H_3$, where

$$H_1 = -\frac{\hbar^2}{2m}\nabla_1^2 + \left(\frac{e^2}{r_{12}} - \frac{2e^2}{r_1}\right),$$

$$H_2 = -\frac{\hbar^2}{2m}\nabla_2^2 + \left(\frac{e^2}{r_{12}} - \frac{2e^2}{r_2}\right),$$

$$H_3 = -\frac{e^2}{r_{12}}.$$

If we multiply H_1 by $|\psi(\mathbf{r}_2)|^2$ and integrate over all space with respect to $\mathbf{r}_2$, we obtain the operator

$$-\frac{\hbar^2}{2m}\nabla_1^2 + V(r_1).$$

But this operator, operating on ψ_1, gives $E\psi_1$, and hence the term H_1 makes a contribution of just E to the whole integral. The term H_2 yields an exactly equal amount, so that the total energy of the two electrons is

$$W = 2E - e^2 \iint \frac{1}{r_{12}} |\psi(\mathbf{r}_1)|^2 |\psi(\mathbf{r}_2)|^2 \, d\tau_1 \, d\tau_2. \tag{7}$$

The total energy of the atom is always *less* than $2E$. The expression $2E$ includes the interaction between the electrons *twice*.

Solutions of Hartree's equations have been obtained for various atoms and ions in the following papers: Hartree, *Proc. Camb. Phil. Soc.* **24** (1928), 89, theory and methods; p. 111, applications to Li, Rb^+, Na^+, Cl^-. *Proc. Roy. Soc.* A, **141** (1933), 283, Cl^-, Cu^+; ibid. **143** (1933), 506, Cu^+, K^+, Cs; ibid. **150** (1935), 96, Al^{+3}, Fe^-, Rb^+; *Phys. Rev.* **46** (1934), 738, Hg. Hartree and Black, *Proc. Roy. Soc.* A, **139** (1933), 311, oxygen in various states of ionization. D. R. Hartree and W. Hartree, ibid. **149** (1935), 210, Be, Ca, Hg; ibid. **150** (1935), 9, Be with exchange. McDougall, ibid. **138** (1932), 550, calculation of terms in the optical spectrum. Torrance, *Phys. Rev.* **46** (1934), 388, C in ground state and with electron configuration $1s^2 2s 2p^3$; Kennard and Ramberg, ibid. **46** (1934), 1034, Na. Fock and Petrushen, *Phys. Zeits. d. Sowjetunion*, **8** (1935), 547, Li with exchange.

1.3. *Fock's equation.* Slater[†] and Fock[‡] have deduced Hartree's equation from a variational principle, and have also shown how a more accurate equation may be obtained. Their method is as follows: If Schrödinger's equation is written in the form (2), then the characteristic solutions Ψ_n^* are those for which the integral

$$\int \Psi_n^* H \Psi_n \, d\tau$$

† *Phys. Rev.* **36** (1930), 57.

‡ *Zeits. f. Phys.* **61** (1930), 126. Cf. *Handb. d. Phys.* **24/1** (1933), 349. A detailed account of Fock's and Hartree's equations and the relation between them has been given by Brillouin, *Actualités Scientifiques*, iv, Hermann & Cie, Paris (1934).

is a minimum for any small variation in Ψ_n' consistent with the normalizing condition

$$\int \Psi_n^* \Psi_n \, d\tau = \text{const.}$$

Now for helium, as we have seen, the exact wave function is a function of the positions of both electrons, $\Psi = \Psi(\mathbf{r}_1, \mathbf{r}_2)$. Slater and Fock show that if we set for Ψ the (necessarily approximate) expression

$$\Psi = \psi_1(\mathbf{r}_1)\psi_2(\mathbf{r}_2), \tag{8}$$

and seek to make the expression

$$\iint \psi_1^*(\mathbf{r}_1)\psi_2^*(\mathbf{r}_2)H\psi_1(\mathbf{r}_1)\psi_2(\mathbf{r}_2) \, d\tau_1 \, d\tau_2$$

a minimum, we obtain Hartree's equations. *Thus the wave functions obtained by this method are the best that can possibly be obtained, as long as we use for the wave function of the whole atom the simple approximation* (8).

The *exact* wave function of the helium atom must be either symmetrical (parhelium) or antisymmetrical (orthohelium) in the space coordinates of the electrons. Therefore, unless the two electrons are in the same state ($\psi_1 = \psi_2$), the approximate wave function (8) bears no resemblance to the true wave function. We may, however, form symmetrical or antisymmetrical wave functions from (8) by writing

$$\Psi(\mathbf{r}_1, \mathbf{r}_2) = \psi_1(\mathbf{r}_1)\psi_2(\mathbf{r}_2) \pm \psi_1(\mathbf{r}_2)\psi_2(\mathbf{r}_1). \tag{9}$$

If for ψ_1, ψ_2 we use the Hartree wave functions in (9), we obtain a certain fairly good approximation. The Hartree functions are not the best that can possibly be chosen, however; to find these we must make $\iint \Psi^* H\Psi \, d\tau_1 \, d\tau_2$ a minimum, using for Ψ the expression (9). One obtains thereby a more complicated set of equations (Fock's equations), which may be written (in atomic units)

$$\left[\tfrac{1}{2}\nabla^2 + \frac{2}{r} + W - H_{22} - G_{22}(\mathbf{r})\right]\psi_1 = \pm[H_{12} + G_{12}(\mathbf{r})]\psi_2$$

$$\left[\tfrac{1}{2}\nabla^2 + \frac{2}{r} + W - H_{11} - G_{11}(\mathbf{r})\right]\psi_2 = \pm[H_{12} + G_{12}(\mathbf{r})]\psi_1, \tag{10}$$

where

$$H_{ik} = \int \psi_i^*\left(-\tfrac{1}{2}\nabla^2 - \frac{2}{r}\right)\psi_k \, d\tau,$$

$$G_{ik}(\mathbf{r}_1) = \int \frac{1}{r_{12}}\psi_i^*(\mathbf{r}_2)\psi_k(\mathbf{r}_2) \, d\tau_2.$$

These equations have at the present time been solved only for beryllium.†

An approximate solution for the states of helium where one electron is in an excited state of high quantum number has been given by L. P. Smith.‡

Fock‖ has given a method of treating the exchange interaction between the valence electrons of an atom and the core without solving the complete Fock equations. We suppose that the wave functions of the core electrons have been obtained (by Hartree's method or otherwise), and are, without spin coordinates, $\psi_i(\mathbf{r})$. We then write

$$\rho(\mathbf{r}, \mathbf{r}') = \sum_i \psi_i^*(\mathbf{r})\psi_i(\mathbf{r}'),$$

where the summation is over all the states of the core, i.e. over *half* the number of electrons. Fock then obtains for the Schrödinger equation of a valence electron

$$\frac{\hbar^2}{2m}\nabla^2\psi + [E - V(r) + A]\psi = 0, \tag{11}$$

where $V(r)$ is the ordinary (self-consistent) potential of the core, and A is the operator defined by

$$A\psi(\mathbf{r}) = e^2 \int \frac{1}{|\mathbf{r} - \mathbf{r}'|}\rho^*(\mathbf{r}, \mathbf{r}')\psi(\mathbf{r}')\, d\tau'.$$

1.4. *The statistical method of Thomas*†† *and Fermi.*‡‡ This is a method for finding the density of electrons in an atom or molecule, which is a good approximation under the following conditions:

(*a*) The system as a whole is in its lowest quantum state.

(*b*) In a volume so small that within it the change in the potential energy is small compared with the mean total energy of an electron, the number of electrons is large.

The condition (*b*) is approximately fulfilled only for heavy atoms.

It follows from (*b*) that the mean wave-length of the electrons is small compared with the distance within which the potential changes appreciably; this is the condition that the electrons may be treated by classical mechanics rather than by quantum mechanics. The laws of quantum mechanics are only used in the assumption that the electrons obey the exclusion principle (Fermi-Dirac statistics).

---

† Hartree (refs. on p. 46).
‡ *Phys. Rev.* **42** (1932), 176. See also *Handb. d. Phys.* **24**/1 (1933), 352.
‖ *Zeits. f. Phys.* **81** (1933), 195.
†† *Proc. Camb. Phil. Soc.* **23** (1927), 542.   ‡‡ *Zeits. f. Phys.* **48** (1928), 73.

We denote by· Φ the electrostatic potential in the system, and by

$$E_{\max} = -e\Phi_0$$

the maximum energy of any electron. Then, if $\mathbf{p}$ is the momentum of any electron at the point (x, y, z),

$$\frac{p^2}{2m} - e\Phi \leqslant -e\Phi_0,$$

and hence $p \leqslant \sqrt{\{2me(\Phi-\Phi_0)\}}.$

The right-hand side will be denoted by $p_{\max}$. Now, according to the Fermi-Dirac statistics, each volume of phase space $d\mathbf{p}d\mathbf{r}$ will contain $(2/h^3)\,d\mathbf{p}d\mathbf{r}$ electrons. The volume of momentum space corresponding to points which are occupied is $\frac{4\pi}{3}p_{\max}^3$, and hence the number of electrons per unit volume at the point (x, y, z) is, substituting for $p_{\max}$,

$$N = \frac{2}{h^3}\frac{4\pi}{3}[2me(\Phi-\Phi_0)]^{\frac{3}{2}}.$$

We obtain a differential equation for Φ by means of Poisson's equation

$$\nabla^2\Phi = -4\pi\rho = 4\pi Ne,$$

whence we obtain

$$\nabla^2\Phi = \alpha(\Phi-\Phi_0)^{\frac{3}{2}}, \qquad \alpha = 2^{1\frac{3}{2}}\pi^2 m^{\frac{3}{2}}e^{\frac{5}{2}}/3h^3. \tag{12}$$

For the tabulated solution of this equation, the reader is referred to the original papers.†

2. The electrostatic field in a metal

A metal contains electrons which are in some sense free to move, as we know from the most characteristic property of metals, their high electrical and thermal conductivities. Both for the alkali metals and for the monovalent metals silver and gold, there is direct experimental evidence‡ that the number of electrons contributing to the conductivity is of the order of magnitude of one per atom. These electrons form a kind of gas inside the metal; but this gas will be very different from a perfect gas, since the electrons must interact strongly with each other and with the positive ions from which they

---

† For a discussion of the Fermi-Thomas model as applied to the metallic state, cf. Lennard-Jones and Woods, *Proc. Roy. Soc.* A, **120** (1928), 727 (a two-dimensional metal); Slater and Krutter, *Phys. Rev.* **47** (1935), 559; and Feinberg, *Phys. Zeits. d. Sowjetunion*, **8** (1936), 416. These authors show that the use of this approximation does not lead to a satisfactory description of metallic cohesion.

‡ From the optical properties, cf. Chap. III, §§ 7 and 8.

have been stripped. It would seem at first sight that a gas of this kind must be too complicated for simple mathematical treatment, because of this very large interaction. In quantum mechanics, however, the effect of one electron on all the others can, to a large extent, be *averaged*; one can treat each electron as moving in the field of the positive ions, and in the *average* field of all the other electrons. This average field may be obtained, for instance, by the 'self-consistent field' method of Hartree.† Let the electrons have wave functions $\psi_1(\mathbf{r}), \psi_2(\mathbf{r}),...$, where $\mathbf{r} = (x, y, z)$. In calculating the self-consistent field, one treats each electron as though it produced a negative charge distribution $-e|\psi(x, y, z)|^2$ throughout the metal, the total charge distribution being obtained by summing over all the electrons. This charge distribution, together with the positive ions, produces the field in which the electrons are supposed to move. We denote the potential energy of an electron in this field by $V(x, y, z)$. The wave functions $\psi(\mathbf{r})$ are then solutions of Schrödinger's equation

$$\nabla^2\psi + \frac{2m}{\hbar^2}[E-V]\psi = 0. \tag{13}$$

In a crystalline metal in which the ions are at rest in their positions of equilibrium, the potential $V(x, y, z)$ is clearly periodic with the

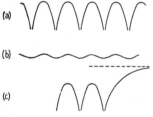

period of the crystal. If a line were drawn passing through the nuclei of atoms which were nearest neighbours in the crystal and V plotted against position along this line, the curve obtained would resemble that of Fig. 19 (a). If V were plotted against position along a parallel line which did not pass through the nuclei, a curve of the type (b) would be obtained. At the surface of the metal the periodicity would be broken; the potential would look, perhaps, as in Fig. 19 (c).

FIG. 19. Potential energy of electron in a metal.

(a) Along a line passing through the centres of the atoms.

(b) Along a line parallel to (a).

(c) At the surface of the metal.

This method, in which the electrons are treated as moving freely through a periodic static field, has often been criticized on the ground that it neglects an essential feature of the problem, namely the 'collisions' between the electrons. The answer to this criticism is as follows: A metal, just like any other solid, must be treated in

† Cf. § 1.2.

quantum mechanics as a gigantic molecule; to determine its properties one must find a wave function for the whole system of about 10^{23} electrons which are responsible for its electrical and chemical properties. It is impossible to find this wave function exactly, but a fairly good approximation to the true wave function is obtained by forming an antisymmetrical determinant (cf. Chap. VI, § 7) from wave functions of the type $\psi_1(\mathbf{r})$. The necessary improvements to this approximate wave function, which do in effect take account of 'collisions', are discussed in Chapter IV, and the effect of such collisions on the electrical conductivity in Chap. VIII, § 9.

3. The free-electron gas

3.1. *The model of Sommerfeld.* In the last section we have proposed a model for the discussion of the free electrons in a metal; we assigned to each electron a wave function $\psi(x, y, z)$, these wave functions being solutions of Schrödinger's equation for a particle moving in a certain field, which is periodic in x, y, z with the period of the lattice (cf. Fig. 19).

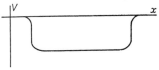

FIG. 20. Idealized potential of electron in metal.

This model was first applied to problems of conductivity by Sommerfeld,[†] who, to simplify the problem, assumed the average field acting on an electron at a given point to be zero, *and therefore $V(x, y, z)$ to be constant within the metal.* The potential of the field in which an electron moves plotted along a line through the metal will therefore appear as in Fig. 20, the potential rising sharply at the surface, but being constant within the metal.

This model is not, of course, accurate for any real metal, since there must in fact be a singularity in the potential at each nucleus. However, for the alkalis[‡] at any rate, the non-constant part of the potential has very little effect on certain of the physical properties.

We wish to emphasize again that the use of Sommerfeld's model does not mean, necessarily, that we neglect the interaction between the electrons; we neglect, firstly, that part of it which cannot be represented by a time average, as in § 1.2, and, secondly, we assume that the time average of the field at a point can be taken as zero over most of the space within a metal.

† *Zeits. f. Phys.* 47 (1928), 1.   ‡ Cf. Appendix I.

3.2. *One-dimensional metal.* Let us consider, as an illustration of the principles involved, a metal in which the electrons are free to move only in one direction, along the x axis, and in which the potential is as shown in Fig. 21, being zero within the metal and rising sharply to some finite value D at the boundaries. If for the moment we take D infinite, the wave functions $\psi(x)$ must vanish† at the boundaries $x = 0$ and $x = L$. The solutions of Schrödinger's equation

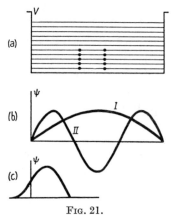

$$\frac{d^2\psi}{dx^2} + \frac{2m}{\hbar^2} E\psi = 0 \qquad (14)$$

which satisfy these boundary conditions are‡

$$\psi = \sqrt{\frac{2}{L}} \sin \frac{\pi n x}{L} \quad (n = 1, 2, 3, ...).$$

The wave functions for $n = 1$ and 3 are shown in Fig. 21 (b), curves I, II. The corresponding energy levels are given by

$$\frac{2m}{\hbar^2} E_n = \frac{\pi^2 n^2}{L^2}.$$

FIG. 21.

(a) Potential in metal; the horizontal lines show energy levels and the dots electrons.

(b) Wave functions of electron in metal ($D = \infty$).

(c) Wave function for a finite value of D.

Energy levels are shown by horizontal lines in Fig. 21 (a). These energy levels are, of course, extremely close together. If L is equal to 1 cm., the energy of the ground state is only 3×10^{-15} electron volts, and thus corresponds to an electron practically at rest. It is only when L is of atomic dimensions that the separation between the levels becomes appreciable.

The great advance made by Sommerfeld (loc. cit.) was the application of Pauli's exclusion principle, or its wave-mechanical form, the Fermi-Dirac statistics, to the calculation of the energy of free electrons in a metal. According to this principle, not more than *two* electrons (one with each spin direction) can ever be in the same state; that is to say, not more than two can have the same wave function. In an atom, only two electrons can be in the K level, two in each of the four L levels, and so on. Similarly, the free electrons in our one-

† If D is finite, ψ must tend to zero outside the metal, as shown in Fig. 21 (c): none of our conclusions are altered.

‡ The factor $\sqrt{(2/L)}$ is to ensure normalization.

dimensional metal cannot all go into the lowest state, in which they would be practically at rest. Only two can go into that state, and two into the next state, and so on. Thus some of the electrons will have considerable kinetic energy. At the absolute zero of temperature, when the metal has no thermal energy, they will all be in the lowest states allowed by the exclusion principle. If, therefore, there are N electrons (N being even), the states 1, 2, 3,..., $\frac{1}{2}N$ will be occupied (two electrons in each) and the states $\frac{1}{2}N+1$, $\frac{1}{2}N+2$ unoccupied. This is illustrated in Fig. 21, where the lower states are shown occupied and the upper states empty.

The wave-length corresponding to the highest occupied state is equal to $2L/N$. If the number of atoms is equal to the number N of electrons, and they are supposed equally spaced along the x-axis, then $2L/N$ is equal to twice the distance between the atoms. Now it is fundamental in wave mechanics that, if an electron has a wavelength of atomic dimensions, it has also a kinetic energy of atomic order of magnitude, i.e. of several electron volts. Thus the kinetic energy is much greater than the thermal energy of an atom at room temperature ($3kT \sim 0.071$ e.v.).

We have therefore reached the following conclusion: the electrons in a (one-dimensional) metal are not at rest, but move with kinetic energies lying between zero and a certain maximum of the order of magnitude of several electron volts. This maximum kinetic energy will be denoted by $E_{\max}$. The *mean* kinetic energy of the electrons will be called the 'mean Fermi energy'.

3.3. *Three-dimensional metal.* We consider now a piece of metal cut into the shape of a cube of side L. Schrödinger's equation for the wave functions $\psi(x,y,z)$ is, corresponding to (14),

$$\nabla^2\psi + \frac{2m}{\hbar^2}E\psi = 0, \tag{15}$$

and the boundary conditions are† that ψ must vanish on the planes $x = 0, x = L; y = 0, y = L; z = 0, z = L$. The wave functions are therefore

$$\psi = \sin\frac{\pi l_1 x}{L}\sin\frac{\pi l_2 y}{L}\sin\frac{\pi l_3 z}{L}, \tag{16}$$

where l_1, l_2, l_3 are integers. The corresponding energy values are

$$E = \frac{\hbar^2}{2m}\frac{\pi^2}{L^2}(l_1^2 + l_2^2 + l_3^2).$$

† Cf. p. 52, footnote.

It will be noticed that all these wave functions extend throughout the volume of the crystal.

As before, not more than two electrons can exist in any one state; but at the absolute zero of temperature all the electrons will be in the lowest states allowed by the exclusion principle. We require to know the *maximum* kinetic energy that any electron will have, i.e. the energy of the highest occupied state.

If $E_{\max}$ denotes this energy and N the number of electrons, then $\frac{1}{2}N$ (the number of occupied states) is equal to the number of sets of positive integers (l_1, l_2, l_3) such that

$$\frac{\hbar^2}{2m}\frac{\pi^2}{L^2}(l_1^2+l_2^2+l_3^2) < E_{\max}, \tag{17}$$

because each set of three integers satisfying (17) specifies an occupied state. The number of such sets of positive integers is equal to the volume of the eighth part of a sphere of radius

$$\sqrt{\left(\frac{2m}{\hbar^2}\frac{L^2}{\pi^2}E_{\max}\right)},$$

so that we have
$$\frac{1}{2}N = \frac{\pi}{6}\left[\frac{2m}{\hbar^2}\frac{L^2}{\pi^2}E_{\max}\right]^{\frac{3}{2}}. \tag{18}$$

Hence for $E_{\max}$, the maximum kinetic energy of any electron, we obtain, writing Ω, the volume, for L^3,

$$E_{\max} = \left(\frac{3}{\pi}\right)^{\frac{2}{3}}\frac{\pi^2\hbar^2}{2m}\left(\frac{N}{\Omega}\right)^{\frac{2}{3}}. \tag{19}$$

As explained above, this maximum energy is numerically equal to several electron volts. It depends only on the number of electrons per unit volume, N/Ω. If Ω_0 is the atomic volume in cubic Å. U. and n_0 the number of 'free' electrons per atom, formula (19) gives

$$E_{\max} = 36 \cdot 1(n_0/\Omega_0)^{\frac{2}{3}} \text{ electron volts.} \tag{19.1}$$

The following table gives the values for some metals calculated from (19), and assuming *one* electron per atom:

Element	Li	Na	K	Rb	Cs	Cu	Ag	Au
$E_{\max}$ (in electron volts)	4·74	3·16	2·06	1·79	1·53	7·10	5·52	5·56

The number of electronic states with energy less than any quantity E is, by an argument identical with that used in obtaining (18) above,

$$\frac{\Omega}{6\pi^2}\left[\frac{2m}{\hbar^2}E\right]^{\frac{3}{2}}.$$

By differentiating this expression with respect to E, we find that the number of states $N(E)\,dE$ with energies between E and $E+dE$ in a volume Ω is†

$$N(E)\,dE = \frac{\Omega}{4\pi^2}\left(\frac{2m}{\hbar^2}\right)^{\frac{3}{2}}\sqrt{E}\,dE. \tag{20}$$

The maximum number of *electrons* in the energy range dE is $2N(E)\,dE$, *two* electrons being in each state. The *total* (kinetic) energy of all the N electrons in a volume Ω is thus

$$2\int_0^{E_{\max}} N(E)E\,dE = \tfrac{3}{5}NE_{\max}. \tag{21}$$

The energy $\tfrac{3}{5}E_{\max}$ will be called the 'mean Fermi energy'.

The thermal energy of the electrons in a metal is shown in Chap. VI, § 2 to be small at ordinary temperatures, of order per electron $(kT)^2/E_{\max}$, and may for many purposes be neglected.

3.4. *Introduction of running waves.* In the preceding section we have described each electron by a standing wave reflected from the sides of the crystal; in the case of one dimension and a crystal of side L (cf. § 3.2) the wave function is

$$\sin kx \quad (k = \pi n/L),$$

where n is a positive integer. In our subsequent chapters it will be more convenient to use a wave function of the type

$$\psi = e^{ikx}, \tag{22}$$

representing motion in a definite direction. Actually such a wave function is only appropriate if our one-dimensional metal is bent round into a ring, so that a current can flow continuously. In this case, if the circumference of the ring is L, L must of course be equal to an integral number of wave-lengths, so that

$$|k| = 2\pi n/L;$$

k may, however, have positive or negative values, so that the number of states for an electron having *energy* in a given range is the same as before.

The number of states for which k lies between k' and $k'+dk'$ is clearly

$$L\,dk'/2\pi.$$

It will be seen that the wave function (22) is periodic with period

† This quantity is equal to the volume of phase-space, $4\pi\Omega p^2\,dp/h^3$, where p is the momentum.

L. In three dimensions it is convenient to take for the electronic wave functions

$$\psi = e^{i(\mathbf{kr})}, \tag{23}$$

and to assume ψ to be periodic along the three cubic axes with period L, where L^3 is equal to Ω, the volume of the metal. This assumption does not correspond to any physical property of the wave functions, but is convenient for obtaining the number of states within a given interval $d\mathbf{k}$. The number of states for which the vector $\mathbf{k}$ lies within the limits given by

$$k'_x < k_x < k'_x + dk'_x$$
$$k'_y < k_y < k'_y + dk'_y$$
$$\cdot \quad \cdot \quad \cdot \quad \cdot \quad \cdot \quad \cdot \quad \cdot$$

is easily seen to be

$$\frac{\Omega}{8\pi^3} dk'_x dk'_y dk'_z. \tag{24}$$

Each of these states can accommodate *two* electrons. Thus, if a metal contains N electrons per unit volume, the maximum value $k_{\max}$ of k which corresponds to an occupied state is given by

$$\tfrac{1}{2}N = \frac{1}{8\pi^3} \frac{4\pi}{3} k^3_{\max}. \tag{25}$$

It will easily be seen that the exclusion principle, expressed in the form (24), is equivalent to the statement that $2h^3 d\omega$ electrons may occupy each volume element $d\omega$ of *phase space*, the volume element being defined by

$$d\omega = dp_x dp_y dp_z dx dy dz,$$

where (p_x, p_y, p_z) is the momentum and (x, y, z) the position of an electron.

4. Motion of electrons in a periodic field

In Sommerfeld's simple treatment discussed in the last section, it is assumed that the time average of the field in which an electron moves is zero at any point within the metal. Each electron has, therefore, a wave function of the type (23), which represents motion parallel to the vector $\mathbf{k}$.

It is a better approximation to assume that the electrons move in the 'self-consistent field', of which the potential $V(x, y, z)$ is illustrated in Fig. 19. This field is periodic in x, y, z, with the periodicity of the crystal lattice. The wave functions will no longer have the simple form (23), but they will be modified by the field. The study of these modifications leads to an understanding of the reason why some crystals are insulators and others conductors, and of other properties of solids.

The wave function ψ of each electron satisfies the Schrödinger wave equation (13). Bloch has shown[†] that the solutions of this equation are of the form

$$\psi = e^{i(\mathbf{kr})} u_k(x, y, z), \tag{26}$$

where u is a function, depending in general on $\mathbf{k}$, *which is periodic in* x, y, z, *with the period of* V, i.e. with *the period of the lattice*. The wave function therefore represents, as before, a plane wave of wavelength $2\pi/k$ travelling in the direction of $\mathbf{k}$, but the wave is now modulated by the periodic field of the lattice.

A proof[‡] of Bloch's theorem will now be given.

4.1. Theorem of Bloch. We confine ourselves to the case of motion in one dimension. The Schrödinger equation then becomes

$$\frac{d^2\psi}{dx^2} + \frac{2m}{\hbar^2}(E-V)\psi = 0, \tag{27}$$

and, if a is the distance between the atoms, V is periodic with period a. Then, if $\psi(x)$ is a solution, $\psi(x+a)$ is clearly also a solution.

Now let $f(x)$, $g(x)$ be two independent real solutions of (27); then, since $f(x+a)$, $g(x+a)$ are also solutions, we may write

$$f(x+a) = \alpha_1 f(x) + \alpha_2 g(x)$$
$$g(x+a) = \beta_1 f(x) + \beta_2 g(x),$$

where α_1, α_2, β_1, β_2 are real functions of E. It follows that, if $\psi(x)$ is any other solution of (27) defined by

$$\psi(x) = Af(x) + Bg(x),$$

where A and B are constants (not necessarily real), then

$$\psi(x+a) = (A\alpha_1 + B\beta_1)f(x) + (A\alpha_2 + B\beta_2)g(x).$$

If now we choose the ratio $A : B$ so that

$$\begin{aligned} A\alpha_1 + B\beta_1 &= \lambda A, \\ A\alpha_2 + B\beta_2 &= \lambda B, \end{aligned} \tag{28}$$

where λ is a constant, then the function $\psi(x)$ so defined has the property

$$\psi(x+a) = \lambda\psi(x). \tag{29}$$

[†] *Zeits. f. Phys.* **52** (1928), 555; see also Kramers, *Physica*, **2** (1935), 483.

[‡] The proof is almost the same as that given in Whittaker and Watson, *Modern Analysis*, 3rd ed., p. 412 (Floquet's theorem).

If we eliminate A and B from equations (28), we obtain the quadratic

$$(\alpha_1-\lambda)(\beta_2-\lambda) = \alpha_2\beta_1, \tag{30}$$

which has two roots λ_1, λ_2. To each of these corresponds a value of the ratio $A : B$, and there are therefore *two*, and only two, independent functions $\psi_1(x)$, $\psi_2(x)$ having the property (29); for these

$$\psi_1(x+a) = \lambda_1\psi_1(x),$$
$$\psi_2(x+a) = \lambda_2\psi_2(x).$$

We now assume that $V(x) = V(-x)$; this will always be the case for fields in crystals if the origin is suitably chosen. It then follows from Schrödinger's equation (27) by writing $-x$ for x that $\psi_1(-x)$ is a solution. But from (29), writing $x-a$ for x,

$$\psi_1(x-a) = \frac{1}{\lambda_1}\psi_1(x),$$

and hence, writing $-x$ for x,

$$\psi_1(-(x+a)) = \frac{1}{\lambda_1}\psi_1(-x).$$

Hence $\psi_1(-x)$ has the property (29), with $\lambda = 1/\lambda_1$. But since only two solutions have this property, it follows that $\psi_1(-x)$ is a multiple of $\psi_2(x)$, and that

$$\lambda_1\lambda_2 = 1. \tag{31}$$

Now the roots of (30) will be real for some ranges of E, and complex for others; in the former case, by (31), we may write

$$\lambda_1 = e^{\mu a}, \qquad \lambda_2 = e^{-\mu a},$$

where μ is a real constant. In the latter case, since the coefficients of the equation are real, it follows that the roots are complex conjugates, and hence, by (31), we may write

$$\lambda_1 = e^{ika}, \qquad \lambda_2 = e^{-ika},$$

where k is a real number.

It follows from (29), therefore, that two solutions of (27) exist with the form either

$$e^{\pm\mu x}u_k(x) \quad \text{or} \quad e^{\pm ikx}u_k(x),$$

where $u(x)$ is periodic with the period a of the lattice. The solutions of the former type are not bounded; therefore they do not correspond to stationary states of electrons in the lattice. Thus there exist ranges of the energy E for which no electronic state exists. These forbidden ranges will be discussed further in the subsequent sections. Solutions of the second type, on the other hand, do correspond to stationary states, and are of the form demanded by Bloch's theorem.

It is clear that, for a given wave function, k is not uniquely defined, since k may be replaced by $k + 2\pi n/a$ without destroying the periodicity of $u(x)$.

It is often convenient to set $k = 0$ for any wave function which has the periodicity of the lattice; there will be a series of such wave functions which we may denote by $u_n(x)$. Then the general wave function will be of the form

$$\psi_{nk}(x) = e^{ikx}u_{nk}(x) \quad (-\pi/a \leqslant k \leqslant \pi/a),$$

where the functions $u_{nk}(x)$ are periodic with period a.

It will be noticed that, if $U(x)$ is any periodic function with the period of the lattice, the quantity

$$\lim_{L\to\infty} \frac{1}{L} \int_0^L \psi_{nk}^* U \psi_{n'k'} \, dx$$

will vanish unless $k = k'$. States for which $k = k'$ are said to 'combine'.

In three dimensions a proof of Bloch's theorem may be given on similar lines. It is obvious that the formula (24) for the number of states in the volume element $dk_x dk_y dk_z$ will still hold.

In the following sections we discuss the form of the wave functions under the following conditions:

I. The potential V is small compared with the total kinetic energy of the electrons; we call this the approximation of 'nearly free electrons'.

II. The atoms are a long way apart, so that the interaction between them is small; we call this the approximation of 'tight binding'.

III. The intermediate case, applicable to most real metals. This can be treated by a method due to Wigner and Seitz.

4.2. *Approximation of nearly free electrons.* In order to obtain a better understanding of the behaviour of electrons in a periodic field, we shall consider the case when V is everywhere numerically small compared with the average kinetic energy of the electrons. This case does not correspond to any real metal, since V becomes infinite at each nucleus; however, for the alkali metals it is probably true that $|V|$ is small over *most* of the volume of the metal.

We shall first consider a *one-dimensional* crystal—i.e. we shall suppose the electron to move along the axis of x only, and the wave function $\psi(x)$ to satisfy equation (27) above, where V is periodic with period a. We shall take our zero of energy so that the mean value of V, i.e. $\int_0^a V(x)\, dx$, vanishes. In the absence of the crystalline field ($V = 0$) the wave function can be written

$$\psi = e^{ikx} \quad (-\infty < k < \infty).$$

In the presence of the field Bloch's theorem tells us that it will have the form

$$\psi = e^{ikx}u(x), \tag{32}$$

where u is periodic with the period a of the lattice, and can therefore be expanded by Fourier's theorem in the form†

$$u(x) = \sum_{n=-\infty}^{\infty} A_n e^{-2\pi inx/a}. \tag{33}$$

If we insert (33) into (27) we obtain, on dividing by e^{ikx},

$$\sum_{n=-\infty}^{\infty} \left[-k_n'^2 + \frac{2m}{\hbar^2}(E-V) \right] A_n e^{-2\pi inx/a} = 0, \tag{34}$$

where

$$k_n'^2 = (k - 2\pi n/a)^2.$$

If V is small compared with E, we should expect that $u(x)$ would be nearly independent of x, and that all the coefficients A_n in (33) would thus be small compared with A_0. If we *assume* this to be the case, we may solve (29) approximately by neglecting small quantities of the second order, i.e. those involving the product of V and A_n ($n \neq 0$). Neglecting these terms, (34) becomes

$$\sum_{n=-\infty}^{\infty} \left[-k_n'^2 + \frac{2mE}{\hbar^2} \right] A_n e^{-2\pi inx/a} = \frac{2m}{\hbar^2} A_0 V.$$

Multiplying both sides of this equation by $e^{2\pi inx/a}$ and integrating from 0 to a, we obtain

$$\left[-k_n'^2 + \frac{2mE}{\hbar^2} \right] a A_n = \frac{2m}{\hbar^2} A_0 \int_0^a V e^{2\pi inx/a}\, dx.$$

If n is put equal to zero, this gives us for the energy of the state k

$$E = \hbar^2 k^2/2m,$$

the term on the right vanishing. The energy is therefore unchanged to the first order in V. For the nth coefficient we have

$$A_n = \frac{2mV_n}{\hbar^2} \frac{A_0}{k^2 - (k - 2\pi n/a)^2}, \tag{35}$$

† The minus sign in the exponential is used for convenience of notation in the later developments.

where
$$V_n = \frac{1}{a} \int\limits_0^a V(x) e^{2\pi i n x/a} \, dx. \tag{36}$$

The second approximation to E may easily be obtained by the usual methods† and gives

$$E = E_0 + \sum_{n \neq 0} \frac{|V_n|^2}{E_0 - E_n}, \tag{37}$$

with
$$E_0 = \hbar^2 k^2/2m, \qquad E_n = \hbar^2 k_n'^2/2m. \tag{38}$$

The assumption on which this calculation is based, that $A_n \ll A_0$ for all values of n other than zero, is seen from (37) to be justified if V is small, unless, for any positive or negative value of n, the denominator in (37) becomes small. If, however, k is equal or nearly equal to $\pi n/a$ for any integral value of n, the corresponding A_n will *not* be small compared with A_0, and the calculation is therefore invalid. In order, therefore, to obtain approximate values of the energy and approximate forms for the wave function in this case, we shall set for the wave function ψ

$$\psi = e^{ikx}(A_0 + A_n e^{-2\pi i n x/a}),$$

thereby neglecting all the other terms which really are small. Writing as before $k_n = k - 2\pi n/a$, the wave function becomes

$$\psi = A_0 e^{ikx} + A_n e^{ik_n x}.$$

Substituting this into the wave equation (27) above, we obtain

$$A_0 e^{ikx}\left[-k^2 + \frac{2m}{\hbar^2}(E-V)\right] + A_n e^{ik_n x}\left[-k_n^2 + \frac{2m}{\hbar^2}(E-V)\right] = 0.$$

If we now multiply this equation either by e^{-ikx} or by $e^{-ik_n x}$, and integrate from 0 to a, we obtain in the two cases, making use of (38),

$$\begin{aligned} A_0(E-E_0) - A_n V_n^* &= 0, \\ -A_0 V_n + A_n(E-E_n) &= 0, \end{aligned} \tag{39}$$

where V_n as before is defined by (36), and the mean value V_0 of V is taken to be zero. Eliminating A_0 and A_n from these equations, we obtain
$$(E-E_0)(E-E_n) - V_n V_n^* = 0,$$

a quadratic equation with the solutions

$$E = \tfrac{1}{2}[E_0 + E_n \pm \sqrt{\{(E_0-E_n)^2 + 4V_n V_n^*\}}]. \tag{40}$$

If k is not in the neighbourhood of $n\pi/a$, so that $|E_0 - E_n|$ is not small,

† Sommerfeld and Bethe, loc. cit. 386.

the square root may be expanded to give for (40)

$$\frac{1}{2}\left[E_0+E_n\pm\left(|E_0-E_n|+\frac{2V_nV_n^*}{|E_0-E_n|}-\dots\right)\right].$$

Since in this case we know that E is approximately equal to E_0, i.e. to $\hbar^2k^2/2m$, it follows that we must take the *negative* value of the square root when $E_0 < E_n$, i.e. when $k < n\pi/a$, and the positive value when $E_0 > E_n$.

If then we use equation (40) to plot E against k, a *discontinuity* in E occurs at $k = n\pi/a$ (cf. Fig. 22); since, for this value of k, $E_0 = E_n$,

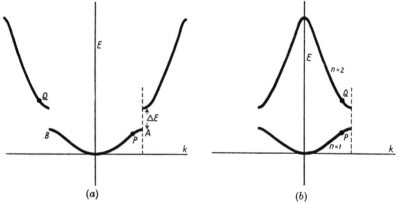

FIG. 22. Energy of an electron in a crystal plotted against the wave number k.
(a) With the variable k defined in the range $-\infty < k < \infty$.
(b) With k defined in the range $-\pi/a \leqslant k \leqslant \pi/a$ and a quantum number n to define the band.

the jump in the energy is given by

$$\Delta E = 2|V_n|. \tag{41}$$

Energies lying between the values $\dfrac{\hbar^2}{2m}\left(\dfrac{n\pi}{a}\right)^2 \pm |V_n|$ are thus impossible; solutions of the wave equation (27) with these values of the energy and with real k do not exist.

The value of A_n/A_0 may be obtained at once from (39). A_n/A_0 is small except in the neighbourhood of the critical wave-length. It will be noticed that, at the critical wave-length, $|A_0|$ and $|A_n|$ are equal, so that the wave function takes the form

$$A\cos(kx+\alpha) \quad (k = n\pi/a),$$

where α is a constant. The wave function therefore represents a *standing wave.* Further, it is easily seen from (39) that, as k passes through the critical wave-length, α changes by $\frac{1}{2}\pi$.

It will be noticed that the values of k for which the bands of forbidden energy occur are simply those for which Bragg reflection from the lattice takes place. The energy gaps ΔE determine the range of energies for which it is impossible for an electron wave to penetrate the lattice, so that, if the wave is incident from outside, *total reflection* occurs. The solution of the wave equation for energies in the forbidden range have actually *complex* values of k (cf. § 4.1).

In the foregoing description each wave function is described by a variable k which ranges from $-\infty$ to ∞. In § 4.1 we saw that an alternative mode of description is possible; we may describe the state by a variable k which ranges from $-\pi/a$ to π/a, and a suffix n to denote to which band of allowed energy values the state belongs. The energy plotted against this new variable k is shown in Fig. 22 (b). In the second zone ($n = 2$), if we write k_1 for the variable k in Fig. 22 (b), we have

$$k_1 = k \pm 2\pi/a.$$

Thus the points Q in the two figures represent the same state.

The states P and Q (in either figure) combine with each other in the sense of p. 59.

Note that the points A and B represent the same state.

One-dimensional motion in a periodic field which is not small

Various authors have investigated the solution of Schrödinger's equation in certain potential fields of special forms. Kronig and Penney[†] have considered the field

$$V = \text{const.} \quad na-b < x < na+b$$
$$= 0 \quad \text{elsewhere}$$

where a and b are constants such that $2b < a$. Morse[‡] has considered the case when

$$V = A \sin x,$$

in which case the solutions are Mathieu functions.[||] In all cases it is found that only for certain zones of E has the Schrödinger equation a solution of the form

$$\psi = e^{ikx}u(x)$$

with real k, k being complex in the forbidden zones.

Motion in three dimensions. In this section we shall confine ourselves to the discussion of a simple cubic lattice; the crystal structures of real metals are discussed in Chapter V.

† *Proc. Roy. Soc.* A, **130** (1931), 499.
‡ Morse, *Phys. Rev.* **35** (1930), 1310.
|| Cf., for example, Whittaker and Watson, *Modern Analysis*, Chap. XIV; Goldstein, *Trans. Camb. Phil. Soc.* **23** (1927), 303.

The wave functions may be written

$$\psi = e^{i(\mathbf{k} \mathbf{r})} u(x, y, z),$$

where u is periodic and may therefore be expanded as

$$u(x, y, z) = \sum_n A_n e^{2\pi i (\mathbf{n} \mathbf{r})/a},$$

and where $\mathbf{n}$ stands for the integers n_1, n_2, n_3 and the summation is over all positive and negative values of these integers. As for the case of one dimension (formula (35)), we find that $A_n \ll A_0$ unless

$$k^2 \simeq (\mathbf{k} - 2\pi \mathbf{n}/a)^2. \tag{42}$$

If, however, (42) is fulfilled, we obtain, by the same methods as before (cf. equation (40)),

$$E = \tfrac{1}{2}[E_0 + E_n \pm \sqrt{\{(E_0 - E_n)^2 + 4 V_n V_n^*\}}],$$

where $E_0 = \hbar^2 k^2 / 2m, \qquad E_n = \hbar^2 (\mathbf{k} - 2\pi \mathbf{n}/a)^2 / 2m,$

$$V_n = \iiint e^{2\pi i (\mathbf{n} \mathbf{r})/a} V \, dx dy dz \Big/ \iiint dx dy dz,$$

the integrations being over the unit cell. Therefore, just as for the case of one dimension, there is a discontinuity in E for certain values of $\mathbf{k}$, i.e. those given by the formula

$$k^2 = (\mathbf{k} - 2\pi \mathbf{n}/a)^2. \tag{43}$$

(43) may be written $(\mathbf{nk}) = \pi n^2/a,$ (44)

or, writing $\mathbf{k} = (k_x, k_y, k_z)$,

$$k_x n_1 + k_y n_2 + k_z n_3 = \pi (n_1^2 + n_2^2 + n_3^2)/a. \tag{45}$$

It will be noticed that the values of $\mathbf{k}$ satisfying (43) are the values for which Bragg reflection of the wave takes place. For the direction cosines of the normal to the set of parallel planes in the crystal with Miller indices (n_1, n_2, n_3) are n_1/n, n_2/n, n_3/n, where $n = \sqrt{(n_1^2 + n_2^2 + n_3^2)}$.

These are also the direction cosines of the normal to the plane (44). The condition for a Bragg reflection in the first order of a wave of wave-length λ, whose direction of propagation makes an angle θ with the normal to the planes (n_1, n_2, n_3), is

$$2a \cos \theta = n\lambda, \tag{46}$$

FIG. 23.

the distance between successive planes of the set (n_1, n_2, n_3) being a/n (cf. Fig. 23). Since $\cos \theta = (\mathbf{nk})/nk$, and the wave-length of the wave specified by $\mathbf{k}$ is

$$\lambda = 2\pi/k,$$

it follows that (45) and (46) are equivalent.

4.3. *The Brillouin zones.* It is convenient to introduce the idea of 'k-space'. We take Cartesian coordinates k_x, k_y, k_z; then any point in 'k-space' represents a state of the electron.

k-space is divided up into zones by planes given by equation (44) across which the energy is discontinuous. These zones have been discussed in detail for cubic structures by Brillouin,[†] and we shall call them 'Brillouin zones'. We show in Fig. 24 the first few zones for the two-dimensional cubic lattice. In Chapter V we discuss the zones for other structures.

We note that points such as P, Q in Fig. 24 are equivalent, i.e. that they have the same wave function. The zones cut off by the triangles BEC, AFD therefore form part of the same zone; one can go from any point in one zone to any point in the other without crossing a plane of energy discontinuity. By writing $k'_x = k_x \pm 2\pi/a$ respectively in the two half zones, one may describe the state of the electron by a new variable $\mathbf{k'}$ which varies continuously within the zone (Fig. 24(b); cf. pp. 63, 72).

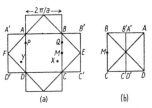

Fig. 24. Brillouin zones for a simple cubic lattice (two-dimensional).

(a) k-space.   (b) k'-space.

It is of interest to know the energy as a function of $\mathbf{k}$ near the bottom of the zones. At the bottom of the first zone

$$E = \frac{\hbar^2}{2m}(k_x^2 + k_y^2 + k_z^2).$$

At the bottom of a zone such as BEC in Fig. 24, i.e. near M, it follows from (40) and (41) that

$$E = \frac{\hbar^2}{2m}(\alpha k''^2_x + k_y^2 + k_z^2), \qquad \alpha = 1 + \frac{4E}{\Delta E}, \tag{47}$$

where ΔE is the energy gap and $k''_x = k_x - \pi/a$. This formula for α is correct to the first order in ΔE.

4.4. *Distance between the atoms large (approximation of tight binding).* In the preceding section we have assumed the potential V of the field to be small in comparison with the kinetic energy of the electrons, an assumption which is certainly not valid in any real metal, since at each nucleus the field becomes infinite. In this section, therefore, we shall make the opposite assumption; we shall

† *Die Quantenstatistik*, Berlin (1931).

consider a field of the type illustrated in Fig. 19, in which the atoms are relatively far away from each other, so that the behaviour of an electron in the neighbourhood of any one atom is influenced only to a small extent by the field of the other atoms. We shall find, just as for the case when V is small, that forbidden bands of energy may occur.

We consider first an electron moving in the field of an isolated atom in which its potential is $U(\mathbf{r})$. The Schrödinger equation is then

$$\nabla^2\phi + \frac{2m}{\hbar^2}(E-U)\phi = 0. \tag{48}$$

We consider first a solution $\phi(r)$ which corresponds to an s state, and is therefore spherically symmetrical. Let E_0 be the corresponding energy, and let the state be non-degenerate, so that no other wave functions have the same energy.

Consider now the wave function $\psi_k(\mathbf{r})$ of an electron in the field of the crystal as a whole. Let $\mathbf{r}_l$ denote the position of any atom; then since the influence of one atom on another is small, the wave function in the neighbourhood of the atom at $\mathbf{r}_l$ will be approximately $\phi(\mathbf{r}-\mathbf{r}_l)$, which is just the unperturbed wave function of an atom with its centre at $\mathbf{r}_l$. We therefore set for ψ

$$\psi = \sum_l c_l \phi(\mathbf{r}-\mathbf{r}_l) \tag{49}$$

where the coefficients c_l may be determined from the theorem of Bloch (§ 4.1), that ψ is the product of a periodic function and a factor† $e^{i(\mathbf{kr})}$; this gives $c_l = e^{i(\mathbf{kr}_l)}$, and hence for a state $\mathbf{k}$

$$\psi_k = \sum_l e^{i(\mathbf{kr}_l)}\phi(\mathbf{r}-\mathbf{r}_l), \tag{50}$$

the summation being over all lattice points of the crystal.

If H denotes the Hamiltonian for an electron in the crystal, given by

$$H = -\frac{\hbar^2}{2m}\nabla^2 + V,$$

then the energy of the electron with wave function ψ_k is

$$E = \int \psi_k^* H \psi_k \, d\tau \Big/ \int \psi_k^* \psi_k \, d\tau. \tag{51}$$

To evaluate these integrals we proceed as follows: we have from (50)

$$H\psi_k = \sum_l e^{i(\mathbf{kr}_l)} H\phi(\mathbf{r}-\mathbf{r}_l), \tag{52}$$

---

† The wave function (50) is not exactly of the required form, but is the best approximation to it of the form (49). Cf. Sommerfeld and Bethe, loc. cit. 394 et seq.

and, in each of the terms of the summation on the right-hand side, we may separate H into two terms, firstly the Hamiltonian of a free atom at the point $\mathbf{r} = \mathbf{r}_l$,

$$H_l = -\frac{\hbar^2}{2m}\nabla^2 + U(\mathbf{r}-\mathbf{r}_l),$$

and secondly the remaining terms in the potential, namely

$$H - H_l = V(\mathbf{r}) - U(\mathbf{r}-\mathbf{r}_l). \tag{53}$$

The functions U and V are illustrated in Fig. 25. We shall treat $H - H_l$ as a *small quantity*. Since by (48) $H_l\phi(\mathbf{r}-\mathbf{r}_l) = E_0\phi(\mathbf{r}-\mathbf{r}_l)$, it follows from (52) that

$$H\psi_k = E_0\psi_k + \sum_l e^{i(\mathbf{k}\mathbf{r}_l)}[H-H_l]\phi(\mathbf{r}-\mathbf{r}_l).$$

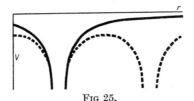

FIG 25.

Full line = potential in a free atom.
Dotted line = potential in a crystal lattice, plotted along a line
passing through two lattice points.

Hence, from (51),

$$E = E_0 + \frac{\int \psi_k^* \sum_l e^{i(\mathbf{k}\mathbf{r}_l)}[H-H_l]\phi(\mathbf{r}-\mathbf{r}_l)\, d\tau}{\int \psi_k^*\psi_k\, d\tau}. \tag{54}$$

Now $\int \psi_k^*\psi_k\, d\tau$ is equal to N, the number of atoms, if the overlap between the atoms is neglected. Thus, since the second term in (54) is itself small compared with E_0, we may write, to the first order in small quantities,

$$E = E_0 + \frac{1}{N}\int \psi_k^* \sum_l e^{i(\mathbf{k}\mathbf{r}_l)}[H-H_l]\phi(\mathbf{r}-\mathbf{r}_l)\, d\tau,$$

or, substituting from (50) for ψ_k^*,

$$E = E_0 + \frac{1}{N}\sum_m \left\{ \sum_l e^{i(\mathbf{k}\cdot\mathbf{r}_m - \mathbf{r}_l)}\int \phi^*(\mathbf{r}-\mathbf{r}_l)[H-H_m]\phi(\mathbf{r}-\mathbf{r}_m)\, d\tau \right\}.$$

The terms in the summation over m are all identical; we have therefore, taking the origin at $\mathbf{r}_m$ in each term, writing $-\mathbf{\rho}_l = \mathbf{r}_m - \mathbf{r}_l$, and substituting for $H-H_l$,

$$E = E_0 + \sum_l e^{-i(\mathbf{k}\rho_l)} \int \phi^*(\mathbf{r}-\boldsymbol{\rho}_l)\{V(\mathbf{r})-U(\mathbf{r})\}\phi(\mathbf{r})\, d\tau. \qquad (55)$$

$\boldsymbol{\rho}_l$ denotes the vector joining an atom at the origin to any other atom l.

We shall neglect all integrals in (55), except those for which the atoms are nearest neighbours; we may then write

$$\left. \begin{aligned} \int \phi^*(\mathbf{r})\{V(\mathbf{r})-U(\mathbf{r})\}\phi(\mathbf{r})\, d\tau &= -\alpha, \\ \int \phi^*(\mathbf{r}-\boldsymbol{\rho})\{V(\mathbf{r})-U(\mathbf{r})\}\phi(\mathbf{r})\, d\tau &= -\gamma, \end{aligned} \right\} \qquad (56)$$

it being clear that, for spherically symmetrical wave functions ϕ, the integral $-\gamma$ is the same for all nearest neighbours. We obtain thus for the energy

$$E = E_0 - \alpha - \gamma \sum_l e^{-i(\mathbf{k}\rho_l)}. \qquad (57)$$

Since $V-U$ is negative, α and γ are *positive*. They may both be of the order of some electron volts.

For the three cubic structures, we have from (57) if a is the side of the cube:

Simple cubic, six nearest neighbours.

$$\boldsymbol{\rho}_l = (\pm a, 0, 0), \quad (0, \pm a, 0), \quad (0, 0, \pm a),$$

$$E = E_0 - \alpha - 2\gamma(\cos k_x a + \cos k_y a + \cos k_z a). \qquad (58.1)$$

Body-centred cubic, eight nearest neighbours.

$$\boldsymbol{\rho}_l = (\pm \tfrac{1}{2}a, \pm \tfrac{1}{2}a, \pm \tfrac{1}{2}a),$$

$$E = E_0 - \alpha - 8\gamma \cos \tfrac{1}{2}k_x a \cos \tfrac{1}{2}k_y a \cos \tfrac{1}{2}k_z a. \qquad (58.2)$$

Face-centred cubic, twelve nearest neighbours.

$$\boldsymbol{\rho}_l = (0, \pm a, \pm a), \quad (\pm a, 0, \pm a), \quad (\pm a, \pm a, 0),$$

$$E = E_0 - \alpha - 4\gamma(\cos \tfrac{1}{2}k_y a \cos \tfrac{1}{2}k_z a + \cos \tfrac{1}{2}k_z a \cos \tfrac{1}{2}k_x a +$$
$$+ \cos \tfrac{1}{2}k_x a \cos \tfrac{1}{2}k_y a). \qquad (58.3)$$

The energy of an electron consists, therefore, of a constant term $E_0 - \alpha$, together with a term which depends on the wave number $\mathbf{k}$. This latter term varies between sharply defined limits, i.e. $\pm 6\gamma$ in the simple cubic case. *Thus for every state of an electron in the free atom there exists a band of energies in the crystal.* The integral γ, and hence the breadth of the band, will be greater the more the wave functions ϕ overlap. The breadth of the band for the inner electrons of the atoms of a crystal will therefore be exceedingly small ($\sim 2 \cdot 10^{-19}$ e.v. for the K electrons of sodium).†

For the valence electrons of real metals, however, the overlap is too great for the method to give accurate results; it is not therefore worth while to work out the integrals explicitly. We can, however, by this method obtain useful qualitative information about the dependence of the energy on $\mathbf{k}$.

For small values of $\mathbf{k}$ the energy for the simple cubic lattice is

$$E \simeq E_0 - \alpha - 6\gamma + \gamma(k_x^2 + k_y^2 + k_z^2)a^2, \qquad (59)$$

with similar forms for the other structures. The energy near the bottom of the band is therefore independent of the direction of motion, as for free electrons (cf. p. 65).

The number of states in the zone which corresponds to any non-degenerate atomic energy level is equal to N, the number of atoms, two electrons filling each state. This is clear from general considerations, or may be seen as follows: the formulae (58.1), (58.2), and (58.3) for the energy are periodic in $\mathbf{k}$, and thus only values of $\mathbf{k}$ lying within a certain polyhedron in k-space will define independent wave functions. The gradient of E normal to the planes bounding this polyhedron will vanish. For the simple cubic the polyhedron is clearly

$$-\pi/a \leqslant k_x \leqslant \pi/a, \quad \text{etc.,}$$

which shows that the polyhedron is a *cube* of volume $8\pi^3/a^3$. Hence, by equation (24), the number of states with wave vectors $\mathbf{k}$ within the cube is N.

In the case of the body-centred cubic, the polyhedron is a dodecahedron bounded by the planes

$$\pm k_x \pm k_y = 2\pi/a, \quad \text{etc.}$$

It is easily seen that the normal derivative of E vanishes on one of these planes; for, taking the plane $k_x + k_y = 2\pi/a$, we have

$$\text{grad}_k E = \frac{1}{\sqrt{2}}\left(\frac{\partial E}{\partial k_x} + \frac{\partial E}{\partial k_y}\right)$$

$$= \frac{a}{2\sqrt{2}} 8\gamma \cos \tfrac{1}{2}k_z a \sin \tfrac{1}{2}(k_x + k_y)a,$$

which vanishes on the plane considered. The dodecahedron is exactly the same as that considered in Chapter V for the case of

† Sommerfeld and Bethe, loc. cit. 398.

nearly free electrons; it has the volume $16\pi^3/a^3$, and since the atomic volume is $1/2a^3$, the number of states is equal to N.

For the face-centred cubic also, $\mathbf{k}$ must lie within the polyhedron considered in Chap. V, § 2.1, as may easily be verified.

We show in Fig. 26 two surfaces of constant energy for the face-centred cubic lattice, calculated from (58.3).

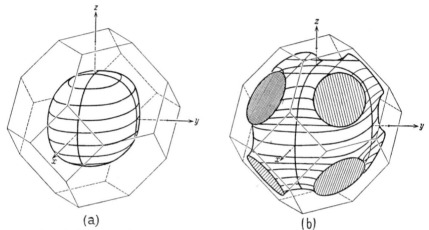

(a)   (b)

FIG. 26. Surfaces of constant energy in k-space, face-centred cubic lattice.†
(a) Zone nearly empty.   (b) Zone nearly full.

p states. We now extend our calculation to atomic p states. The atomic p state is triply degenerate; the wave functions are of the form

$$xf(r),\ yf(r),\ zf(r);$$

we shall denote them by

$$\phi_1(\mathbf{r}),\ \phi_2(\mathbf{r}),\ \phi_3(\mathbf{r}),$$

and the corresponding energy by E_1. We shall confine ourselves to a simple cubic lattice,‡ so that the position of any atom is given by the vector

$$\mathbf{r}_l = a\mathbf{l},$$

where $\mathbf{l} = (l_1, l_2, l_3)$ and l_1, l_2, l_3 are integers. For the simple cubic lattice the wave functions ϕ_1, ϕ_2, ϕ_3 do not combine,‖ so that we may set for the wave function in the field of the whole crystal, similarly to (50),

$$\psi_{nk} = \sum_l e^{ia(\mathbf{lk})}\phi_n(\mathbf{r}-a\mathbf{l})\quad (n = 1, 2, 3).$$

† From Sommerfeld and Bethe, loc. cit. p. 401.
‡ For the other cubic structures the calculation is much more complicated, cf. Sommerfeld and Bethe, loc. cit. 404.
‖ Cf. p. 59.

Equation (55) follows just as before with $\rho_l = al$. On the other hand, since the functions $\phi(\mathbf{r})$ are no longer spherically symmetrical, the integrals for nearest neighbours are no longer equal. We obtain therefore from (55)

$$E = E_1 - \alpha_1 + 2\gamma_1 \cos ak_x + 2\gamma_1'(\cos ak_y + \cos ak_z), \qquad (60)$$

where

$$\left.\begin{aligned}
\alpha_1 &= -\int \phi_1^*(\mathbf{r})[V-U]\phi_1(\mathbf{r})\,d\tau, \\
\gamma_1 &= \int \phi_1^*(x+a,y,z)[V-U]\phi_1(x,y,z)\,d\tau, \\
\gamma_1' &= \int \phi_1^*(x,y+a,z)[V-U]\phi_1(x,y,z)\,d\tau.
\end{aligned}\right\} \qquad (60.1)$$

The signs and orders of magnitude of these integrals are of interest: since $V-U$ is negative (cf. Fig. 25), α_1 is positive. In the integrand in γ_1, midway between the atoms where the overlap is largest, $\phi_1(x+a)$ and $\phi_1(x)$ have opposite signs, and hence γ_1 is positive. In the integrand in γ_1', on the other hand, midway between the atoms, $\phi_1(y+a)$ and $\phi_1(y)$ have the same sign, so that γ_1' is negative. Moreover, both $\phi_1(y)$ and $\phi_1(y+a)$ vanish along a plane passing through both atoms concerned, so that we may assume

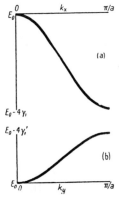

FIG. 27. Energies as functions of $\mathbf{k}$ in the p band.
(a) Plotted against k_x.
(b) Plotted against k_y.

$$|\gamma_1'| < \gamma_1.$$

The energy in the p band is shown in Fig. 27(a) plotted along the k_x-axis, and in Fig. 27 (b) along the k_y-axis. In Fig. 28 we give a contour diagram showing the energies in the $(k_x k_y)$-plane. The energies of the points marked 'max' is $E_0 - 4\gamma_1'$ and of the points marked 'min' $E_0 - 4\gamma_1$. A section in the $(k_z k_x)$-plane is exactly similar to Fig. 28, the k_z- replacing the k_y-axis. In the complete cube bounded by the planes $k_x = \pm\pi/a$, $k_y = \pm\pi/a$, $k_z = \pm\pi/a$, there are therefore four maxima and two minima. The minima of the band occur at $k_x = \pm\pi/a$. The whole of this band was derived from atomic p states whose nodal planes are perpendicular to the x-axis. Similar bands having the same energies arise from the other two p states.

Fig. 27 should be compared with Fig. 22, obtained using the approximation of nearly free electrons. Note that the diagonal planes of energy discontinuity do not arise in this approximation, but would occur if second nearest neighbours were considered. Otherwise the surfaces of constant energy, etc., are similar.

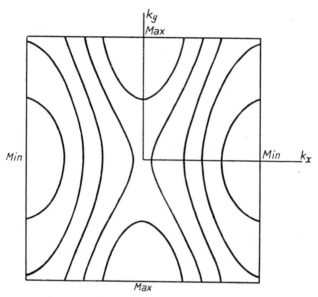

FIG. 28. Lines of constant energy in the p band.

s and p states degenerate. In the preceding sections we investigated the form of the zones arising from the s and p states of an electron in the free atom. The approximations used are clearly valid only if the width, 12γ, of the band is small compared with the separation $E_1 - E_0$ between the atomic s and p states. In this section we investigate what happens if this is not the case, and $E_1 - E_0$ is comparable with γ. We confine ourselves again to the case of the simple cubic.

As before, we denote by $\phi_0(\mathbf{r})$ the wave function for an s state of the atom and by ϕ_1, ϕ_2, ϕ_3 the wave functions for the p states. We form the wave functions

$$\psi_{nk} = \sum_l e^{ia(\mathbf{lk})}\phi_n(\mathbf{r}-a\mathbf{l}) \quad (n = 0, 1, 2, 3)$$

as before. We now set for the wave function of an electron in the whole crystal

$$\Psi_n(\mathbf{k}; \mathbf{r}) = \sum_m a_{nm}(\mathbf{k})\psi_{mk}(\mathbf{r}). \tag{61}$$

We then have, multiplying the equation $(H-E)\Psi_n = 0$ by each of the functions ψ_{nk}^* in turn, integrating over all $\mathbf{r}$ and eliminating the coefficients $a_{nm}(\mathbf{k})$,

$$|H_{nm} - E\delta_{nm}| = 0, \tag{62}$$

where

$$H_{nm} = \int \psi_n^* H\psi_m \, d\tau \, / A_n A_m,$$

$$A_n^2 = \int \psi_n^* \psi_n \, d\tau.$$

This is the usual secular equation of quantum mechanics.

The diagonal elements $(n = m)$ have already been evaluated (equations (56), (60.1)); we write them

$$H_0, \; H_1, \; H_2, \; H_3.$$

The non-diagonal elements are small in the sense of our approximation; we may therefore put $A_n^2 = A_m^2 = N$ (number of atoms) and obtain

$$H_{nm} = 0 \text{ if neither } n \text{ nor } m \text{ is zero,}$$

$$H_{0n} = \frac{1}{N} \sum_{ll'}' e^{-ia(\mathbf{k} \cdot \mathbf{l} - \mathbf{l'})} \int \phi_0^*(\mathbf{r} - a\mathbf{l})[H - H_l]\phi_n(\mathbf{r} - a\mathbf{l'}) \, d\tau.$$

Neglecting all but nearest neighbours, substituting from (53) for $H - H_l$, and using the equations

$$\int \phi_0^*(x+a, y, z)[V - U]\phi_1(x, y, z) \, d\tau$$
$$= -\int \phi_0^*(x-a, y, z)[V - U]\phi_1(x, y, z) \, d\tau$$

and

$$\int \phi_0^*(x+a, y, z)[V - U]\phi_n(x, y, z) \, d\tau = 0 \quad (n \neq 1),$$

we obtain

$$H_{0n} = 2i\beta \sin k_n a \quad (n = 1, 2, 3), \tag{63}$$

where

$$\beta = \int \phi_0^*(x+a, y, z)[V - U]\phi_1(x, y, z) \, d\tau.$$

We may expect β to have approximately the same numerical value as γ.

The secular equation (62) for the energy becomes

$$\begin{vmatrix} H_0(\mathbf{k}) - E & 2i\beta \sin k_x a & 2i\beta \sin k_y a & 2i\beta \sin k_z a \\ -2i\beta \sin k_x a & H_1(\mathbf{k}) - E & 0 & 0 \\ -2i\beta \sin k_y a & 0 & H_2(\mathbf{k}) - E & 0 \\ -2i\beta \sin k_z a & 0 & 0 & H_3(\mathbf{k}) - E \end{vmatrix} = 0, \tag{64}$$

which is a quartic for E. The solutions will give the energies in the s and the three p bands as functions of $\mathbf{k}$.

We consider first the energies for states on the k_x-axis $(k_y = k_z = 0)$. Equation (64) becomes

$$2E = H_0 + H_1 \pm [(H_0 - H_1)^2 + 16\beta^2 \sin^2 ak_x]^{\frac{1}{2}}. \tag{65}$$

Now H_0 and H_1 are the energies corresponding to an s or p wave function already calculated; these are shown in Fig. 29 (a) and (b) by the dotted lines. Two cases may arise; either the curves do not cross, Fig. 29 (a), or else they cross, Fig. 29 (b). In either case the full lines show the energy E calculated from formula (65) with a suitable value of β. It should be noticed, firstly, that the curves do not cross, and, secondly, that they lose their symmetrical form, and approach much more nearly the form given by the theory of nearly free electrons (Fig. 22).

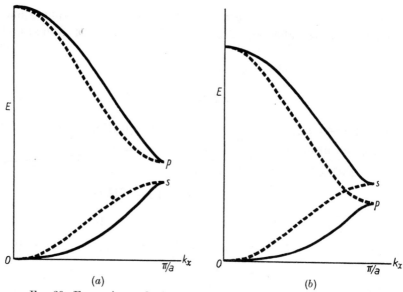

(a)   (b)

FIG. 29. Energy in s and p bands; full lines with interaction, dotted lines without interaction.

In case (b) it is noteworthy that the highest state of the lower band is built completely of p wave functions. Considering again the states with $k_y = k_z = 0$, a short calculation gives from equations (61), (63), (65) for the normalized wave functions in the lower and upper bands

$$\Psi_0^c(k_x, 0, 0) = \cos \tfrac{1}{2}\theta\,\psi_{0k} + i\sin \tfrac{1}{2}\theta\,\psi_{1k},$$
$$\Psi_1^c(k_x, 0, 0) = \sin \tfrac{1}{2}\theta\,\psi_{0k} - i\cos \tfrac{1}{2}\theta\,\psi_{1k},$$

where $\cot \theta = [H_1(\mathbf{k}) - H_0(\mathbf{k})]/4\beta \sin k_x a.$

In the case shown in Fig. 29 (a) we see that $\cot \theta$ remains positive throughout the whole range $\pi/a \geqslant k_x \geqslant 0$. When

$$k_x = 0, \quad \cot \theta = +\infty;$$

it then falls to some minimum value greater than zero for some value of k_x between $\frac{1}{2}\pi/a$ and π/a, and then rises to $+\infty$ when $k_x = \pi/a$. Thus $\sin\frac{1}{2}\theta$, which is the coefficient for the p wave function in the wave function for the lower zone, increases from zero to some maximum value and becomes zero again at π/a. In the case (b), however, $\cot\theta$ varies from $+\infty$ to $-\infty$, becoming equal to zero at the point where the two curves in Fig. 29 cross; thus $\sin\frac{1}{2}\theta$ varies from 0 to 1. The wave function therefore changes uniformly from an s to a p function within a given band.

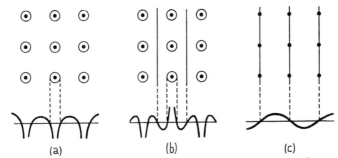

(a)   (b)   (c)

Fig. 30. Nodes and amplitude of wave functions of a metallic electron.

Near the bottom of the first band it is clear that the coefficient of the p wave function is

$$\sin\tfrac{1}{2}\theta = \text{const.}\, k$$
$$= \text{const.}\, \sqrt{E}. \tag{66}$$

It is interesting to consider the form of the wave functions at certain special points in the zones. We shall consider the case where the principal quantum number is 2 (e.g. lithium).

We show in Fig. 30 each wave function in two ways:

(1) by plotting ψ along a line in the (100) direction passing through the nuclei.

(2) by showing, in a (100) plane passing through the nuclei, the nodes of the wave function.

We show:

(a). The lowest s state ($k_x = k_y = k_z = 0$).

(b) and (c). The two states for $k_x = \pi/a$, $k_y = k_z = 0$. As already emphasized, we cannot say, without a detailed calculation of their energies, which is the highest state of the first zone and which the lowest state of the second zone. In the case (*a*) of Fig. 29, 30 (*b*) will have the lower energy; in the case 29 (*b*), 30 (*c*) will be the lower.

4.5. *Exact wave functions; the method of Wigner and Seitz.* Wigner and Seitz† have shown how to calculate, with fair accuracy, the wave function corresponding to the state of lowest energy in any metal for which the field of the ion is known. The method has at present been applied only to metals having the face-centred and body-centred cubic structures. In lattices with these structures we can fill up the whole of space with polyhedra, one surrounding each

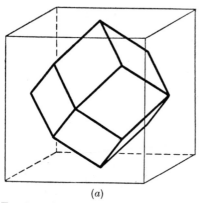
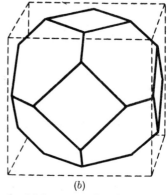

(*a*)   (*b*)

Fɪɢ. 31. Atomic polyhedra surrounding an atom, for (*a*) face-centred and (*b*) body-centred cubic structures. The cubes show the unit cells in either case.

atom, in the following way: for the face-centred structure we must draw planes bisecting the lines joining each atom to its nearest neighbours, obtaining thus a dodecahedron surrounding each atom (Fig. 31 (*a*)); for the body-centred structure we bisect the lines joining an atom to its nearest and next nearest neighbours, and obtain thus a truncated octahedron (Fig. 31 (*b*)). We call these 'atomic polyhedra'.‡

Near the boundary of each atomic polyhedron the field will be small; near the middle it will be spherically symmetrical. We shall take it to be spherically symmetrical within the whole of each polyhedron, within which we shall denote the potential energy of an electron by $V(r)$, r being the distance from the centre of the polyhedron. For reasons to be discussed in Chapter IV, it is a good

---

† *Phys. Rev.* **43** (1933), 804; **46** (1934), 509, referred to as loc. cit. I and II. References to further work are given on p. 134.

‡ Wigner and Seitz call them '*s*-polyhedra'.

approximation to take for $V(r)$ the potential of the free singly charged positive *ion*.

Now the wave function ψ for the lowest state is periodic in (x, y, z) with the period of the lattice (§ 4.1); further (cf. Fig. 30), it is symmetrical about any nucleus. Hence, on the boundary of any atomic polyhedron, it will satisfy the condition

$$\frac{\partial \psi}{\partial n} = 0, \qquad (67)$$

$\partial/\partial n$ denoting differentiation normal to the bounding plane. Since, however, the polyhedra approximate closely to spheres, it will be a good approximation to apply the boundary condition (67) over the surface of a sphere of equal volume. We call such spheres 'atomic spheres'; if the radius of such a sphere is r_0, we have

$$\frac{4\pi}{3} r_0^3 = \Omega_0 = \text{atomic volume.}$$

The boundary condition (67) thus becomes

$$\left(\frac{\partial \psi}{\partial r}\right)_{r=r_0} = 0, \qquad (68)$$

and, to obtain the wave function within each sphere, we have to solve the Schrödinger equation

$$\frac{1}{r^2} \frac{d}{dr}\left(r^2 \frac{d\psi}{dr}\right) + \frac{2m}{\hbar^2}[E - V(r)]\psi = 0 \qquad (69)$$

subject to the boundary condition (68). E is the corresponding energy of an electron in its lowest state in the lattice field.

If equation (69) is integrated numerically for given E, the value r_0 of r for which $\partial \psi/\partial r$ vanishes may be determined. Calculations have been carried out on these lines by Wigner and Seitz[†] for sodium, by Seitz[‡] for lithium, by Fuchs[||] for copper and silver. In the calculations of Fuchs for copper and silver, $V(r)$ was taken to include the exchange interaction of the valence electron with the inner shells, and therefore denotes an operator (cf. § 1.3).

Fig. 32 shows the wave functions for silver obtained for different values of E. In the free atom the valence electron is in the $5s$ state; the wave function has therefore four zeros, of which the three outermost are shown. In the metal the wave function must also have four zeros; thus we may only consider values of $\partial \psi/\partial r$ at distances

---

† Loc. cit.   ‡ *Phys. Rev.* **47** (1935), 400.
|| *Proc. Roy. Soc.* A, **151** (1935), 585.

from the origin greater than that of the fourth zero. As the figure shows, there will at first be two values r_0 of r for which $\partial\psi/\partial r$ vanishes;

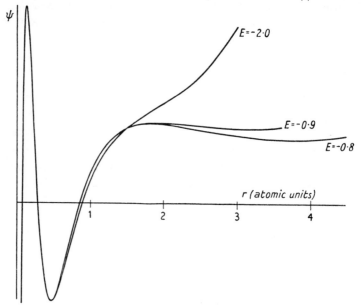

Fig. 32. Wave functions in the field of a silver ion for different values of the energy E (in Rydberg units).

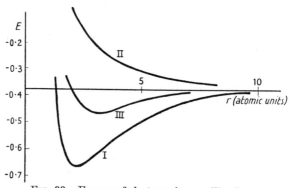

Fig. 33.   Energy of electrons in metallic silver.
I. Energy E_0 of lowest state.
II. Mean Fermi energy E_F.
III. Mean energy E_0+E_F of electrons.
All energies in Rydberg units (13·52 e.v.).

as the energy is decreased they move closer together and finally disappear. The energy E_0 of the electron in its lowest state, plotted against r_0, therefore appears as in Fig. 33, curve I.

The mean energy of the electrons in the lattice is not of course equal to E_0, because only two electrons can be in the lowest state. To obtain the mean energy we must add the 'Fermi energy' E_F (cf. § 3). Methods of calculating E_F are given below. In Fig. 33 E_F and the mean energy $E_0 + E_F$ are shown also.

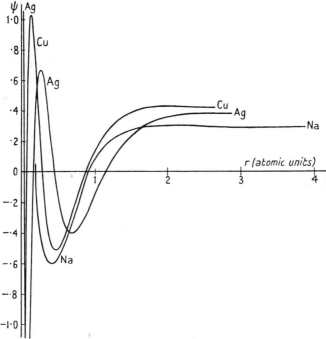

FIG. 34. Wave functions of electrons in lowest states in metals; the wave functions are normalized so that $\int \psi^2 r^2\, dr = 1$. The values of r_0 for which these are calculated are Cu 2·65, Ag 2·9, Na 3·88. (atomic units, 0·54 A).

In Chapter IV it will be shown that $E_0 + E_F$ is approximately equal to the binding energy per atom of the metal. Thus, for the crystal in equilibrium, the value of the atomic radius r_0 will be given approximately by the minimum of the curve III in Fig. 33. It will be noticed that this is not far from the minimum of curve I. Hence, for the atomic radius of the actual crystal, the maximum and minimum of ψ are close together. *The wave function ψ, therefore, for the actual atomic radius is rather flat except in the middle of the atom.* This is illustrated in Fig. 34, which shows the actual wave functions for sodium, copper, and silver. This flatness of the wave function is the reason why the approximation of § 3 (neglect of the periodic field) gives good results in certain cases. Moreover, since the volume within which the wave function is not flat is relatively small, the charge density in the flat region is almost exactly e/Ω_0, where Ω_0 is the atomic volume. This is illustrated in Fig. 35 for lithium.

It is of interest to discuss further the reason for the minimum in the energy curve I of Fig. 33. We may write the energy

$$E = \frac{\hbar^2}{2m} \int |\mathrm{grad}\,\psi|^2 \, d\tau + \int \psi^* V \psi \, d\tau, \qquad (70)$$

the first term representing the kinetic energy of the electron and the second term the potential energy. As the atomic radius is decreased, the potential

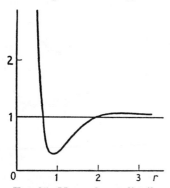

energy decreases, because the electron comes nearer to the positively charged nucleus. On the other hand, if the atomic radius is decreased too much, the 'flat' part of the wave function (Fig. 34) occupies relatively less volume, which means that the kinetic energy increases. The rise in the energy is thus due to increasing kinetic energy.

In the neighbourhood of the minimum the dependence of the two terms on r_0 can be estimated very roughly as follows: a change in r_0 has very little effect upon $\partial\psi/\partial r$ in the middle of the atom, as Fig. 32 shows; thus the only reason for a variation with r_0 of the *kinetic energy* is that a change in the atomic volume alters the absolute magnitude of ψ, since $\int \psi^* \psi \, d\tau$ over the atomic volume is con-

FIG. 35. Mean charge distribution in units of e/Ω_0 within a given atomic sphere for lithium (r in atomic units).

stant. Since, moreover, ψ is constant over most of the volume, the kinetic energy will be proportional to $1/r_0^3$. On the other hand, for the same reason, the potential energy will be a multiple of $-e^2/r_0$.

The method given above can be used to obtain only the wave function for the lowest state, for which the wave vector $\mathbf{k}$ is zero. Denoting this wave function by $\psi_0(r)$ and its energy by E_0, a fair approximation to the wave function for higher states will be, within any one atomic sphere,

$$\psi_k(\mathbf{r}) = e^{i(\mathbf{k}\mathbf{r})}\psi_0(r), \qquad (71)$$

provided that $\mathbf{k}$ lies within the first Brillouin zone not too near its boundaries. Using the approximation (71) for the wave function, we may calculate the energy from (70), and obtain

$$E_k = E_0 + \hbar^2 k^2/2m; \qquad (72)$$

to this approximation the extra energy of the state $\mathbf{k}$ is just the same as it would be in the absence of a periodic field. These approximate formulae will be used frequently in discussing the experimental evidence in the theory of conductivity, and elsewhere.

From (72) we have for the mean Fermi energy,

$$E_F = \int \frac{\hbar^2 k^2}{2m} k^2 \, dk \bigg/ \int k^2 \, dk,$$

the integrations being from zero to the value k_{max} of the wave number of the highest occupied state. This gives

$$E_F = \frac{3}{5} \frac{\hbar^2}{2m} k_{max}^2, \tag{73}$$

or, from (25), in terms of r_0 (cf. p. 77),

$$E_F = \frac{3}{10} \left(\frac{9\pi}{4}\right)^{\frac{2}{3}} \frac{\hbar^2}{mr_0^2}. \tag{74}$$

This formula is used in plotting E_F in Fig. 33.

More exact formulae for the energy E_k as a function of the wave number **k** may be obtained in several ways. For $k = 0$, $\partial E/\partial k$ vanishes; we shall write

$$\alpha = \frac{m}{\hbar^2} \left(\frac{\partial^2 E}{\partial k^2}\right)_{k=0},$$

so that, for small k, E_k may be written

$$E_k = E_0 + \alpha \hbar^2 k^2 / 2m. \tag{75}$$

It may be shown (cf. Chap. III, § 6) that

$$\alpha = 1 + \frac{2}{m} \sum_i \frac{|p_{i0}|^2}{E_i - E_0}, \tag{76}$$

where

$$p_{i0} = \frac{\hbar}{i} \int \psi_i^* \operatorname{grad} \psi_0 \, d\tau.$$

Here the summation is over all states i (with energy E_i and wave function ψ_i) which combine with the lowest state ψ_0. Since ψ_0 is spherically symmetrical, the vector $\operatorname{grad} \psi_0$ may be taken in any direction.

Wigner and Seitz† have transformed (76) into

$$\alpha = \Omega_0 |\psi_0(r_0)|^2 - \frac{\hbar}{m} \sum_i \frac{p_{i0}}{E_0 - E_i} \int x \psi_0 \frac{\partial \psi_i}{\partial n} \, dS,$$

where the integral is over the surface of the atomic polyhedron. The advantage of this form is that the second term gives virtually no contribution from the states with quantum number lower than the

† Loc cit. II.

ground state (i.e. the X-ray levels). For sodium Wigner and Seitz find the second term to be small, so that

$$\alpha \simeq \Omega_0 |\psi_0(r_0)|^2. \tag{77}$$

For lithium Seitz has calculated α from the exact formula (76). For copper and silver Fuchs has calculated it from (77). For these metals the values of α are as follows:

Metal	r_0 (atomic units)	α	r_0 (observed)
Lithium	3·21	0·744	3·21
	3·62	0·810	
Sodium	3·67	1·08	3·9
	4·05	0·99	
	4·74	0·89	
Copper	2·65	1·10	2·53 (3·12 calc.)
	3·14	0·983	
Silver	2·91	1·192	3·03
	3·30	1·083	

Formulae of the type (75) for the energy are probably a fair approximation in the whole of the first Brillouin zone, except near the planes of energy discontinuity. On the first plane of energy discontinuity, at the point nearest to the origin (cf. Fig. 24), the wave function will take the form illustrated in Fig. 30(c), with nodes passing through the origin. As an approximation to this wave function, Slater[†] and Millman[‡] take within any atomic polyhedron

$$\psi = f(r)\cos\theta,$$

where f satisfies

$$\frac{d^2f}{dr^2} + \frac{2}{r}\frac{df}{dr} + \frac{2m}{\hbar^2}\left(E - V - \frac{2}{r^2}\right)f = 0,$$

and the boundary condition $f'(r_0) = 0$. They call the corresponding state 'the lowest p state'.

Fig. 36 shows the energies of the lowest s and p states calculated for Li and Na; the highest states are also shown, i.e. those whose wave functions vanish for $r = r_0$. For lithium the energy interval E_B between the lowest s and p states turns out to be only 4 e.v. For completely free electrons it would be $h^2/8md^2$, where d is the distance between the (111) planes in lithium; this gives 6·5 e.v. In lithium, therefore, the density of states is greater than for free electrons;

† Phys. Rev. **45** (1934), 794; Rev. Mod. Phys. **6** (1934), 210.
‡ Phys. Rev. **47** (1935), 286.

this corresponds to the fact that the values of α in the above table are less than unity.

Krutter† has carried out similar calculations for d electrons, and has applied them to copper. He finds that the d band overlaps the s band, but that in copper all the ten d states are occupied, giving one electron per atom in the s band.

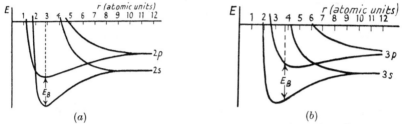

FIG. 36. Energies of states in crystal lattices; (a) lithium, (b) sodium.

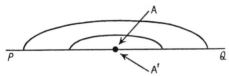

FIG. 37. Energy surfaces at the bottom of a higher zone.

A convenient formula may be obtained for the energy as a function of $\mathbf{k}$ near the bottom of any higher zone which is separated from the next zone below it by a *small* energy gap ΔE. In Fig. 37 let PQ represent a plane of energy discontinuity in k-space, and A the point of lowest energy in the zone lying above PQ. If the x-axis is perpendicular to the plane PQ, we may write for the energy in the neighbourhood of A

$$E = \frac{\hbar^2}{2m}\{\alpha_1 k_x^2 + \alpha_2(k_y^2 + k_z^2)\}.$$

Now formula (76) may be applied to calculate α_1, giving

$$\alpha_1 = 1 + \frac{2}{m}\sum_i \frac{|(p_x)_{i0}|^2}{E_i - E_0},\tag{78}$$

where E_0 refers to the state A and E_i to all other states which combine with it. Now, if ΔE, the energy gap from A' to A, is small, all terms in the summation in (78) may be neglected except that which refers to the pair of states A', A, since all others will have a much

† *Phys. Rev.* **48** (1935), 664.

larger value of $|E_i - E_0|$. Moreover, the wave functions at A' and at A will have the forms

$$u(\mathbf{r})e^{i(k_y y + k_z z)} \frac{\sin}{\cos}\left(\frac{\pi x}{d}\right),$$

where d is the distance between the successive reflecting planes in the crystal which give rise to the plane of energy discontinuity considered, and $u(\mathbf{r})$ is periodic in $\mathbf{r}$ with the period of the lattice. It will probably not introduce any serious error if we take $u(\mathbf{r})$ constant over most of the volume of the crystal (cf. Fig. 34). We thus obtain

$$(p_x)_{i0} = h/2di,$$

and hence

$$\alpha_1 = 1 + h^2/2md^2\,\Delta E = 1 + 4E_0/\Delta E, \tag{79}$$

where $E_0 \; (= h^2/8md^2)$ is the energy which a *free* electron would have in the state A. We note that (79) is the formula obtained on p. 65 from the approximation of nearly free electrons.

On the other hand, since $(p_y)_{i0} = 0$, α_2 and α_3 will be of the order of magnitude unity. Thus, if ΔE is small, the surfaces of constant energy near the bottom of the zone are very eccentric ellipses.

4.6. *Density of states.* For the discussion of many properties of metals, such as the specific heat, paramagnetism, etc., it is necessary to know the number of electronic states per unit volume of the metal with energy between E and $E + dE$. We denote this by $N(E)\,dE$; two electrons may occupy each state. A formula for $N(E)$ has already been given for the case where the energy is the same function of the wave number as for free electrons (formula (20)).

If the energy is given by the formula $E = \alpha\hbar^2 k^2/2m$, then clearly

$$N(E)\,dE = \frac{1}{4\pi^2}\left(\frac{2m}{\alpha\hbar^2}\right)^{\frac{3}{2}}\sqrt{E}\,dE. \tag{80}$$

Thus small α (large effective mass[†]) gives large density of states. Further, if the energy is given by the formula

$$E = \frac{\hbar^2}{2m}(\alpha_1 k_x^2 + \alpha_2 k_y^2 + \alpha_3 k_z^2),$$

it may easily be shown that

$$N(E)\,dE = \frac{1}{4\pi^2}\left(\frac{2m}{\hbar^2}\right)^{\frac{3}{2}}\frac{\sqrt{E}\,dE}{\sqrt{(\alpha_1\,\alpha_2\,\alpha_3)}}. \tag{81}$$

† Cf. Chap. III, § 3.

If E is given by the formula

$$E = E_0 - \alpha\hbar^2 k^2/2m,$$

then, similarly,

$$N(E)\,dE = \frac{1}{4\pi^2}\left(\frac{2m}{\alpha\hbar^2}\right)^{\frac{3}{2}}\sqrt{(E_0-E)}\,dE.$$

In any case where $N(E)$ is of the form $C\sqrt{E}$, the number N of electrons per unit volume is $\frac{4}{3}CE_{\max}^{\frac{3}{2}}$, and hence

$$N(E_{\max}) = (\tfrac{3}{4}NC^2)^{\frac{1}{3}},$$

which gives the variation of $N(E_{\max})$ with the number of electrons.

The variation of $N(E)$ within a band is also of interest; we shall therefore calculate it from formulae (58) for the s band, using the approximation of tight binding.

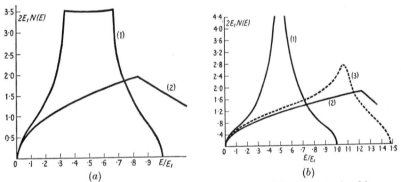

FIG. 38. Density of states $N(E)$; (a) simple cubic, (b) body-centred cubic.

Simple cubic structure. The boundaries of the first Brillouin zone in k-space, which in this case is a cube, are given by the planes

$$k_x = \pm\pi/a, \quad k_y = \pm\pi/a, \quad k_z = \pm\pi/a.$$

It follows directly from the definition of $N(E)$ that

$$N(E) = \frac{1}{8\pi^3}\iint\frac{1}{|\mathrm{grad}\,E|}\,dS, \tag{82}$$

where the integration is over the surface in k-space on which the energy is constant and has the value E. When the energy has the special form given by (58.1), it is possible to reduce the integral in (82) to an integration over a single variable, which can then be evaluated numerically. The result is shown graphically in curve 1, Fig. 38 (a). The energy is plotted in units of the maximum energy in the band E_1, and the area under the curve 1 is equal to the total number of states in the band, namely 2 per atom. The curve is symmetrical about the mid-point, $E = \frac{1}{2}E_1$, and for small values of E it behaves like the corresponding curve for free electrons, which varies as $\sqrt{E}$. Curve 2 shows the value of $N(E)$, calculated on the assumption that the energy surfaces are spheres or portions of spheres at all points inside the zone, thus giving the $N(E)$ curve for *free electrons* in states within this zone.

The two curves represent the $N(E)$ values in two limiting cases: (1) for very large energy gaps across the zone boundaries (tight binding), and (2) for very small energy gaps. The energy surface which first touches the boundary in case (1) encloses $\frac{1}{6}$ of the total volume of the zone, whilst in case (2) the first sphere to touch the boundary contains 0·524 of the total volume.

Body-centred cubic structure. The expression for the energy in this case is given by (58.2), and here again the integral in (82) can be reduced to a single integral which is easily evaluated numerically. Fig. 38 (b), curve 1, shows the form of the $N(E)$ curve. $N(E)$ becomes infinite at the point $E = \frac{1}{2}E_1$, but in such a way that the area under the curve remains finite. The surface $E = \frac{1}{2}E_1$ is a cube which just touches the boundary of the zone and contains half the total volume. Curve 2 shows $N(E)$ for free electrons in the same units. The break in the curve at $E/E_1 = 1\cdot234$ occurs where the spherical energy surface just touches the boundaries of the zone. This sphere contains 0·740 of the total volume of the zone. Just as for the simple cubic structure, curves 1 and 2 of Fig. 38 (b) represent the limiting curves of very large and very small energy gaps respectively. The probable form for a real metal is shown by curve 3.†

5. Crystalline field in alloys

If a foreign atom is dissolved in a pure metal, there will at that point be a break in the periodicity of the field. The question then arises whether it is legitimate to speak of zones for a disordered alloy. The question has not been cleared up yet in a satisfactory way, but it seems fairly certain from the evidence of Chapter V that the break in the periodicity may be considered not to affect the zone structure very much. The discussion of the conductivity of alloys in Chapter VII is also relevant.

A rough calculation‡ may be made, by the method of Thomas and

---

† See note on p. 88.   ‡ Mott, *Proc. Camb. Phil. Soc.* **32** (1936), 281.

Fermi,† of the field surrounding a dissolved atom of different valency from the solvent metal. We suppose the solvent metal to be copper, silver, or gold, with one electron outside a closed d shell, and the dissolved metal to have $z+1$ electrons outside a closed shell, so that, for instance, z is one for Zn and two for Al.

We treat the solvent metal by the method of § 2; that is to say, we suppose the positive charge to be uniformly distributed throughout the metal. If N_0 is the number of atoms per unit volume, $\rho_+ = N_0 e$ is the density of positive charge‡. The dissolved atom will be treated as a positive charge ze. Our problem is to find the density of negative charge round it.

Let $\Phi(r)$ be the electrostatic potential at a distance r from the dissolved atom; the boundary conditions for Φ are then

$$\Phi \sim ze/r \quad (r \to 0)$$
$$\to 0 \qquad (r \to \infty). \tag{83}$$

The electron density in the metal, by the usual assumption of the Thomas-Fermi method, is

$$N(r) = \frac{8\pi}{3h^3}[2me(\Phi+\Phi_0)]^{\frac{3}{2}},$$

where $e\Phi_0$ is the maximum kinetic energy of any electron in the degenerate Fermi gas, given by equation (19), which we write

$$e\Phi_0 = \frac{h^2}{2m}\left(\frac{3N_0}{8\pi}\right)^{\frac{2}{3}}.$$

The density of negative charge is hence $\rho(r) = -eN(r)$, and the total charge density $e(N_0-N)$. Poisson's equation gives therefore

$$\nabla^2\Phi = -4\pi e(N_0-N).$$

Substituting for N and N_0, we have

$$\nabla^2\Phi = \alpha[(\Phi+\Phi_0)^{\frac{3}{2}}-\Phi_0^{\frac{3}{2}}], \tag{84}$$

where $\qquad \alpha = 2^{\frac{13}{2}}\pi^2 m^{\frac{3}{2}}e^{\frac{5}{2}}/3h^3.$

A solution in terms of known functions can only be given if Φ is treated as small compared with Φ_0; expanding (84) we obtain

$$\nabla^2\Phi = \tfrac{3}{2}\alpha\Phi_0^{\frac{1}{2}}\Phi = q^2\Phi, \tag{85}$$

where $\qquad q^2 = 4me^2(3N_0/\pi)^{\frac{1}{3}}/\hbar^2.$

† Cf. § 1.4. $\qquad\qquad\qquad\qquad$ ‡ e in this section is positive.

The solution satisfying the boundary conditions (83) is†

$$\Phi = +\frac{ze}{r}e^{-qr}.$$

We see, therefore, that the isolated point charge is screened, the electrons being drawn nearer to its position than they would be in its absence.

Values of $1/q$ calculated from (85) are

	Cu	Ag	Au
$1/q$ (Å. U.)	0·55	0·58	0·58

† The potential round a point charge is similar to that obtained in the Debye-Hückel theory of electrolytes. In our case, however, q is independent of the temperature.

Note on recent development.

Curve 2 of Fig. 38 (a) has been given by Stoner, *Proc. Leeds Phil. Soc.*, **3** (1936), 120. A calculation of the density of states in the *d* band has been given by Slater, *Phys. Rev.*, **49** (1936), 537.

III

MOTION OF ELECTRONS IN AN APPLIED FIELD

1. Conductors, insulators, and semi-conductors

ONE of the greatest successes of the quantum theory of metals is the explanation which it has given for the sharp distinction existing in nature between metallic conductors on the one hand and insulators and semi-conductors on the other. While most substances have at least a small conductivity, it is only for metals that the conductivity is greatest for low temperatures and for the purest specimens; for insulators and semi-conductors the resistance becomes greater under these conditions.

We must emphasize that the distinction between conductors and insulators cannot depend *directly* on the electric field surrounding each atom; we must not imagine that in insulators the electrons are held to their respective atoms so firmly that they cannot escape. We know that, according to quantum mechanics, an electron can pass fairly easily through a potential barrier a few volts high and 2–3 Å.U. broad; differences in the potential barriers could only account for a factor of 10 or 100 in the ratio between the conductivities of different substances. The reason for the factor of about 10^{24} between the conductivities of, say, silver and fused quartz must therefore be sought elsewhere.

We have seen in Chapter II that the possible states of an electron in the lattice may be divided into zones (the Brillouin zones). The energies of the states in a given zone form a continuous band, which may be separated by an interval of forbidden energies from the energies corresponding to the next zone. It is easy to see that, *if all the states in a given zone are occupied by electrons, then the total current due to these electrons is always zero.* For in such a zone, for every electron moving to the right there will be another electron moving with exactly equal velocity to the left. The resultant current therefore vanishes. In a zone which is only partly full, on the other hand, it may happen that more electrons are moving in one direction than in the other, so that the total current does not vanish.

We see, therefore, that *an insulating crystal†* is one in which all the Brillouin zones which contain any electrons at all are full. The fact

† Liquids are considered in Chap. VII, § 10.

that no current is then possible is due essentially to the Pauli exclusion principle, and thus to the same principle as that which ensures that the total orbital current (and hence the orbital magnetic moment) shall vanish in a closed shell of an atom. A conductor, on the other hand, is a crystal in which one or more zones are only partly full.

We must now consider under what conditions there will exist zones which are only partly full. In the first place, such zones will exist for all monovalent metals which form a cubic lattice—i.e. for the alkalis and for copper, silver, and gold. For we have seen (Chap. II, § 4)

(a)   (b)

FIG. 39. Possible forms in k-space of the surface of the Fermi distribution; (a) monovalent metal, (b) divalent metal.

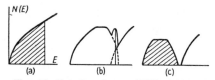

(a)   (b)   (c)

FIG. 40. Density of states $N(E)$; (a) monovalent metal, (b) divalent metal, (c) semiconductor. In (a) and (c) occupied states are shaded.

that for such lattices the first Brillouin zone contains $2N$ possible states, where N is the number of atoms, the 2 being due to the two possible spin directions. The first zone cannot therefore be filled by N electrons. The surface in k-space which separates occupied and unoccupied states is shown in Fig. 39 (a).

If the atom has two valence electrons, there are just enough electrons to fill the first zone, and the crystal will be an insulator, unless the first zone overlaps the second. Since the divalent metals are conductors, we must assume that this overlap does in fact occur. The surface of the Fermi distribution will thus be as shown in Fig. 39 (b), and the density of states† $N(E)$ as in Fig. 40 (b). When a field is applied, only the electrons in states corresponding to the unshaded regions in Fig. 39 (b) are free to move; in the shaded parts there will be as many electrons moving parallel to the field as in the opposite direction. Thus the actual number of electrons free to move in a divalent metal will be considerably less than two per atom. Actually most divalent metals are worse conductors than most monovalent metals.‡

† Cf. § 9.1 for experimental evidence that $N(E)$ actually does look like this.
‡ Cf. Chap. VII, § 3.

For three, four, or five electrons per atom it is impossible to say without a detailed investigation whether a crystal will be a conductor or insulator.

In some monovalent metals it is quite possible that the surface of the Fermi distribution touches the sides of the first Brillouin zone, as in Fig. 26 (b). For instance, this will be the case with one electron per atom if we use for the energy formula (58) of Chapter II. This will clearly *lessen* the current produced in a given time by a given field.†

We may emphasize here that a perfectly periodic lattice has no resistance; the wave functions ψ_k describe an electron with definite momentum, and in a perfectly periodic lattice there is no reason why the electron should change its momentum. Resistance is due to the departure from periodicity in the lattice due to thermal motion; this is described in Chapter VII.

It is convenient to define a quantity N_{eff} which we call the 'effective number of free electrons' per unit volume in a metal. N_{eff} is defined as follows: if a field F acts for a time δt on a metal, the current δj produced is

$$\delta j = \frac{N_{\text{eff}} e^2 F}{m} \delta t. \tag{1}$$

δt is of course assumed to be so small that the distance travelled by the electrons is short compared with the mean free path.

We shall denote the quantity N_{eff}/N by n_{eff}. These quantities are calculated in certain special cases in § 4.

We shall now consider certain insulating crystals. Crystals of the type of rock salt consist of ions each of which contains six outer p electrons forming a closed shell. From the point of view of the Bloch theory, corresponding to the levels of the Na+ and of the Cl− ions there will exist in the crystal bands of levels, which will be completely filled. The crystals will be conductors only if the p band of the Na+ or Cl− ions overlaps with the next highest band.

Diamond and some other crystals are discussed in Chap. V, § 2.

A crystal in which a full zone and an empty zone are separated by a very small energy gap ΔE, as in Fig. 40 (c), would be an insulator at the absolute zero of temperature, but as the temperature was raised a few electrons would come up into the empty zone, and the crystal would conduct. The conductivity would increase with

† Cf. Chap. VII, p. 275, and also Appendix I.

temperature, so that the crystal would behave like a *semi-conductor*. The behaviour of such crystals has been discussed by A. H. Wilson;[†] according to Wilson, in most semi-conductors the full bands from which the electrons come are due to impurities or imperfections in the lattice rather than to the periodic lattice itself.

2. The velocity of an electron in a given state in the lattice

For one-dimensional motion the state of an electron in the lattice is specified by the wave vector k. For free electrons (vanishing lattice field) the velocity of the electron, i.e. the group velocity of the waves, is given by

$$v = \hbar k/m. \tag{2}$$

For non-vanishing lattice fields we use the well-known formula for the group velocity of any waves:

$$v = \frac{d\nu}{d(1/\lambda)}.$$

Since the frequency ν of any de Broglie wave is equal to E/h, and since $1/\lambda = k/2\pi$, this gives us for the velocity

$$v = \frac{1}{\hbar} \frac{\partial E}{\partial k}. \tag{3}$$

We deduce for the current contributed by an electron in the state k

$$j = -ev = -\frac{e}{\hbar} \frac{\partial E}{\partial k}, \tag{4}$$

an equation which may easily be generalized for three-dimensional motion (cf. equation (11)).

An alternative proof may be given, starting from the well-known formula[‡] for the velocity $\mathbf{v}$ (in the absence of a magnetic field),

$$\mathbf{v} = \frac{\hbar}{2mi} \int (\psi^* \operatorname{grad} \psi - \psi \operatorname{grad} \psi^*) \, d\tau \bigg/ \int \psi^* \psi \, d\tau, \tag{5}$$

where the integration is over the whole of space. We shall transform this expression in the following way: If we differentiate the Schrödinger wave equation, Chap. II, equation (13), partially with respect to k_x, we obtain

$$\nabla^2 \frac{\partial \psi}{\partial k_x} + \frac{2m}{\hbar^2} \left[\frac{\partial E}{\partial k_x} \psi + (E - V) \frac{\partial \psi}{\partial k_x} \right] = 0. \tag{6}$$

---

† *Proc. Roy. Soc. A,* **133** (1931), 458; **134** (1931), 277. *Actualités Scientifiques et Industrielles,* No. 82, Paris (1933).

‡ Cf., for instance, Condon and Morse, *Quantum Mechanics,* p. 30, New York (1929).

Also, writing ψ in the form $e^{i(\mathbf{kr})}u(\mathbf{r})$ (cf. Chap. II, § 4.5), we obtain

$$\frac{\partial\psi}{\partial k_x} = ix\psi + e^{i(\mathbf{kr})}\frac{\partial u}{\partial k_x}. \tag{7}$$

Operating on (7) with ∇^2, we obtain

$$\nabla^2\frac{\partial\psi}{\partial k_x} = 2i\frac{\partial\psi}{\partial x} + ix\nabla^2\psi + \nabla^2\left(e^{i(\mathbf{kr})}\frac{\partial u}{\partial k_x}\right). \tag{8}$$

Substituting from (8) in (6), we obtain

$$2i\frac{\partial\psi}{\partial x} + \frac{2m}{\hbar^2}\frac{\partial E}{\partial k_x}\psi + \left[\nabla^2 + \frac{2m}{\hbar^2}(E-V)\right]\frac{\partial u}{\partial k_x}e^{i(\mathbf{kr})} = 0. \tag{9}$$

We now multiply by ψ^* and integrate over all space; the last term may then be shown to vanish, for, by Green's theorem,

$$\int \psi^*\nabla^2\left(\frac{\partial u}{\partial k_x}e^{i(\mathbf{kr})}\right)d\tau - \int \frac{\partial u}{\partial k_x}e^{i(\mathbf{kr})}\nabla^2\psi^*\,d\tau$$

is equal to a surface integral over the boundary of the volume of integration, which, since all the terms occurring are periodic, may be taken to be zero. We have, therefore,

$$\int \psi^*\left[\nabla^2 + \frac{2m}{\hbar^2}(E-V)\right]\frac{\partial u}{\partial k_x}e^{i(\mathbf{kr})}\,d\tau$$
$$= \int \frac{\partial u}{\partial k_x}e^{i(\mathbf{kr})}\left[\nabla^2 + \frac{2m}{\hbar^2}(E-V)\right]\psi^*\,d\tau = 0, \tag{10}$$

since ψ^* satisfies the same Schrödinger equation as ψ.

We therefore obtain from (9)

$$i\int \psi^*\frac{\partial\psi}{\partial x}\,d\tau = -\frac{m}{\hbar^2}\frac{\partial E}{\partial k_x}\int \psi^*\psi\,d\tau,$$

whence it follows from (5) that

$$v_x = \frac{1}{\hbar}\frac{\partial E}{\partial k_x}, \tag{11}$$

which is the result we require.

We see therefore from Fig. 22 that the velocity of an electron tends to zero at the bottom *and at the top* of a band.

In the special case of vanishing lattice field, where the energy is given by $E = \hbar^2 k^2/2m$, (11) reduces to the trivial result (2).

3. Acceleration of the electrons by an external electric field

We now require an expression for the acceleration of an electron in an external field. We confine ourselves first to motion in one dimension. If we represent the electron by a wave function $\psi_k(x)$

extending right through the crystal, the position of the electron is entirely undefined, and the acceleration therefore difficult to visualize. We shall therefore take for the wave function of the electron a *wave packet*†

$$\psi = \int c(k')\psi_{k'}(x)\,dk'.$$

$c(k')$ is a function which vanishes except in the range

$$k - \Delta k \leqslant k' \leqslant k + \Delta k,$$

where Δk is small compared with k; it follows that the volume of wave packet must extend over many atoms, but we may consider it to be small compared with the dimensions of the whole crystal. Bearing in mind that we are dealing always with a wave packet, we may give a purely classical derivation of the formula for the acceleration.

We denote the electric field by F, the velocity of the electron by v, and its energy by E_k, measured from the lowest state in the zone considered.‡ We consider an interval of time δt, and suppose the wave number k of the electron to change by δk, and the energy therefore by

$$\delta E = \frac{\partial E}{\partial k}\delta k. \tag{12}$$

But by the conservation of energy

$$\delta E = eFv\,\delta t. \tag{13}$$

We have seen (§ 2) that $v = \dfrac{1}{\hbar}\dfrac{\partial E}{\partial k}$, and hence, equating (12) and (13), we obtain

$$\delta k = eF\,\delta t/\hbar,$$

or

$$\frac{dk}{dt} = \frac{eF}{\hbar}. \tag{14}$$

It follows that, under the influence of an electric field, the wave vector in the direction of the field increases uniformly with the time. For the case of vanishing lattice field, when $k = mv/\hbar$ and $E = \frac{1}{2}mv^2$, the result (14) reduces to Newton's third law of motion, namely $m\dot{v} = eF$.

In three dimensions (14) is to be understood as a vector equation

$$\dot{\mathbf{k}} = e\mathbf{F}/\hbar. \tag{14.1}$$

The proof is similar.

The proof given here of the important result (14) needs amplifying in several respects. Firstly, as k increases, the energy of the electron

---

† Cf., for instance, Mott and Massey, *The Theory of Atomic Collisions*, Chap. I, § 9, Oxford (1933).

‡ For electrons in no periodic field E_k would be the kinetic energy.

will increase until it comes to a band of forbidden energies (cf. Fig. 41). What happens then has been discussed by Zener,† who finds that, if k is increasing in the direction shown by the arrow, there is a very small probability that the electron will make a transition from one band to the next. In general, however, the point in k space which represents the electron travels from A' to A and then reappears at A'; it must of course be remembered that A' and A represent the *same* state of the electron.

The probability that the electron 'jumps' from A to B is found by Zener to be, per unit time,

$$\frac{eFa}{h}\exp\left\{\frac{-\pi^2 ma(\Delta E)^2}{h^2|eF|}\right\}, \quad (15)$$

where F is the external field, a the lattice constant, and ΔE the energy gap AB.

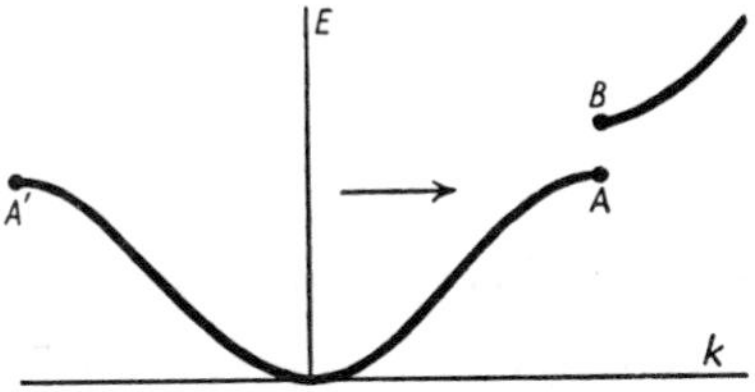

FIG. 41. Energy E of an electron plotted against wave number.

Secondly, the proof given above is of course semi-classical; a proof based more fully on quantum mechanics has been given by Jones and Zener.‡

From equation (14) we obtain, for one-dimensional motion, the following expression for the acceleration of an electron in an external field F:

$$\frac{dv}{dt} = \frac{1}{\hbar}\frac{d}{dt}\left(\frac{dE}{dk}\right) = \frac{d^2E}{dk^2}\frac{eF}{\hbar^2}. \quad (16)$$

Comparing this with the classical expression eF/m for the acceleration, we may call $\hbar^2 / \left|\dfrac{d^2E}{dk^2}\right.$ the 'effective mass' of an electron in the lattice, though the expression is in some ways unfortunate, since the quantity is often negative. We note that for a narrow band d^2E/dk^2 is numerically small and hence the effective mass large. In other words, the acceleration produced by a given field is small. This is easy to understand, since a band is narrow if the atoms are a long way from each other, and under these conditions an electron will require a relatively long time to jump from one atom to the next.

The dimensionless quantity

$$\frac{m}{\hbar^2}\frac{d^2E}{dk^2} = \frac{\text{electronic mass}}{\text{'effective mass'}}$$

is also called the 'oscillator strength of frequency zero' (cf. § 5).

† *Proc. Roy. Soc.* A, **145** (1934), 521.   ‡ *Proc. Roy. Soc.* A, **144** (1934), 101.

For motion in three dimensions, equation (16) is only valid if E may be written in the form $Ak_x^2 + Bk_y^2 + Ck_z^2$. In the general case, denoting by v_s, k_s, F_s ($s = 1, 2, 3$) the three components of $\mathbf{v}$, $\mathbf{k}$, $\mathbf{F}$, we have

$$\frac{dv_s}{dt} = \frac{e}{\hbar^2} \sum_t \frac{\partial^2 E}{\partial k_s \partial k_t} F_t \tag{16.1}$$

or in vector notation

$$\frac{d\mathbf{v}}{dt} = \frac{e}{\hbar^2} \operatorname{grad}_k(\mathbf{F} \operatorname{grad}_k E).$$

Thus a field $\mathbf{F}$ may change the velocity in directions other than that of $\mathbf{F}$.

The quantity $\hbar^2 \left(\dfrac{\partial^2 E}{\partial k_s \partial k_l} \right)^{-1}$ has been called the 'mass tensor'.[†]

4. The effective number of free electrons

To obtain the rate of change of the current in a metal under the influence of an external field F we must multiply (16) or (16.1) by the electronic charge and sum over all occupied states; in the one-dimensional case, since the number of states per unit volume in the range dk is $2dk/2\pi$, we obtain

$$\frac{dj}{dt} = \frac{e^2 F}{\hbar^2} \frac{2}{2\pi} \int \frac{d^2 E}{dk^2}\, dk, \tag{17}$$

the integration being over *occupied* states and hence between $\pm k_{\max}$. Comparing this with the expression for vanishing lattice field

$$\frac{dj}{dt} = \frac{e^2 F}{m} N \tag{17.1}$$

we may call $$\frac{1}{\pi} \frac{m}{\hbar^2} \int \frac{d^2 E}{dk^2}\, dk = N_{\text{eff}}, \text{ say,} \tag{18}$$

the 'effective number of free electrons' in a metal[‡] per unit volume.

Integrating (18), we obtain

$$N_{\text{eff}} = \frac{1}{\pi} \frac{m}{\hbar^2} 2 \left(\frac{dE}{dk} \right)_{k=k_{\max}}. \tag{19}$$

It follows that, since dE/dk vanishes at the top of a band (Fig. 41), *for a full zone the effective number of electrons is zero*, as we should expect from the considerations of §1. Further, for a narrow band, N_{eff} will be small compared with N.

† Blochinzev and Nordheim, *Zeits. f. Phys.* **84** (1933), 168.
‡ Cf. § 1.

We note also that the free electron number depends only on the state of affairs at the surface of the Fermi distribution.

In the three-dimensional case we obtain similarly, from (16.1) instead of (16),

$$\frac{d\mathbf{j}}{dt} = \frac{2e^2}{\hbar^2} \frac{1}{8\pi^3} \int \int \int \mathrm{grad}_k(\mathbf{F}\,\mathrm{grad}_k\,E)\,dk_x\,dk_y\,dk_z.$$

If we require only the current in the direction of the field, this may be written

$$\frac{dj_x}{dt} = \frac{2e^2 F_x}{\hbar^2} \frac{1}{8\pi^3} \int \int \int \frac{\partial^2 E}{\partial k_x^2}\,dk_x\,dk_y\,dk_z.$$

This may be transformed into a surface integral over the boundary in k-space of the occupied states: we obtain

$$\frac{dj_x}{dt} = \frac{2e^2 F_x}{\hbar^2} \frac{1}{8\pi^3} \int \left(\frac{\partial E}{\partial k_x}\right)^2 \frac{dS_k}{|\mathrm{grad}\,E|}, \tag{20}$$

dS_k denoting an element of surface in k-space, the integration being over the whole surface of the Fermi distribution where it does not touch a plane of discontinuity in the energy.

The effective number of free electrons is thus

$$N_{\text{eff}} = \frac{m}{\hbar^2} \frac{1}{4\pi^3} \int \left(\frac{\partial E}{\partial k_x}\right)^2 \frac{dS_k}{|\mathrm{grad}\,E|}. \tag{21}$$

We shall denote by $n_{\text{eff}} = N_{\text{eff}}/N_a$ the effective number of electrons per *atom*, N_a denoting the number of atoms per unit volume.

It is of interest to work out n_{eff} in a number of cases:

(a) If the electrons are free, i.e. if

$$E = \hbar^2 k^2/2m, \tag{22}$$

then n_{eff} is equal to the *actual* number of free electrons per atom (n_0).

(b) If the energy is proportional to k^2,

$$E = \alpha\hbar^2 k^2/2m, \tag{23}$$

then $n_{\text{eff}} = \alpha n_0$. In this case it is especially convenient to refer to m/α as 'effective mass' of the electrons in the metal.

(c) If the energy near all or part of the surface of the Fermi distribution has the form

$$E = \frac{\hbar^2}{2m}(\alpha_1 k_x^2 + \alpha_2 k_y^2 + \alpha_3 k_z^2), \tag{24}$$

the 'effective number of free electrons' depends on the direction of the current. In the direction of the x-axis it may easily be shown that

$$n_{\text{eff}} = \alpha_1 n_0. \tag{25}$$

This formula may be applied to the case of divalent metals, where the Fermi surface overlaps into higher zones, near the bottom of each of which the energy is given by a formula of the type (24). By n_0 we must understand the actual number of electrons per atom in the zone considered.

Referring to formula (79) of Chap. II and to Fig. 37, we see that in the case considered there $n_{\text{eff}} \sim n_0$ for motion parallel to the y- and z-axes, but $n_{\text{eff}} \gg n_0$ for motion parallel to the x-axis.

(d) In the limiting case of tight binding, for a body-centred cubic lattice the energy within an s band as a function of $\mathbf{k}$ is given by (cf. Chap. II, equation (58.2))

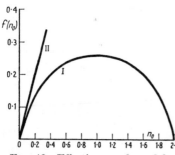

FIG. 42. Effective number of free electrons plotted against the number n_0 of electrons per atom: I, for a single zone (case of tight binding); II, for free electrons.

$$\tfrac{1}{2}E_1(1 - \cos\tfrac{1}{2}ak_x \cos\tfrac{1}{2}ak_y \cos\tfrac{1}{2}ak_z),$$

where E_1 is the breadth of the energy band, and a is the lattice constant. With this form for the energy the double integral in (21) can be reduced to a single integral, which may be evaluated numerically. This gives N_{eff} as a function of the energy of the limiting surface, and hence, since $N(E)$ is known as a function of E (cf. Chap. II, § 4.6), we can compute the number of electrons enclosed within a given energy surface and thus obtain N_{eff} in terms of the number of electrons n_0 per atom. The result may be expressed as follows:

$$N_{\text{eff}} = \frac{ma^2 E_1}{8\hbar^2} f(n_0).$$

Fig. 42 shows $f(n_0)$ plotted against n_0. In the special case when the width of the band is given by

$$E_1 = 8\hbar^2/ma^2,$$

the energy in the immediate neighbourhood of the origin of k-space is the same function of $\mathbf{k}$ as for free electrons. In this case N_{eff} is given directly by $f(n_0)$, and, as we should expect, near the origin $f(n_0)$ approaches the straight line which represents N_{eff} for free electrons.[†]

† See note on p. 131.

5. Periodic electric fields; light waves†

In the preceding sections we have discussed the motion of the electrons in a metal in an electrostatic field; we now discuss their behaviour under the influence of a *periodic* field (light wave).

In any solid, conductor or insulator, a light wave will set up an alternating current to which is due the dispersion and refraction of the light; for certain ranges of wave-length, moreover, the electrons can absorb energy from the light by a process in which an electron jumps from one zone to another. We call this process 'internal photoelectric absorption'. We consider first this absorption process.‡

We consider a light wave of frequency ν travelling in the direction of the z-axis with its electric vector F parallel to the x-axis; if the wave-length λ is large compared with the lattice constant a, we may take, in a volume large compared with a but small compared with λ,

$$F = F_0 \sin 2\pi\nu t,$$

and, for the vector and scalar potentials $\mathbf{A}$, A_0,

$$A_x = \frac{cF_0}{2\pi\nu} \cos 2\pi\nu t, \qquad A_y = A_z = 0, \qquad A_0 = 0.$$

The Schrödinger equation for an electron moving in the lattice under the influence of the light wave is therefore

$$i\hbar \frac{\partial \psi}{\partial t} = -\frac{\hbar^2}{2m} \nabla^2 \psi + V\psi + \frac{e\hbar}{mci}(\mathbf{A}\,\mathrm{grad}\,\psi), \tag{26}$$

where V, as usual, denotes the potential energy of the electron in the field of the lattice. The perturbing term

$$\frac{e\hbar}{mci}(\mathbf{A}\,\mathrm{grad}) = \frac{eF_0\hbar}{2\pi i m\nu} \cos 2\pi\nu t \frac{\partial}{\partial x}$$

is periodic‖ in (x, y, z), and therefore the selection rules discussed in Chap. II, § 4.1, are valid; *an electron in a state with given wave number* $\mathbf{k}$ *in a given band* n_0 *can make a transition only to one state in any other band* n, namely, the state with the same wave number. The absorption spectrum of a single electron is thus a line absorption spectrum.

---

† Kronig, *Proc. Roy. Soc.* A, **133** (1931), 255; Wilson, ibid. **151** (1935), 274.

‡ The absorption process discussed in this section is different from the *surface* photoelectric effect, which is responsible for the ordinary photoelectric emission.

‖ It may easily be seen that the operator 'grad' is periodic in the sense of Chap. II, § 4.1.

Fig. 43 shows the states in two bands n_0, n between which transitions are possible; an electron in any state in the lower band can ònly make transitions to states vertically above it.

To find the probability of such a process we proceed as follows: We denote the wave function and energy of the initial state n_0 by $\psi_0(\mathbf{r})$, E_0. At any subsequent time we expand the wave function ψ of the electron

$$\psi = \psi_0(\mathbf{r})e^{-iE_0 t/\hbar} + \sum_n B_n(t)\psi_n(\mathbf{r})e^{-iE_n t/\hbar}. \tag{27}$$

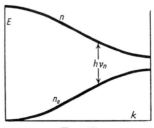

FIG. 43.

$|B_n|^2$ then denotes the probability that the electron is in the state n at the time t. To obtain B_n we substitute (27) in (26), multiply by ψ_n^*, and integrate over all space; we thus find in the usual way,[†] after integrating with respect to the time,

$$|B_n(t)|^2 = \left(\frac{eF_0}{4\pi m\hbar\nu}\right)^2 |p(n0)|^2 2\left[\frac{1-\cos 2\pi(\nu_n-\nu)t}{\{2\pi(\nu_n-\nu)\}^2} + \frac{1-\cos 2\pi(\nu_n+\nu)t}{\{2\pi(\nu_n+\nu)\}^2}\right],$$

where $\qquad p(n0) = \dfrac{\hbar}{i} \displaystyle\int \psi_n^* \dfrac{\partial}{\partial x}\psi_0 \, d\tau, \qquad h\nu_n = E_n - E_0.$

If we put ν equal to one of the frequencies ν_n, the transition probability $|B|^2$ increases with the *square* of the time; this difficulty arises in all quantum-mechanical solutions of this type, and is due to the neglect of the finite breadth of an absorption line. In the theory of line absorption by an atom, it is avoided by considering non-mono-chromatic radiation and integrating over all frequencies ν; in our case, since the energies of the electrons and hence the frequencies ν_n form practically a continuum, it is more convenient to integrate over all the electrons in initially occupied states. Since $2d\mathbf{k}/(2\pi)^3$ is the number of states per unit volume of metal with their wave vectors in the volume element $d\mathbf{k}$, the probability that after a time t

† Cf., for example, Mott and Massey, *The Theory of Atomic Collisions*, p. 258.

a quantum of radiation has been absorbed is

$$\frac{2}{(2\pi)^3} \int |B|^2 \, d\mathbf{k}. \tag{28}$$

The frequency ν_n for any two bands n_0, n is, of course, a function of $\mathbf{k}$, as Fig. 43 shows. The value of the integral (28) increases with the time if, and only if, the integration extends over one of the resonance points. Let us first consider the case of one-dimensional motion. Let k_ν be the wave number for which the resonance takes place, defined therefore by

$$\nu_n(k_\nu) = \nu.$$

Writing $\nu_n - \nu = \xi$, and taking in the square bracket only the first term, which gives the resonance, (28) becomes†

$$\frac{2}{2\pi}\left(\frac{eF_0}{4\pi m\hbar\nu}\right)^2 \int |p(n0)|^2 2\left[\frac{1-\cos 2\pi\xi t}{(2\pi\xi)^2}\right]\left\{\frac{dv_n}{dk}\right\}^{-1} d\xi. \tag{29}$$

For νt large the function within the square brackets has a strong maximum at $\xi = 0$; since

$$2\int_{-\infty}^{\infty} \frac{1-\cos 2\pi\xi t}{(2\pi\xi)^2}\, d\xi = t,$$

(29) gives finally

$$\frac{2}{2\pi}\left(\frac{eF_0}{4\pi m\hbar\nu}\right)^2 |p(n0)|^2 \bigg/ \left(\frac{dv_n}{dk}\right)_{\nu_n=\nu} \tag{30}$$

for the number of quanta absorbed per unit volume after a time t equal to unity.

We may deduce the so-called 'conductivity for frequency ν', $\sigma(\nu)$, defined‡ as the energy absorbed per cm.³ per sec. divided by the mean square of the electric vector, $\frac{1}{2}F_0^2$. This is

$$\sigma(\nu) = \frac{e^2}{2\pi m^2 h\nu}|p(n0)|^2 \bigg/ \frac{dv_n}{dk}. \tag{31}$$

The case of three dimensions is more complicated, and will be understood best by reference to Fig. 44, which

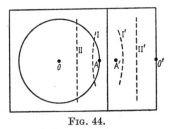

FIG. 44.

shows the first and part of the second Brillouin zones for a simple cubic lattice. The full line shows the surface of the Fermi distribution and the dotted lines show states between which transitions are

---

† 2π is substituted for the term $(2\pi)^3$ occurring in (28), since we are dealing with one-dimensional motion.

‡ Cf. § 7; $\sigma(\nu) = nk\nu$, where n and k are the optical constants.

possible for given frequency; thus electrons can jump from any point on the line I to the corresponding point on the line I' absorbing a frequency ν_I, say, or from II to II' absorbing a greater frequency ν_{II}. To obtain the absorption coefficient for given frequency ν_I, we must integrate (30) for all points lying on the line (surface) I and corresponding to occupied states. Such a calculation has in fact been carried out by Wilson;† we shall here discuss some of his results.

The minimum frequency for absorption ν_1 will correspond to the transition AA' (Fig. 44). Clearly $h\nu_1$ will be greater or equal to the energy gap between the first and second zones. One may obtain a very rough estimate of ν_1 by assuming that the approximation of free electrons (Chap. II, § 3) is valid and that the energy gap vanishes. In that case, if $(n_1 n_2 n_3)$ are the Miller indices of the first plane of Bragg reflection of the crystal, we have

$$h\nu_1 = [(2k_n - k_{max})^2 - k_{max}^2]\hbar^2/2m,$$

where k_{max} is the maximum wave number in the Fermi distribution (Chap. II, equation (25)) and

$$E_n = \frac{\hbar^2 k^2}{2m} = \frac{h^2}{8md^2}(n_1^2 + n_2^2 + n_3^2),$$

d being the lattice constant; E_n is thus the energy of an electron which would suffer reflection for normal incidence on the $(n_1 n_2 n_3)$-plane.

For the three cubic structures we have, therefore:

	Simple cubic	Body-centred	Face-centred
First plane of reflection . .	(100)	(110)	(111)
$8md^2E_n/h^2 = n_1^2+n_2^2+n_3^2$ . .	1	2	3
Number of atoms in unit cell .	1	2	4
$8(\pi/3)^{2/3}md^2E_{max}/h^2$ . . .	1	$2^{2/3}$	$4^{2/3}$
$8md^2h\nu_1/h^2$	0·062	0·96	1·18

We may express these formulae in terms of the atomic volume Ω_0:

$$\lambda_1 = c/\nu_1 = 5{\cdot}32 \times 10^{10}\Omega_0^{2/3} \qquad \text{body-centred cubic lattice}$$
$$= 7{\cdot}10 \times 10^{10}\Omega_0^{2/3} \qquad \text{face-centred cubic lattice,}$$

whence we deduce for λ_1 the values shown below:

Theoretical wave-lengths at which photoelectric absorption begins

	Li	Na	K	Rb	Cs	Cu	Ag	Au
λ_1, cm. $\times 10^4$	0·41	0·62	0·91	1·1	1·23	0·37	0·47	0·47

† Loc. cit.

These results may be correct only as to the order of magnitude, but they do suggest that the alkalis should absorb in the infra-red, and noble metals in the visible or ultra-violet. For more detailed comparison with experiment cf. § 8, where it is shown that the noble metals actually have absorption edges between 0.3 and 0.6×10^{-4} cm.

The shape of the absorption band near the low-frequency limit. For given frequency the absorption coefficient depends upon the number of occupied states for which the transition is possible (i.e. the area cut off by the Fermi surface from the surface represented by the dotted line in Fig. 44). If the surface of the Fermi distribution does not touch the first plane (in k-space) of energy discontinuity (as shown in Fig. 44), then it may easily be seen that, near the low-frequency limit ν_0, the absorption coefficient $nk\nu$ (p. 107) is given by

$$nk\nu \propto \text{const.}(\nu-\nu_0); \tag{32}$$

on the other hand, if it does touch, as in Fig. 26, then a short calculation gives

$$nk\nu \propto \text{const.}\sqrt{(\nu-\nu_0)}. \tag{33}$$

Formula (33) gives also the shape of the absorption band due to a fully occupied zone.†

6. Dispersion

To obtain the current in the lattice due to the light wave, the procedure is exactly the same as for a single atom; the current due to a single electron in the state $n_0, \mathbf{k}$ is given by the ordinary formula of the Kramers-Heisenberg dispersion theory‡

$$j_x = \frac{e^2 F_0}{2\pi m\nu}\left[1-\frac{2}{hm}\sum_n \frac{\nu_n|p(n0)|^2}{\nu_n^2-\nu^2}\right]\sin 2\pi\nu t,$$

where $F_0 \cos 2\pi\nu t$ is the electric vector of the light wave and $p(n0), \nu_n$ are defined in the last section. As may easily be verified, we may write this

$$j_x = \frac{e^2 F_0}{2\pi m}\left[\frac{f}{\nu}-\sum_n \frac{\nu f_n}{\nu_n^2-\nu^2}\right]\sin 2\pi\nu t, \tag{34}$$

where

$$f = 1-\frac{2}{hm}\sum_n \frac{|p(n0)|^2}{\nu_n}, \tag{35}$$

$$f_n = \frac{2}{hm}\frac{|p(n0)|^2}{\nu_n}, \tag{36}$$

and where the terms in $n = 0$ are to be omitted from the summations.

† See note on p. 131.
‡ Cf., for example, Sommerfeld, *Wave Mechanics*, p. 168, London (1930).

If we compare this expression with the current that would be produced by a harmonic oscillator with charge ef, mass m, and natural frequency ν_0, viz.

$$-\frac{e^2\nu f}{2\pi m(\nu_0^2-\nu^2)}\,F_0\sin 2\pi\nu t,$$

we see that *the electron behaves like a system of such oscillators, one of them having the frequency zero (no restoring force) and strength f and the others the frequencies ν_n and strengths f_n.*

The difference between the behaviour of an electron in an atom and in a crystal is as follows: for an atom, the quantity f vanishes by the sum rule due to Thomas[†] and Reiche[‡]; for an electron in a periodic field, however, owing to the different boundary conditions[||] satisfied by the wave function, this is not the case; it may in fact be shown from (35) that[††]

$$f=\frac{m}{\hbar^2}\frac{\partial^2 E}{\partial k_x^2}. \tag{37}$$

We call this theorem the 'modified f-sum rule'.

This result should be compared with (16); if in equation (34) we make $\nu \to 0$, we obtain

$$j_x = e^2 f F_0 t/m,$$

which is just the result (16) obtained for steady fields.

We may therefore sum up the results of this paragraph as follows: the current due to a single electron in the field of a lattice is the same as that which would be produced by a free electron of effective mass independent of frequency, together with a series of oscillators of natural frequency ν_n and oscillator strength f_n.

We note from (35) and (36) that

$$f+\sum f_n = 1, \tag{38}$$

which is the form taken by the sum rule for the electrons in a metal. It follows that, if the frequency ν of the incident light is large compared with the frequencies ν_n of all transitions for which f_n is not small, the current is

$$j_x = e^2 F_0\sin 2\pi\nu t/2\pi m\nu.$$

[†] *Zeits. f. Phys.* **33** (1925), 408.

[‡] Ibid. **34** (1925), 510.

[||] The wave function in an atom tends to zero at infinity, but that in a crystal lattice remains finite.

[††] This seems first to have been proved by Bethe, *Handb. d. Phys.* **24**/2 (1933), 378. An alternative proof has been given by Wilson, loc. cit., Appendix.

In other words, for high frequencies, the electron behaves as though it were free.

To obtain the current in a metal, we must sum (34) over all the electrons. The term in f, independent of frequency, has already been discussed. For frequency ν the corresponding current will be $N_{\text{eff}} F_0 e^2 \sin 2\pi\nu t / 2\pi m \nu$, where N_{eff} is defined by equation (21).

The oscillators (36) may be treated as follows: the metal behaves as though it contained a series of oscillators of frequencies ν_n lying between two limits ν_1, ν_2; for unit volume of metal the strength of such oscillators having frequencies in the range $d\nu_n$ is (for one-dimensional motion)

$$df_n = \frac{1}{\pi} \frac{d\nu_n}{d\nu_n/dk} \frac{2}{hm} \frac{|p(n0)|^2}{\nu_n}. \tag{39}$$

We may express this in terms of the absorption coefficient $\sigma(\nu)$ by (31)

$$df_n = \frac{4m}{e^2} \sigma(\nu_n)\, d\nu_n, \tag{40}$$

a formula which is true also in three dimensions. Thus the total current in the metal is

$$j_x = \left[\frac{N_{\text{eff}} e^2}{2\pi m \nu} - \frac{2\nu}{\pi} \int_{\nu_1}^{\nu_2} \frac{\sigma(\nu_n)\, d\nu_n}{\nu_n^2 - \nu^2} \right] F_0 \sin 2\pi\nu t.$$

The polarization $P \ (= \int j_x\, dt)$ is thus

$$P = \left[-\frac{N_{\text{eff}} e^2}{4\pi^2 m \nu^2} + \frac{1}{\pi^2} \int \frac{\sigma(\nu_n)\, d\nu_n}{\nu_n^2 - \nu^2} \right] F_0 \cos 2\pi\nu t. \tag{41}$$

7. Optical properties of metals; theory

7.1. *Definition of the optical constants.* The optical constants n and k are defined by the form of a light wave in a medium, viz. for E or H

$$e^{-k\omega x/c} \frac{\cos}{\sin} \omega\left(\frac{nx}{c} - t \right), \tag{42}$$

where $\omega/2\pi$ is the frequency. The constant n is called the refractive index, and k the extinction coefficient.† The notation $k = n\kappa$ is often used. If we use a complex form for the light wave *in vacuo*

$$E = E_0\, e^{i\omega(x/c - t)},$$

† Throughout §§ 7 and 8, k denotes the extinction coefficient, and not the wave vector of an electron, as elsewhere in this book.

then in the medium we have

$$E = E_0 e^{i\omega(\mathbf{n}x/c - t)}, \tag{43}$$

where $$\mathbf{n} = n + ik. \tag{44}$$

We shall refer to $\mathbf{n}$ defined by (44) as the 'complex refractive index'.

The intensity of light reflected from the surface of a medium may be obtained in terms of n and k, by making use of the boundary conditions that the tangential components of E and H shall be continuous. The reflection coefficient is, for normal incidence,

$$R = \frac{(n-1)^2 + k^2}{(n+1)^2 + k^2}. \tag{45}$$

For oblique incidence the reader is referred to any text-book on optics.

7.2. *Method of calculation.* A theoretical calculation of the optical constants of any isotropic substance, whether conductor or insulator, consists of two parts. The atomic model assumed must first be used to calculate the current j that will be produced by a light wave of frequency $\omega/2\pi$. This current will not in general be in phase with the electric vector of the light; if we write $E = E_0 \sin \omega t$, the current will have the form

$$j = E_0(A \cos \omega t + B \sin \omega t).$$

The rate of loss of energy Ej is proportional to B, since the term in $\cos \omega t$ is 90° out of phase with E. In a transparent medium B will vanish, and the term $A \cos \omega t$ represents the current due to the polarization of the constituent atoms.

In this chapter we shall use the complex form

$$E = E_0 e^{-i\omega t} \tag{46}$$

and shall obtain from our atomic model a relation between the current and the field of the type

$$j = aE, \tag{47}$$

where a is a *complex* function of the frequency.†

We shall also define the polarization P by

$$dP/dt = j, \tag{48}$$

where the constant of integration is chosen so that the time average of P is zero. When a is found, we must use Maxwell's equations to

---

† Only for $\omega = 0$ is a equal to the conductivity.

derive n and k. Maxwell's equations for an isotropic body containing no space charge are

$$c\operatorname{curl}H = \frac{dE}{dt} + 4\pi j, \tag{49 a}$$

$$c\operatorname{curl}E = -\frac{dH}{dt}, \tag{49 b}$$

$$\operatorname{div}E = 0, \tag{49 c}$$

$$\operatorname{div}H = 0. \tag{49 d}$$

The values of the field vectors occurring in these equations denote the *average* values of E and H taken over a volume large compared with atomic dimensions but small compared with the wave-length of light. This form of the theory cannot therefore be used for radiation of wave-length comparable with atomic dimensions.

It will be noticed that in equation (49 a) the electric intensity E, and not the displacement D, is used. This is because the polarization of the medium is already included in the expression j for the current. One may write instead of (49 a)

$$c\operatorname{curl}H = \frac{dD}{dt} = \frac{dE}{dt} + 4\pi\frac{dP}{dt}; \tag{49 e}$$

by virtue of (48), equations (49 a) and (49 e) are identical. If we write $P/E = A$, equation (49 e) may be written

$$c\operatorname{curl}H = (1+4\pi A)\frac{dE}{dt}. \tag{49 f}$$

These equations are thus the usual Maxwell equations for a medium with dielectric constant $1+4\pi A$, the only new point being that A is complex; we obtain at once, if the light wave has the form (46),

$$\mathbf{n}^2 = 1+4\pi A$$
$$= 1+4\pi P/E = 1-4\pi j/i\omega E. \tag{50}$$

The optical constants n and k may be deduced at once from (47).

The real part of $\mathbf{n}^2$, namely n^2-k^2, will be referred to as the dielectric constant and denoted by $\epsilon(\nu)$. The imaginary part, $2ink$, is related to the rate of loss of energy. The rate of loss per unit volume is

$$\tfrac{1}{2}(Ej^*+E^*j), \tag{51}$$

which by (50) reduces to $nk\nu EE^*$.

Hence $$nk\nu = \frac{\text{rate of loss of energy}}{\text{square of electric vector}}.$$

nkv will therefore be called the 'conductivity for frequency ν', and denoted by $\sigma(\nu)$.

The *absorption coefficient* (energy absorbed per unit thickness of material divided by the energy incident) is thus

$$2nk\omega/c.$$

7.3. *Long wave-lengths; the Hagen-Rubens relation.* For electro-magnetic waves of sufficiently long wave-length—in practice greater than 10^{-3} cm.—one may assume that the current is almost in phase with the electric vector. Equation (47) then becomes

$$j = \sigma_0 E, \tag{52}$$

where σ_0 is the conductivity measured for static fields. This leads, by (50), to

$$n^2 - k^2 + 2ink = \mathbf{n}^2 = 1 + 4\pi i\sigma_0/\omega,$$

and hence, with $2\pi\nu = \omega$,

$$n^2 - k^2 = 1,$$

$$nk = \sigma_0/\nu. \tag{53}$$

As we shall see below, $\sigma_0/\nu \gg 1$ for the wave-lengths for which the approximation of this section is valid, and hence, approximately,

$$n = k = \sqrt{(\sigma_0/\nu)}. \tag{54}$$

From (45) we see that the reflecting power is

$$R = 1 - 2\sqrt{(\nu/\sigma_0)} \tag{55}$$

and from (42) that the intensity of the radiation (which is proportional to E^2) falls off as e^{-x/x_0}, where

$$1/x_0 = 4\pi k\nu/c = 4\pi\sqrt{(\sigma_0\nu)}/c. \tag{56}$$

The formula (55) for the reflecting power is known as the Hagen-Rubens relation, and has been compared with experiment† for infrared radiation and for various metals and temperatures. The formula is in general in fair agreement with experiment for

$$\lambda \geqslant 10\mu \quad (1\mu = 10^{-4} \text{ cm.}).$$

This fact will be discussed further below (§ 8.3). For short wave-lengths, on the other hand, the formulae (54) and (55) are not even approximately in agreement with experiment. The reason for this is the non-validity of the assumption (52). The current j will be in phase with the electric field, as there assumed, only if the time of

† Hagen and Rubens, *Ann. d. Physik*, **11** (1903), 873. Cf. also Schaefer and Matossi, *Das Ultrarote Spectrum*, Berlin (1930).

relaxation† of the electron is small compared with the period $2\pi/\nu$ of the light, so that, during the time taken by an electron to traverse its mean free path, the field may be taken as constant. If this is not so, the current will be out of phase with the field, and this is the case which we must now consider.

7.4. *Short wave-lengths; formula of Zener.* It is instructive to consider the opposite extreme to the case treated in the preceding section, namely, an ideal metal in which the mean free path and time of relaxation are infinite, so that there is no possibility of absorption of energy from the electrons by the metal. The calculations of this section are based on the classical theory; but it will easily be seen from § 3 that the results follow also from quantum mechanics, provided the influence of the periodic field of the crystal lattice on the motion of the electrons may be neglected.

We consider a plane polarized light wave, and denote its electric vector by
$$E = E_0 e^{-i\omega t}.$$

If x is the coordinate of the electron parallel to E, we have for the equation of motion of the electron‡

$$m\frac{d^2x}{dt^2} = eE = eE_0 e^{-i\omega t}. \tag{57}$$

By integrating this equation we obtain
$$x = -eE/m\omega^2,$$
and since the polarization P is equal to eNx, where N is the number of electrons per unit volume,
$$\frac{P}{E} = -\frac{Ne^2}{m\omega^2}.$$

By (50), therefore, we have for the *complex* refractive index **n**

$$\mathbf{n}^2 = 1 + \frac{4\pi P}{E} = 1 - \frac{4\pi Ne^2}{m\omega^2}. \tag{58}$$

The behaviour of the metal depends on whether the quantity $4\pi Ne^2/m\omega^2$ is greater or less than unity. In the latter case (short wave-lengths) the refractive index is real but less than unity. The metal is therefore transparent to normally incident light, but a

---

† The time of relaxation is twice the mean free path divided by the mean velocity (cf. Chap. VII, § 2).

‡ This assumes that the force on an electron is E and not $E + \frac{4\pi}{3}P$; cf. p. 116.

critical angle of incidence exists beyond which total reflection at the metal surface takes place.

If $4\pi Ne^2/m\omega^2$ is greater than unity (long wave-lengths), $\mathbf{n}$ is imaginary, and total reflection takes place at the surface for all angles of incidence.

The effect of introducing a finite mean free path or damping coefficient is as follows: the electrons vibrate freely for a finite time only, and then give up their energy to the lattice vibrations. For $4\pi Ne^2/m\omega^2$ greater than unity, therefore, the reflecting power becomes slightly less than unity, since some energy is absorbed in the surface layer; for $4\pi Ne^2/m\omega^2$ less than unity, the wave in the metal is

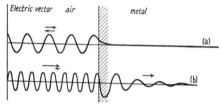

FIG. 45. Electric vector of light wave incident on a metal.
(a) $4\pi Ne^2/m\omega^2 > 1$. (b) $4\pi Ne^2/m\omega^2 < 1$.

damped, so that the metal is opaque, except for relatively thin films. The behaviour of the metal in the two cases is illustrated in Fig. 45.

R. W. Wood[†] has found that the alkalis become transparent in the ultra-violet, and Zener[‡] has pointed out that this simple model is adequate to give, approximately, the wave-length λ_0 at which the transparency begins, by means of the formula

$$4\pi Ne^2/m\omega_0^2 = 1, \qquad \lambda_0 = 2\pi c/\omega_0,$$

as the following table shows:

Wave-length λ_0 in Å. U.

Metal	Cs	Rb	K	Na	Li
λ_0 observed	4,400	3,600	3,150	2,100	2,050
λ_0 calculated	3,600	3,200	2,900	2,100	1,500

This phenomenon is discussed further in § 8.4.

7.5. *Intermediate wave-lengths.* We shall now show in greater detail how to take account of the resistance (non-infinite mean free path), and hence of the absorption of energy by the medium. So

† *Phys. Rev.* **44** (1933), 353.   ‡ *Nature*, **132** (1933), 968.

long as there is no photoelectric absorption, we may assume that the loss of energy is proportional to the square of the current:

$$\frac{\text{rate of loss of energy per unit volume}}{(\text{current})^2} = \frac{1}{\sigma_0}. \tag{59}$$

σ_0 will at any rate be of the order of magnitude of the conductivity for steady currents; if the period of the electric vector is large compared with the time of relaxation, it is not clear that they will be exactly the same; we shall, however, assume this to be the case.[†]

In order to see the modification introduced by (59) into the equations of motion, we consider equation (57), which may be written

$$\frac{m}{Ne^2}\frac{dj}{dt} = E.$$

We have to introduce a damping term into this equation; the equation will become

$$\frac{m}{Ne^2}\frac{dj}{dt} = E - \text{damping term.}$$

Multiplying both sides by j, and taking a time average, we have, since the time average of $j\,dj/dt$ is zero,

$$\overline{Ej} = \text{time average of } j \times \text{damping term.}$$

But the time average of Ej is the rate of loss of energy, and hence, by (59), we must take for our damping term j/σ_0; our new equation becomes therefore

$$\frac{dj}{dt} = \frac{Ne^2}{m}\left(E - \frac{j}{\sigma_0}\right). \tag{60}$$

From (60) we may obtain the current j and polarization P; we have, if E, j, P have the time factor $e^{-i\omega t}$,

$$\left(-i\omega + \frac{Ne^2}{m\sigma_0}\right)j = \frac{Ne^2}{m}E,$$

whence $\qquad P = \dfrac{j}{-i\omega} = E\left(-\dfrac{i\omega}{\sigma_0} - \dfrac{m\omega^2}{Ne^2}\right)^{-1}. \tag{61}$

We introduce the notation

$$\tau = m\sigma_0/Ne^2, \tag{62}$$

so that τ denotes the *time of relaxation*[‡] (twice time between collisions). Hence, from (61), making use of the relation

$$\mathbf{n}^2 = n^2 - k^2 + 2ink = 1 + 4\pi P/E,$$

† Cf. Fujioka, *Zeits. f. Phys.* **76** (1932), 537.   ‡ Cf. Chap. VII, § 1.

we have

$$n^2-k^2 = 1 - \frac{4\pi Ne^2}{m}\left(\omega^2+\frac{1}{\tau^2}\right)^{-1},$$

$$2nk = \frac{4\pi Ne^2}{m}\frac{1}{\omega\tau}\left(\omega^2+\frac{1}{\tau^2}\right)^{-1}. \tag{63}$$

The formulae (63) are the classical formulae for the optical constants which we require, expressed in terms of the frequency $\omega/2\pi$ and the

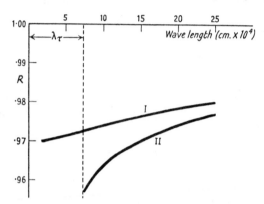

FIG. 46. Reflecting power of platinum in the infra-red.
I. Calculated from (63) and (45) (exact classical formulae).
II. Calculated from the Hagen-Rubens relation (55).
For σ_0 the observed conductivity at room temperature has been taken.

time of relaxation τ. Expressed in terms of the wave-length, they become

$$n^2-k^2 = 1-\left(\frac{\lambda}{\lambda_0}\right)^2\frac{1}{1+(\lambda/\lambda_\tau)^2},$$

$$2nk = \left(\frac{\lambda}{\lambda_0}\right)^2\frac{\lambda}{\lambda_\tau}\frac{1}{1+(\lambda/\lambda_\tau)^2}, \tag{64}$$

where $\lambda_0^2 = c^2\pi m/Ne^2$, $\lambda_\tau = 2\pi c\tau$. The two functions of λ are illustrated in Fig. 47.

If $1/\omega \gg \tau$, i.e. if the period of the light is large compared with the time of relaxation, then from (63)

$$nk = \sigma/\nu \quad \text{and} \quad n^2-k^2 \ll nk.$$

These are the conditions for the validity of the Hagen-Rubens relations. Therefore the formulae (63) tend to those obtained in § 7.3 for sufficiently long wave-length, as we should expect. The reflecting power of a solid, calculated from the formulae (63), in the transition region where $1/\omega \sim \tau$, is shown in Fig. 46.

If, on the other hand, $1/\omega \ll \tau$, the formulae become similar to those of § 7.4, which were obtained by neglecting the damping. In this case we have approximately

$$n^2-k^2 = 1-4\pi Ne^2/m\omega^2, \\ 2nk = 4\pi Ne^2/m\omega^3\tau. \qquad \Big\} \qquad (65)$$

With the above definition of λ_0,

$$n^2-k^2 = 1-(\lambda/\lambda_0)^2.$$

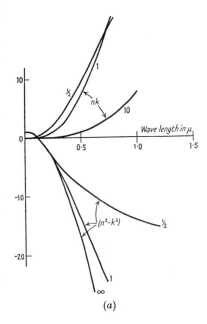

(a)

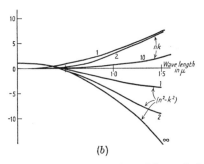

(b)

FIG. 47. Theoretical values of the optical constants according to the classical theory, with different values of the damping constant $\lambda_\tau = 2\pi m\sigma_0 c/Ne^2$ shown in units of 10^{-4} cm.

(a) Mercury: $9 \cdot 5 \times 10^{22}$ atoms per cm.[3] The observed resistivity of $94 . 10^{-6}$ ohms at $0°$ C. corresponds to $\lambda_\tau = 1 \mu$.

(b) Caesium: $0 \cdot 86 \times 10^{22}$ atoms per cm.[3] The observed resistivity of $20 . 10^{-6}$ ohms corresponds to $\lambda_\tau = 40 \mu$.

λ_0 is the wave-length for which the 'ideal' metal (Fig. 45) becomes transparent. *For good conductors the wave-length λ_τ, for which $\omega = 1/\tau$, lies in the far infra-red* (cf. p. 119). Therefore in the visible region formulae (65) are a good approximation.

Note that from (65) the 'conductivity for frequency ν', $\sigma(\nu)$, defined on p. 101, is given by

$$\sigma(\nu) = nk\nu = \frac{1}{4\pi^2}\left(\frac{Ne^2}{m}\right)^2 \frac{1}{\sigma_0\nu^2},$$

so that for short wave-lengths $\sigma(\nu)$ is *inversely* proportional to the conductivity σ_0 for steady currents for a given number of free electrons.

In Fig. 47 we show nk and n^2-k^2, calculated from formulae (63), plotted against the wave-length, for two values of N, those for Hg and for Cs, and for various values of the resistance $1/\sigma_0$.

In Fig. 48 we show the reflecting power of a metal from formulae (45), (65), in the short wave-length region, plotted against λ/λ_0; we show also some experimental points for potassium,[†] fitted with λ_0 put equal to 3,300 Å. U.

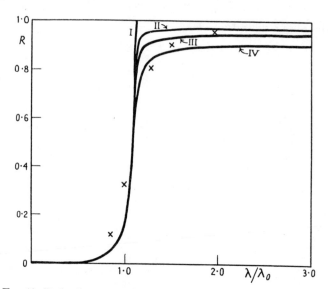

FIG. 48. Reflecting power of a metal calculated on the classical theory.

I. $\lambda_\tau = \infty$ (resistance zero).
II. $\lambda_0/\lambda_\tau = 0{\cdot}01$.
III. $\lambda_0/\lambda_\tau = 0{\cdot}025$.
IV. $\lambda_0/\lambda_\tau = 0{\cdot}05$.

The crosses show the observed reflecting power for solid potassium, with $\lambda_0 = 0{\cdot}33\,\mu$.

The observed resistivity of potassium for steady currents gives $\lambda_0/\lambda_\tau = 0{\cdot}003$; the results thus indicate that the absorption in the surface layer is much greater than can be accounted for by the resistance.

7.6. *Corrections due to the periodic field of the lattice.* The formulae (63) and (64) are valid in the quantum theory as in the classical theory so long as it is assumed that the lattice field acting on an electron is zero (approximation of Chap. II, § 3). We have seen, however, that, in the actual periodic field in a crystal, the acceleration of an electron under the action of an external field is different from, and usually smaller than, the acceleration of a perfectly free electron. Therefore, for the quantity N (number of electrons per

† Frehafer, *Phys. Rev.* **15** (1920), 110.

unit volume) occurring in (63), we must substitute the 'effective' number of free electrons, as shown in §§ 5 and 6. We must thus write instead of N,
$$N_{\text{eff}} = N_a\, n_{\text{eff}},$$
where N_a is the number of atoms per unit volume and n_{eff} the number introduced in § 4, being between 0·5 and 1 for monovalent metals, and about 0·2–0·4 for divalent metals.

Secondly, energy may be transferred from the radiation to the metal by photoelectric absorption. This may occur at the surface of the metal for any wave-length (though for frequencies below the photoelectric threshold the electrons are not ejected). The surface absorption for clean surfaces is, however, small (less than 1 per cent. of the incident light). It may also be much more strongly absorbed (up to 97 per cent.) in the body of the metal, either by ejecting an electron from a closed shell (e.g. the d shell in Cu) to the surface of the Fermi distribution, or by causing a conduction electron to jump from one zone to another, as discussed in § 5. In neither case can a quantum of radiation be absorbed for frequencies less than a certain critical frequency, ν_1. We shall therefore expect, for decreasing wave-length, a sudden increase in the absorption coefficient nk.

Thirdly, to the polarization P of the conduction electrons one must also add the polarization of the atomic cores; this will add a term $(\epsilon-1)_{\text{core}}$ to the expression for the dielectric constant; further, as we have seen in § 6, an electron in a crystal lattice under the influence of a light wave behaves like a series of oscillators with frequencies $\nu_1, \nu_2, \ldots$, and we have shown (equation (41)) how to find the polarization due to these. We call the contribution to the dielectric constant from these oscillators $(\epsilon-1)_{\text{ph}}$. We thus have finally instead of (65)
$$n^2-k^2 = 1-\frac{4\pi N_{\text{eff}}\, e^2}{m\omega^2}+(\epsilon-1)_{\text{core}}+(\epsilon-1)_{\text{ph}}. \tag{66}$$

The last two terms, however, tend to constant values as $\omega \to 0$, and will therefore be unimportant for long wave-lengths.

For high frequencies (considerably higher than those of the absorption band) we obtain, however, from the considerations of p. 104 the classical formula (65), where N is the actual number of electrons outside a closed shell.

We must mention finally the question as to whether it is correct to assume that the force on an electron is $E = D-4\pi P$, as in the preceding sections, or whether a Lorentz-Lorenz correction must be introduced, giving a force

$$E + \frac{4\pi}{3} P.$$

The question is one of some difficulty, and has been discussed by several authors;[†] we quote A. H. Wilson in stating that such a correction is inconsistent with the approximations used in deriving the one-electron wave function used here, and is automatically excluded by our assumptions.

8. Comparison with experiment

8.1. *Experimental methods.* Information about the optical constants of metals can be obtained by polarimetric methods, polarized light being reflected from a polished metal surface,[‡] or by transmission methods. A very large number of metals have been investigated by the polarimetric method;[||] the results obtained by different observers differ widely and depend largely on the method adopted of polishing the specimen. This is not surprising, since various workers have shown that a polished layer is either amorphous[††] or consists of very small crystals, and the layer may be as thick as 500 Å. U., which is of the order of the distance to which the light penetrates. Therefore, the measured optical constants for a heavily polished layer refer to metal in the amorphous rather than the crystalline state. Recent work by Lowery and his co-workers[‡‡] has been directed to obtaining a polished layer of thickness much smaller

---

† Kronig and Groenewold, *Physica*, **1** (1934), 255; Darwin, *Proc. Roy. Soc.* A, **146** (1934), 13; Wilson, loc. cit.

‡ For an account of the methods of determining n and k, see any text-book on physical optics, e.g. Wood, *Physical Optics*, 3rd ed., p. 542, London (1934).

|| Cf. *International Critical Tables* or Landolt-Börnstein's *Tabellen*.

†† The hypothesis is due to Beilby, *Aggregation and Flow in Solids* (1921), who observed through a microscope the consecutive changes occurring when a metal is polished. Electron diffraction experiments by French (*Proc. Roy. Soc.* A, **140** (1933), 637), Darbyshire and Dixit (*Phil. Mag.* **16** (1933), 961), Raether (*Zeits. f. Phys.* **86** (1933), 82), and by Finch, Quarrell, and Roebuck (*Proc. Roy. Soc.* A, **145** (1934), 676) have shown that the surface layer is either amorphous or made up of exceedingly small crystals. A discussion of the experimental evidence is given by Thomson (*Phil. Mag.* **18** (1934), 140). Hopkins and Lees (*Trans. Far. Soc.* **31** (1935), 1095) have investigated the depth of the layer. They find that it varies widely with different methods of polishing; the truly amorphous layer has a depth of from 10–50 Å. U., while down to 150–500 Å. U. the surface layer consists of much smaller crystals than for the unpolished metal. Cf. also a report by Bates (*Science Progress*, **30** (1935), 87).

‡‡ *Phil. Mag.* **13** (1932), 935; **20** (1935), 390.

than 500 Å.U., so that the optical constants shall be determined mainly by crystalline metal. The very striking results obtainable are illustrated in Fig. 49, which shows the absorption coefficient nk of copper as measured by Minor[†] and by Tool[‡] with heavily polished surface, and by Lowery. It appears from these results that the

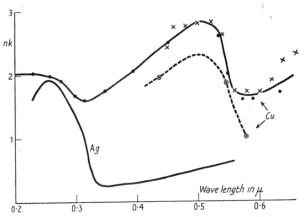

FIG. 49. Absorption coefficient nk for copper and silver.

Copper
{
• Minor.
× Tool.
⊙ Lowery, Bor, and Wilkinson,[||] using lightly polished surface.
}

crystalline metal absorbs less strongly than the amorphous, polished layer, a conclusion which will be referred to below.

We may say, therefore, that it is probably hopeless to try to account for the actual numerical values of n and k in terms of any atomic model, but that 'bumps' in the curves in which nk, n^2-k^2 are plotted against wave-length are likely to have theoretical significance. One might also expect n^2-k^2, which depends on the effective number of free electrons, to be less sensitive to the method of polishing than nk, which depends on the absorption.

In transmission experiments, from which k can be determined, the chief difficulty is to obtain a film of uniform thickness. Important results have, however, been obtained by Wood[††] for the alkali metals, for which at a critical wave-length there is a sudden change of k.

8.2. *Internal photoelectric absorption.* We have seen that, according to the theory of § 5, all metals should have an absorption band

---

[†] *Ann. d. Physik*, **10** (1903), 581.   [‡] *Phys. Rev.* **31** (1910), 1.
[||] *Phil. Mag.* **20** (1935), 390.   [††] Cf. § 7.4.

due to *internal* photoelectric absorption. An absorption band gives rise to large values of nk.

For the alkalis nk is very small in the visible and ultra-violet; the absorption band has not been observed, and is probably in the infra-red.†

For copper, silver and gold the absorption band appears to start at the following wave-lengths:‡

Metal	Cu	Ag	Au
Wave-length (Å. U.) .	5,750	3,100	$\sim$ 5,000
Energy (in e.v.) .	2·1	4·0	2·5

The absorption band may *either* be due to ejection of the 3d, 4d, or 5d electrons—i.e. to the same mechanism as X-ray absorption—*or*

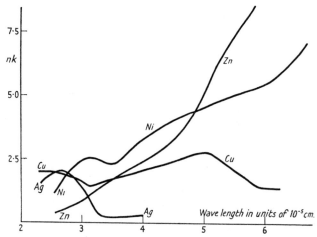

Fig. 50. Absorption coefficient nk for various metals, from Meier, *Ann. d. Physik*, **31** (1910), 1017.

to the transition of an s electron (conduction electron) to a higher zone.‖ We think that probably in Ag the latter process is responsible for the absorption edge at 3,100 Å. U., because, for this always monovalent element, it is unlikely that so small an energy as 4 e.v. is sufficient to ionize the d shell. For Cu the edge at 5,750 Å. U. may be due to ionization of the d shell, and the bump at 3,000 Å. U. (cf. Fig. 49) to the conduction (4s) electron.

Fig. 50 shows nk for a number of metals. Divalent metals (e.g. Zn) appear to have an absorption band in the infra-red; the reason is

† Cf. § 5, and also the calculations referred to in Chap. II, § 4.5.
‡ Meier, *Ann. d. Physik*, **31** (1910), 1017.
‖ Cf. § 5.

not at present clear, but may be connected with the occupation of higher zones.

The transition metals Ni, Pd, Pt also have absorption bands in the infra-red, as Fig. 50 shows, the high value of nk being too great to be accounted for by the same mechanism as electrical resistance. The discussion of these metals in Chap. VI, § 5 shows that the work required to remove an electron from a state in the d band to an empty state in the s band is very small, and we believe the absorption band to be due to such transitions. The electrons in the s band should give an absorption spectrum similar to that of a noble metal, and it is possible that the secondary maximum at 3,000 Å. U. shown in Fig. 50 for Ni may be due to this cause.

Colour of metals. The red colours of copper and gold are thus due to the absorption bands, which cause the reflection coefficient to fall as the wave-length decreases. In silver, the absorption does not begin until the ultra-violet, and in the alkalis also there is no absorption band in the visible. In the transition metals and divalent metals, absorption begins in the infra-red and is fairly uniform in the visible. It seems that one does not get a range of the visible spectrum with no photoelectric absorption unless only the first Brillouin zone is occupied, which is probably the reason why the α- and β-brasses are red or yellow† and γ-brass colourless (cf. Chap. V, §§ 2.5, 2.6).

8.3. *Absorption for long wave-lengths.*

(a) *Far infra-red; the Hagen-Rubens relation.* As shown in § 7.3, the condition for the validity of the Hagen-Rubens formula for the reflecting power is that the period of the light, or more exactly $1/2\pi\nu$, shall be long compared with the time of relaxation τ. The critical wave-length $\lambda_\tau = 2\pi c\tau$, beyond which the Hagen-Rubens relation should be valid, calculated from the observed electric resistance at room temperature for steady fields from the classical formula (1) of Chap. VII, has the following approximate values:

$$\lambda_\tau \simeq 100\mu = 10^{-2} \text{ cm.} \qquad \text{noble metals and alkalis at } 0° \text{ C.}$$

$$= 7\mu \qquad \text{platinum at } 0° \text{ C.}$$

$$= 1\cdot5\mu \qquad \text{platinum at } 1,000° \text{ C.,}$$

† α-brass (Cu–Zn) is more yellow than Cu, which means that the absorption band is farther towards the blue end of the spectrum. If the absorption band is due to the ejection of a $3d$ electron from copper, then in brass the energy required for the transition should be greater than for pure copper, because E_{max}, the energy of the lowest unoccupied state, will increase with the mean number of electrons per atom (cf. Chap. V, § 3).

and in quantum mechanics these should give at least the right orders of magnitude.

Now Hagen and Rubens† have shown that, for radiation of wavelength $12\,\mu$ and the metals copper, silver, and gold, equation (55) is in agreement with experiment for the reflecting power to within 30 per cent. of $(1-R)$. In view of the large time of relaxation and consequent large value of λ_r for these good conductors, this fact is difficult to interpret.

For the poor conductors, platinum, nickel, steel, the agreement between theory and experiment is better, to within 10 per cent.; for platinum, also, Hagen and Rubens have deduced $1-R$ from the emissivity between the temperatures 635° and 1,455° C., and have found agreement with the theoretical formula for $6\,\mu$ and $4\,\mu$ but not for $2\,\mu$, using always for σ_0 in formula (55) the observed electrical conductivity at these temperatures. We must deduce from these measurements that the electrical resistance of the surface layers of these poor conductors is not much greater than for the metal in bulk.

(b) *Near infra-red and visible.* For copper, silver, and gold in the red and near infra-red and for silver in the visible, there appear to be no absorption bands of the type discussed in § 5. There is, however, considerable absorption (cf. Fig. 49), and the observed electrical conductivity for steady fields is too small by a factor of about ten to account for it‡ using formulae (63). We believe that this high absorbing power is due to the fact that the surface layer is amorphous. This conclusion is strengthened by the large decrease in the absorption which can be obtained by preparing a thin polished layer (Fig. 49). The high absorption may be due to a large 'Restwiderstand' in the surface layer, or more probably to surface photoelectric absorption at the boundaries between the small crystals of the polished layer.

Further evidence for the hypothesis that the thermal resistance is not to any large extent responsible for the absorption in the visible is afforded by the fact that the optical constants are in general almost independent of temperature.‖

† Loc. cit. For nickel near the Curie point see also Löwe, *Ann. d. Physik*, **5** (1936) 213.

‡ For actual values, cf. Meier, loc. cit., for the visible region, and Kronig, loc. cit., for the near infra-red, who uses the experimental results of Försterling and Fréedericksz, *Ann. d. Physik*, **40** (1913), 201.

‖ de Selincourt, *Proc. Roy. Soc.* A, **107** (1925), 247, however, finds that the reflecting power of silver near the minimum at 3,100 Å. U. increases with temperature.

8.4. *Dispersion.* It is essential to remember that, for good conductors, the time of relaxation τ is long compared with the period
$(1/\omega = 1/2\pi\nu)$ of light in the visible part of the spectrum. Thus the
wave-length given by the equation

$$\lambda_\tau = 2\pi c \tau$$

is $60\,\mu$ for Na, $140\,\mu$ for Ag, $7\cdot2\,\mu$ for Pt, $0\cdot75\,\mu$ for Hg (liquid), while
the visible spectrum ends at $0\cdot72\,\mu$. Thus, except for bad conductors
like mercury, one may use formulae (65) for the optical constants,
so that n^2-k^2 is independent of the mean free path and depends only
on the position of the absorption band and on the value of the
critical wave-length λ_0 (p. 110), at which, for an ideal metal with
infinite mean free path, $\mathbf{n}^2$ changes sign.

We first note that if the low-frequency limit of the absorption
band is sufficiently high, or if the absorption band is weak, the
optical constants may be used to determine N_{eff}, the effective number
of free electrons per unit volume. Equation (65) or Fig. 47 shows
that, for $\lambda \gg \lambda_0$, $$k \gg n;$$

in this region, therefore, we have approximately from (65)

$$k \simeq \lambda/\lambda_0$$

for long wave-lengths. If therefore k/λ plotted against the wave-
length tends to a constant value for long wave-lengths, we may deduce

$$\frac{k}{\lambda} \to \frac{1}{\lambda_0} = \frac{1}{c}\sqrt{\frac{N_{\text{eff}}e^2}{\pi m}}.$$

For copper, silver, and gold, where the low-frequency limit of the
absorption band lies in the visible, λ_0 and hence N_{eff} may be determined in this way from the polarimetric determinations of Försterling
and Fréedericksz;[†] this has been done by Kronig,[‡] who deduces the
following values for $n_{\text{eff}} = N_{\text{eff}}/N$:

Effective number of free electrons n_{eff} per atom, from polarimetric
measurements

Cu	Ag	Au
0·37‖	0·89	0·73.

† *Ann. der Physik*, **40** (1913), 201.
‡ Loc. cit. Cf. also Sommerfeld and Bethe, p. 583.
‖ The low value for copper is interesting; Dr. Lowery has informed us that k is
less sensitive to the method of polishing than nk; we may thus deduce that the
effective number of free electrons is actually considerably smaller for crystalline copper
than for silver and gold (cf. Appendix I).

For the alkali metals, as already pointed out, the high reflection coefficient and small value of k suggest that the absorption band lies in the infra-red and is weak in the visible. No experiments have been carried out in the infra-red. The polarimetric determinations in the visible are somewhat uncertain; from the values for the visible given in the *International Critical Tables*, we deduce[†] the following values of n_{eff}

Element	Na	K
n_{eff}	0·87	0·75.

Estimations of the optical constants can also be made from the opacity and reflection coefficient for frequencies in the neighbourhood of λ_0. As we have already seen (Fig. 48), for a metal in which there is no photoelectric absorption, there is a rapid change at $\lambda = \lambda_0$ in the extinction and reflection coefficients. To estimate λ_0, we must, however, estimate the contribution to the polarization from the atomic cores. This may be deduced from the known polarizability of the cores: viz.[‡]

	Li^+	Na^+	K^+	Rb^+	Cs^+	Ag^+
$\dfrac{4\pi P}{E} =$	0·017	0·058	0·14	0·195	0·26	1·3.

We have, therefore, e.g. for potassium, the following equation for the wave-length λ_0 for which $\mathbf{n}^2$ changes sign

$$1\cdot14 - N_{\text{eff}}\, e^2 \lambda_0^2/\pi mc^2 = 0.$$

With R. W. Wood's values for λ_0 (p. 110) we obtain

	Li	Na	K	Rb	Cs
$n_{\text{eff}} =$	0·55	1·1	0·97	0·94	0·85.

A determination of n_{eff} for any other metal by this method is not possible, because, owing to the strong absorption bands, the dispersion for frequencies near to ν_0 does not follow the simple formula (65).

At the wave-length $\lambda_0 = c/\nu_0$ at which an alkali metal becomes transparent its reflection coefficient drops also. This drop in the reflection coefficient was shown first by Frehafer,[‖] whose experimental points for potassium are shown in Fig. 48.

[†] Mott and Zener, *Proc. Camb. Phil. Soc.* **30** (1934), 249.

[‡] Cf. Van Vleck, *Electric and Magnetic Susceptibilities*, p. 225, Oxford (1932). The quantity plotted there is $\kappa = \dfrac{4\pi}{3} \dfrac{\text{moment per gram atom}}{E}$; to obtain $4\pi P/E$ we have multiplied κ by $3 \times \text{density/atomic wt.}$

[‖] *Phys. Rev.* **15** (1920), 110.

For Cu, Ag, and Au, in order to discuss the dispersion in the optical region, we must take account of the anomalous dispersion. We shall give here a discussion of silver; copper and gold could be treated in the same way, but the available experimental evidence does not make it worth while.

The polarizability of the cores is[†]

$$(\epsilon-1)_{\text{core}} = 1{\cdot}3 \tag{67}$$

and we shall assume this to be independent of wave-length. For the anomalous dispersion corresponding to the absorption band, we may use formula (41), which gives the dispersion in terms of the absorption. Denoting by $(nk)_{\text{ph}}$ that part of nk which is due to photoelectric absorption we have from (41), for frequency ν,

$$(\epsilon-1)_{\text{ph}} = \frac{4}{\pi} \int (nk)_{\text{ph}} \frac{\nu' \, d\nu'}{\nu^2-\nu'^2}, \tag{68}$$

nk being taken as a function of ν'. We then have for the total dispersion

$$\epsilon-1 = n^2-k^2-1 = (\epsilon-1)_{\text{core}}+(\epsilon-1)_{\text{ph}}-\frac{N_{\text{eff}}e^2}{\pi m \nu^2}. \tag{69}$$

In Fig. 51 we show, for silver:

I. The experimental dielectric constant n^2-k^2.

II. The dielectric constant calculated for perfectly free electrons with $n_{\text{eff}} = 0{\cdot}87$.

III. The difference, $(\epsilon-1)_{\text{core}}+(\epsilon-1)_{\text{ph}}$.

IV. The theoretical value of this quantity, calculated from (67) and (68).

In calculating $(\epsilon-1)_{\text{ph}}$ we took the experimental values of $(nk)_{\text{ph}}$ shown in Fig. 49; it is quite easy to separate approximately the part due to the absorption band from that due to other causes. For wave-lengths less than $0{\cdot}24\,\mu$ a rough extrapolation of nk was made.

The well-known minimum in the reflection coefficient[‡] and maximum in the transparency of silver[||] is thus explained as follows:[††] For long wave-lengths n^2-k^2 is negative, and the extinction coefficient is large, though the absorption coefficient is small. At a certain wave-length n^2-k^2 changes sign, and the distance that the light can penetrate is determined by the absorption (cf. Fig. 45 (b)), as in Wood's

[†] Van Vleck, loc. cit., p. 225.
[‡] de Selincourt, *Proc. Roy. Soc.* A, **107** (1925), 247.
[||] Smakula, *Zeits. f. Phys.* **86** (1933), 185; Fröhlich, ibid. **81** (1933), 297.
[††] Kronig, *Naturwiss.* **21** (1934), 11.

experiments on the alkalis. At a slightly shorter wave-length photo-electric absorption sets in, and the reflection and extinction coefficients increase. The fact that the two wave-lengths are so close together must be a consequence of the particular form of the periodic field, and does not occur for any other metal.

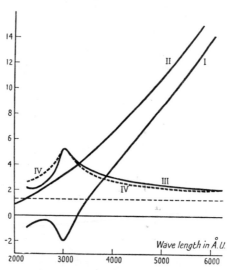

FIG. 51. Dielectric constant of silver.

I. $-(n^2-k^2)$ observed by Meier (*Ann. d. Physik*, **31** (1910), 1017).
II. $-(n^2-k^2)$ calculated for free electrons with 0·87 electron per atom.
III. The difference between I and II.
IV. The theoretical value of this quantity calculated from (67) and (68).
The horizontal dotted line gives the contribution from the cores, given by (67).

8.5. *Liquid metals.* The optical constants of mercury have been determined by Meier† between 3,200 and 6,200 Å. U., and of liquid bismuth, lead, cadmium, and tin between 5,790 and 4,040 Å. U. by Kent.‡ As both authors have pointed out, the results obtained agree with the *classical* formulae (63), (64), using the following values of the parameters n_{eff} (effective number of free electrons per atom) and $1/\sigma_0$ (resistivity):

Metal	Bi	Pb	Cd	Sn	Hg
n_{eff}	5·1	5·1	2·4	4·1	2·1
$1/\sigma_0$ (microhms/cm.) . . .	128	94	33·4	54	87·3
Resistivity, observed (microhms/cm.)	134	98	34	52	94

† Loc. cit.   ‡ *Phys. Rev.* **14** (1919), 459.

In the last row we give also the observed resistivity for steady fields. The agreement, both between n_{eff} (observed) and the number of electrons in the outermost shell of the atom, and between $1/\sigma_0$ and the observed resistivity, is remarkable.

In view of the considerations of Chap. VII, § 10, we believe that the effective number of electrons N_{eff} in the liquid state is about the same as in the solid state (except for Bi and Hg) and hence *less* than one per atom. The large values of n_{eff} obtained seem to us to show, therefore, that the energy $h\nu$ of a quantum of visible light is much greater than the energy gap separating the Brillouin zones; under these conditions, as we saw on p. 104, we obtain for N_{eff} the *actual* number of free electrons.

Experiments to show whether the optical constants are different in the liquid and solid states have not been carried out; if our explanation is correct, the same value of N_{eff} should be obtained in both states, since the energy gap is not likely to be very different in the liquid and solid phases, in any case for close-packed metals.

8.6. *Optical constants of alloys.* Very few reliable determinations exist. Recently Lowery, Bor, and Wilkinson† have investigated the copper-nickel series of alloys, which form the cubic close-packed structure over the whole range. Certain of their results have been interpreted by Mott,‡ assuming that the absorption in nickel is due to the ejection of electrons from the $3d$ shell. It appears from the experimental results that the energy required to eject an electron from a closed d shell containing ten electrons is less than from an incomplete d shell; both are present in nickel (cf. Chap. VI, § 5).

9. X-ray emission and absorption

The study of the emission and absorption of X-rays by metals provides some of the most direct evidence for the conclusions of Chapter II. In particular, we may obtain a direct proof that the energies of the conduction electrons lie in a range of values given approximately by the theoretical formulae, and also that the unoccupied and occupied states are divided into zones, as explained in Chap. II, § 4. We shall discuss first the emission of soft X-rays.

9.1. *Soft X-ray emission.* When a metal is bombarded by electrons of sufficient energy, the K and L levels are ionized by collision. The radiation emitted when the conduction electrons make transitions

---

† *Phil. Mag.* **20** (1935), 390.   ‡ Ibid. **22** (1936) (in press).

to the empty K and L levels does not form a sharp line, but a band; for, although the K level is relatively sharp even in the lightest metal, lithium, the conduction electrons, as we have seen, have all energies between two limits. Thus the breadth (in energy units) of the X-ray emission band gives directly the breadth of the Fermi distribution—i.e. the range of energies of the conduction electrons.

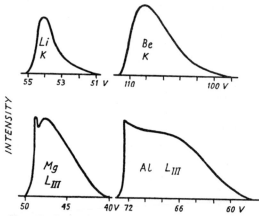

FIG. 52. Intensity of X-ray emission plotted against energy in volts (O'Bryan and Skinner).

Since the breadth of Fermi distribution is, at most, about 15 volts, one must, in order to obtain a band of measurable breadth, use transitions to relatively high levels, giving soft X-rays of energies one or two hundreds of volts. Fig. 52 shows the X-ray K emission bands of lithium, beryllium, magnesium, and aluminium, measured by O'Bryan and Skinner.† The sharp cut-off on the high-energy side represents the 'top' of the Fermi distribution (Fig. 40 (a)); it is due to the fact that the states with kinetic energy greater than E_{max} are *empty*. Magnesium is a divalent metal, and we know, therefore, that the conduction electrons lie in two zones (cf. § 1, and particularly Fig. 40). The kink in the emission band will obviously be connected with this fact.

A detailed theory of the emission has been given by Jones, Mott, and Skinner ‡ The intensity $I(E)$ of the emission band will be proportional to the product of $N(E)$, the density of states for corresponding energy of the conduction electrons, and the transition probability $p(E)$, so that we have

$$I(E) \propto N(E)p(E). \tag{70}$$

† *Phys. Rev.* **45** (1934), 370.   ‡ Ibid. 378.

Consider now the behaviour at the low-energy limit of the emission band; we have $N(E) = \text{const.}\sqrt{E}$. The behaviour of $p(E)$ depends on whether the final state is an s or p state. $p(E)$ is proportional to the square of an integral of the type

$$\int \psi_f^* \frac{\partial}{\partial x} \psi_i \, d\tau, \tag{71}$$

where ψ_i, ψ_f are the initial and final wave functions. ψ_i is symmetrical about any nucleus of the lattice (cf. Fig. 30). Thus if the wave function ψ_f of the X-ray level is an s wave function, the integral vanishes, by symmetry; and it may be shown further that, as E increases, $p(E) \propto E$. We thus have, for small E,

$$I(E) = \text{const.}\, E^{\frac{3}{2}} \quad \text{(final state } s)$$
$$= \text{const.}\, E^{\frac{1}{2}} \quad \text{(final state } p). \tag{72}$$

We note (Fig. 52) that for Li and Be K radiation, where the final state is an s state, that the tail at the low-energy end of the band seems more pronounced than for Mg L_{III} radiation, where the final state is a p state. This is in agreement with formula (72).

We consider now the transition probability for the lowest state in the *second* Brillouin zone. As shown in Chap. II, § 4.4, there are two possibilities: the nodes of the wave function may either pass through the nuclei of the atoms, or mid-way between them (cf. Fig. 30). The experimental results for Be K and Mg L_{III} radiations suggest that the transition probability is small when the final state is s, large when it is p. This will be the case if the nodes pass mid-way between the atoms for both these metals (Fig. 30).

The breadths of the bands for the metals investigated by O'Bryan and Skinner are in surprisingly good agreement with the simple Sommerfeld formula, Chap. II, equation (19), as the following table shows:

Element	Breadth of band (volts)	
	Observed	Calculated
Li . . .	4·2±0·6	4·6†
Be . . .	13·5±2·5	13·8
Na . . .	3·5±1	3·2
Mg . . .	4·0±1·5	7·2
Al . . .	16·0±2	12·0
Si . . .	19·2±2·5	13·0

† With the corrections introduced by Seitz (Chap. II, § 4.5) this becomes 3·5 e.v.

The numbers of electrons assumed in obtaining the theoretical values were: Li and Na, 1; Be and Mg, 2; Al, 3; Si, 4.

Other measurements of soft X-ray emission bands have been made by Siegbahn and Magnusson, *Zeits. f. Phys.* **87** (1934), 291; **88** (1934), 559; **95** (1935), 133; **96** (1935), 1; by Brodi, Glocker, and Kiesig, ibid. **92** (1934), 27; and by Kiesig, ibid. **95** (1935), 555. Of especial theoretical interest are the K emission bands for diamond and graphite obtained by Siegbahn and Magnusson, ibid. **96** (1935), 10; the shapes of the bands are similar to that shown in Fig. 52 for Be, as one would expect for an insulator. The same authors (ibid. **87** (1934), 309) have examined a band from Al_2O_3 and compared it with that from pure Al.

9.2. *X-ray absorption.* Coster and Veldkamp, and others, have observed a fine structure in the X-ray absorption spectrum of metals, extending several hundreds of electron volts on the high-frequency side of the absorption edge (cf. Fig. 54). According to Kronig,† the reason for this is as follows: the X-rays are absorbed by electrons which make transitions from the deep-lying K and L levels to empty levels lying above the Fermi distribution. Now these levels are grouped into zones, just as are the filled levels responsible for the X-ray emission. Thus neither the density of states, $N(E)$, nor the transition probability $p(E)$, is a monotonic function of the energy. The absorption coefficient is proportional to the product of these two, and will therefore show a fine structure.

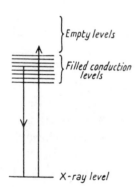

FIG. 53. Illustrating the emission and absorption of X-rays in a metal.

Very near the absorption edge the fine structure will depend on the field within the crystal in a way that has not yet been investigated theoretically. At some distance (say greater than 50 electron volts) from the edge, however, it is possible to treat the electrons as nearly free (in the sense of Chap. II, § 4.2), and hence to ascribe a *unique wave number* to each state of the electron. The fine structure depends, then, essentially on the fact that, for certain wave numbers, the ejected electron suffers Bragg reflection from the lattice. This

† *Zeits. f. Phys.* **70** (1931), 317; **75** (1932), 191. Cf. also *Handb. d. Phys.* **24/2** (1933), 295.

has two consequences: firstly, for a given direction of the ejected electron, there will be ranges of energy which are forbidden, so that absorption cannot take place; secondly, the wave functions, for the critical wave-lengths, are standing waves, and we have seen in the last section that the transition probabilities from X-ray levels may vanish in this case. The transition probabilities depend strongly on whether the initial state is an s or a p state; one would expect a minimum for an s state to correspond to a maximum for a p state.

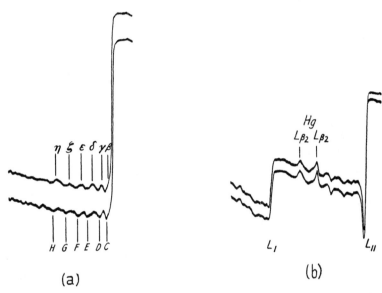

(a)

(b)

FIG. 54. Fine structure of X-ray absorption edges (from Veldkamp, *Physica*, **2** (1935), 25).

(a) L_{III} edge of gold.

(b) L_I and L_{II} edges of tungsten. Note the maximum in the absorption next to the L_{II} edge. No such maximum is observed near the L_I edge.

As a consequence of the theory of Kronig, therefore, it may be predicted:

(a) that, except in the immediate neighbourhood of the absorption edge, the fine structure depends rather on the crystal structure than on the nature of the constituent atoms;

(b) that the fine structure will depend on whether the initial X-ray level is an s or a p level.

The experimental work, carried out for the most part in Groningen, has established the following points:

(a) In monatomic gases there is no fine structure.†

(b) In solids the general character of the fine structure depends on the lattice structure; it is thus similar for the body-centred cubic elements Cr, Fe, but the face-centred elements Cu and Ni show a different type of structure. Similarly, the crystals KCl, KBr, KI have their own type of structure, as do the γ- and ε-brasses.‡

(c) In substances showing the same fine structure, the distance between the absorption edge and a given maximum, measured in

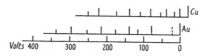

Fig. 55. Maxima (short vertical lines) and minima (long vertical lines) of the absorption coefficient at the Au—L_{III} and Cu—K edges, both taken with an Au–Cu alloy (from Coster and Veldkamp).

energy, or frequency, units, is inversely proportional to the square of the lattice constant, as the theory leads one to expect.‖

(d) A particularly convincing proof of the theory has been given by Coster and Veldkamp,†† who examined the K absorption edge of Cu and the L_{III} edge of Au, in an alloy (solid solution) containing 50 atomic per cent. of Cu and 50 atomic per cent. of Au. They found that both edges showed a similar fine structure with the *same* distance between corresponding maxima, corresponding to the lattice constant of the composite Cu–Au crystal (cf. Fig. 55). On the other hand, the fine structure was displaced relatively to the main L_{III} edge in the gold, compared with the copper K absorption. This may be due to the fact that in one case we are dealing with absorption from an s level, and in the other case from a p level (see above).

(e) It has been possible‡‡ to account quantitatively for the positions of the main maxima and minima for Cu (face-centred) and Fe (body-

† Cf. Coster and van der Tuuk, *Zeits. f. Phys.* **37** (1926), 367, for measurements in argon; Hanawalt, *Phys. Rev.* **37** (1931), 715, for measurements in Kr and Xe and in Zn and Hg vapours.

‡ Smoluchowski, *Zeits. f. Phys.* **94** (1935), 775.

‖ Cf. the summaries by Kronig, *Handb. d. Phys.* **24/2** (1934), 295; Glocker, *Naturwiss.* **20** (1932), 536; Hanawalt, *J. Franklin Inst.* **214** (1932), 569. For the original papers see for instance Coster and Veldkamp, *Zeits. f. Phys.* **74** (1932), 191, or ibid. **82** (1933), 776; *Physica*, **2** (1935), 25, where references to earlier work are given.

†† *Zeits. f. Phys.* **74** (1932), 206.   ‡‡ Kronig, *Zeits. f. Phys.* **75** (1932), 191.

centred), by assuming that for every velocity v of the ejected electron for which perpendicular reflection from a set of crystal planes takes place, a minimum in the absorption occurs. For some ranges of energy there are a large number of such planes, and these produce the minima in the absorption coefficient.

(*f*) For the transition metals W and Ta, Veldkamp† has found a very strong absorption line close to the L_{II} and L_{III} edges, but not the L_I (cf. Fig. 54). We know (Chap. VI, § 4) that there exists a band of unoccupied d states just above the highest occupied state, and that the density of states, $N(E)$, in this band is abnormally large. The large density of states is connected with the high paramagnetism and low electrical conductivity of these metals. It is reasonable to suppose that the absorption line observed by Veldkamp corresponds to transitions to these states, and that its non-occurrence for the L_I edge is due to the fact that the initial level is an s level, so that the transition would be forbidden.

The absorption line was not observed for Pt, perhaps because of the insufficient resolving power of the apparatus.

The most recent results on the emission and absorption of ultra-soft X-rays are described by Skinner, *Reports on Progress in Physics*, **5** (1938), 257.

† *Physica*, **2** (1935), 25.

Notes of recent developments.

Recently Skinner and Johnston have investigated the fine structure of absorption edges in the region of ultra-soft X-rays (100–300 A.), thus obtaining a much higher resolving power (∼ 0·25 e.v.) than in the work at Groningen. Preliminary investigations (*Nature*, **137** (1936), 826) on the L_{23} edge of metallic magnesium show a pronounced variation in the absorption coefficient within 3 e.v. from the edge.

p. 98. Darwin (*Proc. Roy. Soc.* A, **154** (1936), 61) has shown that experiments devised to show the inertia of an electric current will always determine the actual mass of the electron rather than the effective mass.

p. 103. The absorption of light by insulating crystals has recently been discussed by Frenkel (*Phys. Zeits d. Sowjetunion*, **9** (1936), 158). Frenkel shows that a *line* absorption is possible for insulators, though not for metals.

IV

COHESION

1. Types of cohesion

IN this chapter we discuss the calculation of the binding energy of a metal. The purpose of the calculation is to obtain the total energy of the metal, at the absolute zero of temperature, as a function of the atomic volume; one may thus deduce the actual *lattice constant* (i.e. the one for which the energy is a minimum), the *binding energy* (heat of sublimation), and the *compressibility*. Some curves deduced from the experimental data are shown in Chapter I, Fig. 6.

The cohesive forces in solids may be classified roughly as hetero-polar, homopolar, metallic, and van der Waals, though there is no *sharp* distinction between the various types. Of these, heteropolar binding is mainly due to an electrostatic attraction between charged spherical ions; the alkali halides are a typical example. The theory of such crystals has been worked out in great detail by Born[†] and his co-workers and will not be discussed further here.

1.1. *van der Waals attraction.* This is responsible for the cohesion of the rare gases in the solid state, and probably of most molecular lattices. The theory of the van der Waals attraction has been given by London.[‡] The van der Waals attraction is due to the mutual polarization of the atoms; it is the only force between neutral un-excited atoms if their wave functions do not overlap. According to London, the potential energy of a pair of atoms distant R apart is

$$V(R) = -C/R^6,$$ (1)

where C is given by

$$C = \frac{3}{2m^2}(e\hbar)^4 \sum_k \sum_{k'} \frac{f_{0k} f_{0k'}}{(E_k - E_0)(E_{k'} - E_0)(E_k + E_{k'} - 2E_0)},$$ (2)

and where E_0 is the energy of the ground state, E_k of the excited states, and f_{0k} the corresponding oscillator strengths.[‖] If V, E_k are

[†] Cf., for example, Born, *Ergebn. d. exakt. Naturw.* **10** (1931), 387, or Born and Göppert-Mayer, *Handb. d. Phys.* **24/2** (1933).

[‡] *Zeits. f. phys. Chem.* B, **11** (1930) 222; *Zeits. f. Phys.* **63** (1930), 245. Cf. also Polanyi and Cremer, *Zeits. f. phys. Chem.* B, **14** (1931), 435; *Handb. d. Radiologie*, **6** (1934).

[‖] Cf., for example, Bethe, *Handb. d. Phys.* **24** (1933), 431. The oscillator strength is given by

$$f_{0k} = \frac{2m}{\hbar^2}(E_k - E_0) \left| \int \psi_k^* x \psi_0 \, d\tau \right|^2.$$

expressed in electron volts, R in Å. U., the factor to the left of the $\sum$ is 17,600.

In the case where one optical transition with frequency ν has oscillator strength nearly unity, so that by the f sum rule[†] the others may be neglected, this becomes

$$C = \frac{3}{4}\left(\frac{e^2\hbar^2}{m}\right)^2 \frac{1}{(h\nu)^3}$$
$$= \tfrac{3}{4}e^2h\nu\alpha^2, \tag{3}$$

where $\alpha = e^2/4\pi^2m\nu^2$ is the polarizability[‡] of the atom.

We see from these formulae that the van der Waals attraction will be large if the excitation potential of the atoms is small.

London[||] has calculated the energy of face- and body-centred cubic lattices, where the atoms are held together by van der Waals forces. He finds that the energy in ergs per atom of such a lattice is

$$W = -3{\cdot}613C/\Omega^2 \quad \text{face-centred lattice} \left.\begin{array}{c} \\ \\ \end{array}\right\} \tag{4}$$
$$= -3{\cdot}63C/\Omega^2 \quad \text{body-centred lattice,}$$

where Ω is the volume per atom in cm.^{3} This formula neglects the energy of the repulsive forces, and will therefore give numerically too great a value.

References to numerical values of C are given on p. 142.

1.2. *Homopolar binding.* We do not wish to make any sharp distinction between homopolar and metallic binding. For both, two methods of calculation are possible, the method of Bloch, Wigner, and Seitz, and the method of London and Heitler.

The method of London and Heitler[††] has been used by Slater[‡‡] in an early paper to calculate the cohesive forces in sodium, but its development is extremely complicated, and it does not appear to be so well suited to this problem as the method of Wigner and Seitz outlined in the next section. It is, however, probably the most accurate method for calculating the interaction between closed shells.

We give below a summary of some of the more important applica-

---

† The f sum rule states that $\sum_k f_{0k} = 1$; cf. Bethe, loc. cit. 434.

‡ For a table of polarizabilities, cf. van Vleck, *Electric and Magnetic Suscepti-bilities*, p. 225.

|| Loc. cit., first reference; see also Kronig, *Handb. d. Phys.* **24/2** (1933), 285.

†† *Zeits. f. Phys.* **44** (1927), 455.

‡‡ *Phys. Rev.* **35** (1930), 509. Cf. also Rosen, ibid. **38** (1931), 255.

tions of the London-Heitler method which have been made to problems of the solid state:

Taylor, Eyring, and Sherman† have calculated the binding energies of four, five, or six molecules of Na, Cu, and H arranged in various configurations. They find that a small crystal has a tendency to be unstable, and to split up into molecules.

Bleick and Mayer‡ have calculated the exchange repulsion between two ions or atoms having a rare gas electron configuration, and have applied their results to neon. Born and Mayer,‖ Mayer and Helmholz,†† and Huggins and Mayer‡‡ have discussed the repulsive energy between alkali and halogen ions; some of their results are given in § 2·3 of this chapter.

Brück‖‖ has calculated the interaction between s and p shells using hydrogen-like wave functions.

Jensen and Lenz††† have developed a statistical method, similar to that of Thomas and Fermi (cf. Chap. II, § 1.4), for calculating the repulsion between the ions of polar crystals; the method has been applied to the ions of metallic copper (cf. § 3.2).

2. Metallic cohesion; method of Wigner and Seitz

The method which has had by far the greatest success in the calculation of metallic binding is that of Wigner and Seitz;‡‡‡ this will now be discussed. We shall discuss first the 'one-electron' metals, copper, silver, and gold and the alkalis, in which one valence electron per atom‖‖‖ moves in the field of atomic cores which are closed shells.

In the method of Wigner and Seitz the valence electrons and cores are treated separately; the non-electrostatic interaction between the cores will be discussed below; the valence electrons are thought of as moving freely through the lattice in the sense discussed in Chapter II. Each electron will therefore have a wave function $\psi_k(\mathbf{r})$, only two electrons corresponding to each wave function. The method by which the wave functions may be obtained has already been discussed in

---

† *J. Chem. Phys.* **1** (1933), 68. ‡ Ibid. **2** (1934), 252.
‖ *Zeits. f. Phys.* **75** (1932), 1. †† Ibid. **75** (1932), 19.
‡‡ *J. Chem. Phys.* **1** (1933), 693. ‖‖ *Zeits. f. Phys.* **51** (1928), 707.
††† Ibid. **77** (1932), 713, 722; **89** (1934), 713.
‡‡‡ Wigner and Seitz, *Phys. Rev.* **43** (1933), 804; **46** (1934), 509 (Na); Seitz, ibid. **47** (1935), 400 (Li); Fuchs, *Proc. Roy. Soc.* A, **151** (1935), 585 (Cu); Wigner, *Phys. Rev.* **46** (1934), 1002; Kimbal, *J. Chem. Phys.* **3** (1935), 560 (diamond); Slater, *Phys. Rev.* **45** (1934), 794; *Rev. Mod. Phys.* **6** (1934), 210; Gorin *Phys. Zeits. d. Sowjetunion*, **9** (1936), 328 (K).
‖‖‖ For evidence that Cu and Au are monovalent in the metallic state cf. Appendix I.

Chap. II, § 4.5; we must first see how to calculate the energy of the crystal as a whole when the wave functions are known.

The energy of the crystal may be divided up into potential energy and kinetic energy. The potential energy will be made up of the following terms:

(1) The electrostatic interaction between the ions;[†] if $\mathbf{R}_i$ denotes the positions of an ion, this term is

$$H_1 = \sum_i \sum_j \frac{e^2}{|\mathbf{R}_i - \mathbf{R}_j|}, \tag{5}$$

the summation being such that each pair is counted once only.

(2) The electrostatic interaction between the electrons; if $\mathbf{r}_j$ denotes the position of an electron, this term is

$$H_2 = \sum_i \sum_j \frac{e^2}{|\mathbf{r}_i - \mathbf{r}_j|}. \tag{6}$$

(3) The interaction energy between the electrons and the ions. We denote by $V(r)$ the potential energy[‡] of an electron distant r from an ion, and suppose that[||] $V(r) \to -e^2/r$ for large r. Then the interaction energy is

$$H_3 = \sum_i \sum_j V(|\mathbf{R}_i - \mathbf{r}_j|). \tag{7}$$

Let $\Psi(\mathbf{r}_1, \mathbf{r}_2, \ldots)$ denote the wave function of the whole system of electrons in the lattice; this wave function will of course depend on the positions of the cores, which are supposed to be at rest in their positions of equilibrium. Then the total potential energy of the lattice will be

$$U = \int \Psi^*(H_1 + H_2 + H_3)\Psi \, d\tau_1 \, d\tau_2 \ldots, \tag{8}$$

and the kinetic energy of the electrons in the lattice

$$T = \int \Psi^* H_4 \Psi \, d\tau_1 \, d\tau_2 \ldots, \tag{9}$$

where
$$H_4 = -\frac{\hbar^2}{2m} \sum_i \left(\frac{\partial^2}{\partial x_i^2} + \frac{\partial^2}{\partial y_i^2} + \frac{\partial^2}{\partial z_i^2} \right). \tag{10}$$

The total energy of the lattice is the sum of T and U.

† We assume the ions to be singly charged and not to overlap. The energy due to their overlapping, also the van der Waals attraction between them, is considered below.

‡ If we wish to include exchange interaction between valence electron and core, V will be an operator (cf. Fock, *Zeits. f. Phys.* **81** (1933), 195).

|| This assumption is necessary in order that the integrals (8) and (9) should converge.

2.1. *Calculation without exchange.* As a first approximation we replace Ψ by a simple product of one-electron wave functions, each of the type considered in Chap. II, § 4.5,

$$\Psi = \psi_1(\mathbf{r}_1)\psi_2(\mathbf{r}_2)\ldots. \tag{11}$$

We refer to this wave function as representing the 'Hartree approximation'. It neglects the correlation between the positions of the electrons; that is to say, so far as can be deduced from (11), the position of any one electron is independent of where all the other electrons are. Actually, as we shall see later on, any one electron is surrounded by a 'hole', in which other electrons are unlikely to be.

Using the wave function (11), the potential energy term (8) becomes merely the electrostatic energy of the array of positive ions together with a continuous negative charge distribution, of density, at any point,†

$$\rho = -e \sum_i |\psi_i(\mathbf{r})|^2. \tag{12}$$

If now we draw lines connecting nearest and next neighbours, and draw planes bisecting them perpendicularly, we divide the lattice up into a series of polyhedra, one surrounding each atom (cf. Chap. II, § 4.5, and Fig. 31). The potential energy of the lattice may be divided up into:

(1) the interaction between the charges in one polyhedron,

(2) the interaction of the polyhedra with each other.

Since each polyhedron is electrically neutral and for symmetrical structures not very far from spherical, we should expect the second term to be small. That this is in fact the case is shown below. The interaction energy of the polyhedra with one another may thus be treated as a small correction, and we have only to calculate the energy of the charges within an *isolated* polyhedron, which may, to a sufficient degree of approximation, be replaced by a sphere of equal volume.

We must now point out the chief errors that have been introduced by writing the wave function in the form of the simple product (11). With the wave function (11) there is no correlation between the positions of the electrons; i.e. the positions of all the other electrons are independent of where any given electron is. For reasons to be

† This is not the case for an atom when we use the same approximation, because there the number of single-electron wave functions which contribute to the charge density is small, while here it is of the order of the total number of atoms in the crystal, so that each wave function makes a vanishingly small contribution to the total charge density.

discussed below, this is not the case for the true wave function; each electron is surrounded by a sphere in which the other electrons are unlikely to be. In their first paper (1933) Wigner and Seitz make the necessarily rather rough assumption that, as a consequence of this correlation between the positions of the electrons, *if any one electron is in the polyhedron surrounding a given atom, no other electron will be in the same polyhedron*.† On this assumption, therefore, in a monovalent metal each electron moves in the field of the singly charged *ion* in whose polyhedron it happens to be. Hence the total energy of the crystal is the sum of the kinetic energy of each electron, and of its potential energy in the field of its own positive ion; the other terms in H_1, H_2, H_3 above cancel out.

Now this is exactly the energy which we have shown how to calculate in Chap. II, § 4.5. Thus for monovalent metals we may divide up the energy of each electron as follows:

$$E_k = E_0 + \hbar^2 k^2 \alpha / 2m. \tag{13}$$

The second term represents the Fermi energy, due to the motion of the electrons through the lattice; α is a numerical factor (cf. p. 81) constant for small k. E_0 is the energy (kinetic and potential) of an electron at rest in the lattice, and is the characteristic energy of the Schrödinger equation

$$\nabla^2 \psi + \frac{2m}{\hbar^2}(E_0 - V)\psi = 0,$$

subject to the condition that $\partial \psi / \partial n$ should vanish at the boundary of the atomic polyhedron. E_0 is shown in Fig. 33 on p. 78 plotted for sodium against the atomic radius r_0 defined, as elsewhere in this book, by $\frac{4}{3}\pi r_0^3 = \Omega_0 = $ volume per atom.

To obtain the energy of the crystal we must sum (13) over all the electrons in the metal, obtaining, as shown in Chap. II, § 4.5.

$$\left.\begin{array}{l} \bar{E} = E_0 + E_F, \\ E_F = \frac{3}{5}\alpha\hbar^2 k_{max}^2 / 2m \end{array}\right\} \tag{14}$$

for the energy per atom.

2.2. *Correlation between the positions of the electrons.* The energy E thus obtained is something of a hybrid; it is neither the energy obtained from Hartree's equations nor from Fock's,‡ because we have taken account of the correlation between the electrons in an

---

† This, of course, only applies to monovalent metals.   ‡ Cf. Chap. II, § 1.3.

3595.17

incomplete way. Before discussing the solution of Fock's equation, we shall obtain the complete solution of Hartree's equations, i.e. the energy calculated using the simple product (11). We shall not calculate the effect on the wave functions, which is probably small, but only on the energy; we shall in fact assume that the wave functions may be replaced by plane waves, so that the charge distribution throughout the crystal is uniform.

With the simple product (11), each electron is uniformly distributed throughout the crystal; an electron in any atomic polyhedron, therefore, moves in the field not only of the ion but also in that of a uniform negative charge distribution. The potential energy of an electron in any polyhedron is, thus, instead of $V(r)$,

$$V(r) + \frac{3e^2}{2r_0} - \frac{e^2 r^2}{2r_0^3} \quad (r \leqslant r_0).$$

The two additional terms give the potential energy function at a distance r from the centre of a sphere of radius r_0 and with charge $-e$ uniformly distributed throughout its volume. The other polyhedra, as before, make practically no contribution to the potential.

Since $|\psi|^2$ is assumed to be constant, this gives an additional term in the energy

$$E_1 = \frac{1}{2} \int_0^{r_0} \left(\frac{3e^2}{2r_0} - \frac{e^2 r^2}{2r_0^3} \right) 4\pi r^2 \, dr \left/ \frac{4\pi}{3} r_0^3 \right.$$

$$= 0 \cdot 6 e^2 / r_0. \tag{15}$$

The factor $\frac{1}{2}$ enters, because the interactions between each pair of electrons must be counted once only. The addition of the term (15) gives then the energy in the Hartree approximation. To this we must now add the (negative) contribution due to the 'hole' round each electron, which we shall see gives a negative contribution of about the same amount. This will now be considered.

In the preceding section we have used for the wave function a simple product (11); a better approximation will be to write for $\Psi(\mathbf{r}_1, \mathbf{r}_2, ...)$ an antisymmetrical wave function from the functions $\psi_k(\mathbf{r})$ already obtained, and to use it to calculate the energy from formulae (8), (9). If this is done, it is easily seen that the kinetic energy, equation (9), is unaltered, because the ψ_k are orthogonal. The only term affected is that involving H_2, which gives the interaction of the electrons one with another. For consider two electronic wave functions $\psi_k(\mathbf{r}_1)$, $\psi_j(\mathbf{r}_2)$, and let us suppose that the electrons in these states have parallel spins. Then Ψ must be antisymmetrical in $\mathbf{r}_1$ and $\mathbf{r}_2$, and the term in the energy corresponding to the interaction between these electrons is

$$\int\int \frac{e^2}{|\mathbf{r}_1-\mathbf{r}_2|} \tfrac{1}{2}|\psi_k(\mathbf{r}_1)\psi_j(\mathbf{r}_2)-\psi_j(\mathbf{r}_1)\psi_k(\mathbf{r}_2)|^2\, d\tau_1\, d\tau_2. \tag{16}$$

This does *not* represent the electrostatic interaction of two charge densities $|\psi_k|^2$, $|\psi_j|^2$. The second factor in the integrand becomes

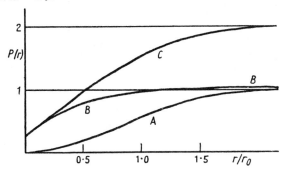

Fig. 56. Hole in charge distribution round a given electron.†

A. Parallel spin.   B. Antiparallel spin.   C. Both spins combined.

$P(r)$ is the relative probability per unit volume that another electron with the spin specified will be found at a distance r from the given electron.

small when $|\mathbf{r}_1-\mathbf{r}_2|$ is small; electrons with parallel spins thus tend to keep away from one another, and the potential energy is therefore smaller than it would otherwise be.

To obtain an estimate of the extent to which the electrons keep apart we proceed as follows: writing $r = |\mathbf{r}_1-\mathbf{r}_2|$, we denote by $P(r)$ the quantity obtained by summing

$$\tfrac{1}{2}|\psi_k(\mathbf{r}_1)\psi_j(\mathbf{r}_2)-\psi_j(\mathbf{r}_1)\psi_k(\mathbf{r}_2)|^2$$

over all $\mathbf{j}$ and averaging over all $\mathbf{k}$; then $P(r)\,d\tau/\Omega$ is the probability that a given electron will be found in the volume element $d\tau$ at $\mathbf{r}_2$, if we know that an electron is at $\mathbf{r}_1$. If ψ_k and ψ_j are replaced by plane waves, the summation can be carried out. We obtain‡

$$P(r) = 1-9(\sin\xi-\xi\cos\xi)^2/\xi^6,$$

where

$$\xi = 3^{\frac{1}{3}}\pi^{\frac{2}{3}}r/\Omega_0^{\frac{1}{3}} = 1\!\cdot\!92r/r_0,$$

so that P is a function of r only. P thus gives the density of charge with parallel spin at a distance r from a given electron. P is shown in Fig. 56.

† From Slater, *Rev. Mod. Phys.* **6** (1934), 228.

‡ Wigner and Seitz, first reference, p. 807.

The extra energy due to the electrons with parallel spins keeping apart is[†]

$$E_p = +\frac{1}{4}\frac{1}{\Omega_0}\int_0^\infty \frac{e^2}{r}[P(r)-1]4\pi r^2\,dr = -0{\cdot}458e^2/r_0. \quad (17)$$

The factor $\frac{1}{4}$ arises because each pair enters once only, and also because only pairs with parallel spins are to be considered.

Electrons with antiparallel spins keep apart also, but not to the same extent, the probability that both electrons occupy the same point being finite, as shown in Fig. 56. The same effect also occurs in many-electron atoms, as shown by the work of Hylleraas[‡] and Bethe[||] on helium. For metals the effect has been investigated by Wigner,[††] to whom the curve shown in Fig. 56 B is due; the correlation is less, the greater the velocity of the electrons. Wigner finds the extra energy to be

$$-be^2/r_0, \quad (18)$$

with b given by

$r_0 =$	1	2	5	10	(atomic units)
$b =$	0·05	0·075	0·13	0·175.	

Wigner estimates these values to be correct to within 20 per cent.

The corrections to the energy E_k, calculated in Chap. II, § 4.5, which are introduced here are thus:

Interaction energy of electron with other electrons in its own cell, with no correlation: $0{\cdot}6e^2/r_0$.

Decrease in energy due to 'Fermi' hole between electrons with parallel spin: $-0{\cdot}458e^2/r_0$.

Decrease due to hole for electrons with antiparallel spin: $-be^2/r_0$.

Total $(0{\cdot}14-b)e^2/r_0 = \Delta E$, say.

Note that ΔE is the increase in the total energy, and thus the decrease in the sublimation energy. The following are the numerical values:

		Li	Na	K	Cu	Ag
At 18° C.	$2r_0$ (Å. U.) .	3·45	4·21	5·23	2·82	3·18
	r_0 (atomic units) .	3·27	3·99	4·95	2·67	3·01
b		0·10	0·11	0·13	0·08	0·09
ΔE (e.v.)		0·3	0·2	0·05	0·6	0·45

† Wigner and Seitz, second reference, p. 512. ‡ *Zeits. f. Phys.* **48** (1928), 469.
|| Ibid. **57** (1929), 815. †† *Phys. Rev.* **46** (1934), 1002.

It will be seen that the sum of the correcting terms is not very large, so that in fact the approximation of Chap. II, § 4.5, is fairly satisfactory.

Application to Ferromagnetism

The result illustrated in Fig. 56, that electrons with parallel spins keep apart from each other more than electrons with antiparallel spins, shows that the electrostatic interaction energy between electrons with parallel spins will be algebraically less than that between electrons with antiparallel spins. These electrostatic forces therefore tend to make the electrons set themselves with their spins parallel, as in a ferromagnet. The opposing tendency is of course the 'Fermi force'; the kinetic energy E_F (equation (14)) would be much greater if the electrons had parallel spins.

Bloch[†] has calculated the tendency to ferromagnetism represented by terms of the type (17); he neglected, however, the correlation between electrons with antiparallel spin, and therefore found too great a tendency to ferromagnetism. The matter has been discussed further by Wigner.[‡] It has not yet proved possible to show with this model under what conditions ferromagnetism will occur.[||] We can, however, say that the correlation forces will in all cases increase the paramagnetism, which will thus be greater than the value obtained by neglecting the correlation (cf. Chap. VI).

Correction to the Fermi energy. It has been assumed above that the Fermi energy corresponding to any electronic state **k** is a function of $|\mathbf{k}|$ only. This is probably a fair approximation for all the occupied states of monovalent metals, and probably does not affect the total energy much. On the other hand, for metals and alloys containing more than one valence electron per atom, the corrections to the Fermi energy due to the gaps for forbidden energy values are very important for determining the crystal structure. This is discussed in Chapter V.

Correction to the potential energy introduced by replacing the atomic polyhedron by a sphere. The energy E_0 of the lowest state has been calculated on the assumption that the atomic polyhedron could be replaced by a sphere. The error in the kinetic energy thus introduced has been estimated by Wigner and Seitz[††] and found to be extremely small. The error in the potential energy may be calculated as follows:[‡‡] near the surface of the polyhedron the charge density is practically uniform. Hence the *difference* ΔE between the potential energy of the real lattice and the potential energy when the

---

[†] *Zeits. f. Phys.* **57** (1929), 545. [‡] Loc. cit. (*Phys. Rev.* **46**).
[||] See, however, Slater, *Phys. Rev.* **49** (1936), 537.
[††] Loc. cit., second reference, Appendix I.
[‡‡] Loc. cit., second reference, Appendix II. Fuchs, *Proc. Roy. Soc.* A, **151** (1935), 585.

polyhedron is replaced by a sphere is equal to the *difference* between the energy of a lattice of positive point charges in a uniform sea of a negative charge, and that of an equal number of positive point charges each surrounded by a uniform sphere of negative charge.

The energy E_s of the uniform sphere is

$$E_s = -\tfrac{9}{10}e^2/r_0, \tag{19}$$

The energy of the lattice of point charges embedded in the negative charge distribution may be calculated by a method due to Ewald,[†] and is given below:

	E_L	$E = E_L - E_s$
Face-centred lattice	-0.89586	0.00414
Body-centred lattice	-0.89593	0.00407

$\left. \right\} \times e^2/r_0$

For sodium and copper the corrections $E_L - E_S$ amount to 0·026 and 0·04 electron volts respectively, and may thus be neglected.

2.3. *Interaction between the closed shells of the atomic cores.* In the preceding work we have considered only the interaction between the ions considered as point charges. We have also to calculate:

(1) The van der Waals attraction between the ions.

(2) The exchange interaction due to overlapping of the closed shells of the ions. This will always give a repulsion.

Methods of calculating the van der Waals attractive energy are discussed in § 1.1. The constant C has been calculated for a number of ions by Mayer and Helmholz,[‡] Mayer,[||] and Mayer and Levy,[††] from the absorption spectrum of the free ions. The values obtained are given below, and also the resulting energy per ion in the metal.

Element	Li[||]	Na[||]	K[||]	Rb[||]	Cs[||]	Cu[††]	Ag[||]
$C \times 10^{60}$ ergs cm.[6] . .	0·073	1·68	24·3	59·4	152	0·41	67
van der Waals energy, e.v. per atom . . .	0·00035	0·0024	0·010	0·016	0·026	0·007	0·54

The 'exchange' repulsion between alkali ions (closed p shells) has been estimated by Born and Mayer,[‡‡] Mayer and Helmholz,[||||] and Huggins and Mayer[†††] from the heats of vaporization and compressi-

† Cf. *Ann. d. Physik*, **64** (1921), 253.
|| *J. Chem. Phys.* **1** (1933), 278, 330.
‡‡ *Zeits. f. Phys.* **75** (1932), 1.
††† *J. Chem. Phys.* **1** (1933), 693.

‡ *Zeits. f. Phys.* **75** (1932), 19.
†† Ibid. 647.
|||| Ibid. 19.

bilities of the alkali halides. These authors find for the mutual potential energy of two ions at a distance r apart

$$W = Ae^{(2r_s-r)/\rho}, \qquad (20)$$

where $\rho = 0.345 \times 10^{-8}$ cm., and the other constants have the values given in the following table:

	Li	Na	K	Rb	Cs
$A \times 10^{12}$ ergs.	2	1·25	1·25	1·25	1·25
$r_s \times 10^8$ cm.	0·475	0·875	1·185	1·320	1·455
Energy of ions in metal $(4W)$, e.v. per atom .	0·01	0·01	0·01	0·005	0·005

The resulting energy is shown also, and is seen to be very small, though owing to the rapid decrease with distance the effect on the elastic constants is not negligible.

For copper the interaction between the ions (closed d shells) has been calculated† by a statistical method using Hartree's wave functions (cf. Chap. II, § 1.2), with the following result, r denoting the distance between the ions:

r (atomic units of length)	4·5	5·0	5·5	5·6	5·7
Energy per ion pair $W(r)$ (Rydberg units $\times 10^3$)	33·0	15·9	2·67	1·51	0·624

3. Results

3.1. *Alkali metals.* In these metals the wave functions of the inner shells overlap very little (cf. Fig. 57), and so, except for the van der Waals attraction between the closed shells, the binding energy of the lattice consists almost entirely of the energy of the valence electrons in the field of the ions.

For lithium and for sodium detailed calculations have been made by the methods of this chapter. The results obtained are as follows:

		Na	Li
Binding energy (kilo-cal. per gm. atom)	obs.	26·9	38·9
	calc.	26·1	33·8
Lattice constant (Å. U.)	obs.	4·23	3·46
	calc.	4·62	3·53
Compressibility (kg. cm.$^2 \times 10^6$)	obs.	10	—
	calc.	16	—

The observed values are extrapolated to apply to the absolute zero of temperature.

The agreement for the binding energy is better than appears from

† Fuchs, loc. cit.

the table, because the binding energy is the *difference* between $-E_0$, the energy of an electron in the lowest state (that of the free atom being taken to be zero), and E_F, the Fermi energy, and, as the following table shows, these quantities are comparable.

	Li	Na	K	Rb	Cs
Binding energy (exp.) kilo-cal. per gm. atom . .	46	30	26·5	25	24
Fermi energy (calc. from formula (19) of Chap. II) .	64·5	44·0	27·8	25·0	21·0

3.2. *Noble metals, Cu, Ag, Au.* The metals differ from the alkalis in that the cores (closed d shells) overlap much more than the closed

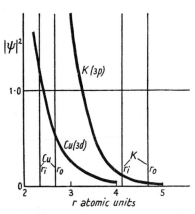

p shells of the alkalis (cf. Fig. 57), so that the repulsive force between the ions, other than the Coulomb force, comes into play. This does not have much effect on the total lattice energy; but owing to the very rapid increase of this force with decreasing distance it has a predominating effect on the lattice constant and compressibility. *One may picture these metals as composed of hard spheres (the ions) held together by the Coulomb attraction between the ions and the valence electrons.* The close-packed structure of these metals may also be accounted for on this model (cf. Chap. V, § 3.2).

FIG. 57. Charge densities of ions of Cu and K, calculated by Hartree.† The unit of $|\psi|^2$ is the reciprocal of the atomic volume in the metallic state; $2r_i$ is the distance between nearest neighbours and r_0 the radius of the atomic sphere.

Fig. 58 shows the energy of the valence electrons, the energy of the ions, and the total energy, as calculated by Fuchs for copper. We give also the compressibility, calculated with and without exchange interaction between the ions.

	Observed	Calculated	
		With ionic interaction	Without ionic interaction
Compressibility of copper (kg. cm.²× 10⁶)	0·70	0·69	2·6

† References are given in Chap. II, § 1.2. The radial extension of Hartree's wave functions is probably too great, and so the overlap shown too large.

Thus, if one does not take into account the exchange repulsion between the ions, the compressibility obtained by the Wigner-Seitz method is much too large. Similar results are obtained for the other elastic constants (cf. § 4).

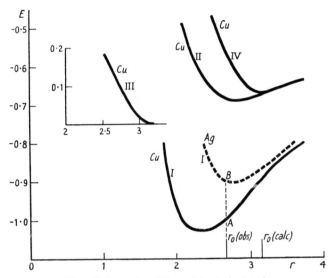

FIG. 58. Energies of Cu and Ag (calculated).

  I. Lowest electronic state.
  II. Total energy of valence electrons.
  III. Repulsive energy of ions.
  IV. Total energy of metal (II+III).

The points marked A, B refer to a discussion of Chapter VII.

3.3. *Transition metals.* No calculations of the binding forces in these metals have yet been made; we believe, however, that for such metals as Co, Ni, Pd, Pt, where the d shells are nearly full, the cohesive forces are of the same nature as in copper, silver, and gold; that is to say, that the attractive forces are due to the electrostatic interaction between the positively charged ions and the 'conduction electrons', i.e. electrons with similar wave functions to those for copper. The repulsive forces, similarly, are due to the 'exchange' repulsion between the nearly closed d shells, the radii of which must therefore determine the interatomic distance. The 'exchange' attractive forces between the positive holes in nearly closed d shells are responsible for the ferromagnetism† in nickel and cobalt, and

† Cf. Chap. VI, §§ 5 and 7.

since the ferromagnetism disappears at a temperature comparable with 1,000°, these forces can only contribute an energy of about $1,000k \sim \frac{1}{10}$ e.v. per atom. They are therefore without appreciable effect on the binding energy.

There are two possible reasons for the smallness of these exchange forces; firstly, the sign of the exchange integral is sometimes positive (Ni) and sometimes negative (Pd); we may therefore expect it to be small numerically in either case; secondly, in a nearly closed shell, the repulsion between electrons in doubly occupied states will prevent any considerable overlap between the few unpaired electrons.†

On the other hand, for transition metals where the d level is less than half full, the d electrons may well play a decisive part in forming the cohesive forces. Evidence in favour of this view may be drawn from the observed values of the constant γ, which measures the rate of change with volume of the vibrational frequency $k\Theta/h$ of the atoms (cf. Chap. I, § 4). The alkalis and alkaline earths have comparatively small values of γ: Li 1·17; Na 1·25; K 1·34; Rb 1·48; Cs 1·29; Ca 1·3; Sr 0·93; Ba 1·1; the noble metals and the platinum triad have much larger values: Cu 1·96; Ag 2·40; Au 3·03; Ni 1·88; Pd 2·23; Pt 2·54. This is to be expected, because in these metals the atoms are held in position by the repulsive 'exchange' forces between the ions, which increase very rapidly with decreasing distance; therefore, when the metals are compressed, the atomic frequencies rise rapidly. Transition metals in which the d shells are *not* nearly full, however, have values of γ more comparable with those of the alkalis, e.g. Mo 1·57; W 1·62 (see also p. 173).

3.4. *Divalent metals.* Under this heading we include the alkaline earths Mg, Ca, Sr, Ba, and the elements Zn, Cd, Hg. No detailed calculations have been made for these elements. The metals Zn,

---

† This does not happen in a nearly closed p shell, such as that in chlorine. A p wave function has one of the forms (cf. Chap. II, § 1.1) $x f(r)$, $y f(r)$, $z f(r)$, and therefore vanishes in a plane and has its largest amplitude in the direction perpendicular to this plane. In a chlorine atom, therefore, which contains five p electrons, there will be two electron pairs with their maximum charge densities in, say, the x and y directions, and one unpaired electron with its maximum density in the z direction. If another chlorine atom approaches from the x or y directions, it will be repelled; but if it approaches from the z direction, and if its unpaired electron has its maximum density in this direction, then the two unpaired electrons may form a bond, resulting in a stable molecule Cl_2. The other electrons, having only a small charge density in the z direction, do not give a large repulsive force.

A d wave function, on the other hand, has the form $xy f(r)$ (cf. Chap. II, § 1.1). It is therefore small in one direction (along the z-axis) and large in a plane.

Cd, and Hg have much smaller binding energy than Cu, Ag, and Au, as the following table shows:

Binding energies in kilo-cal./gm. atom.

Cu	76	Ag	64·5	Au	83
Zn	32·5	Cd	28	Hg	18·5

This may perhaps be attributed to the much greater Fermi energy contributed by *two* valence electrons (cf. Chap. II, § 3), which of course decreases the binding energy.[†] On the other hand, the binding energies of the alkaline earths are greater than those of the corresponding alkalis (cf. Chap. I, § 5).

The problem of the transition between the van der Waals forces and the metallic forces has not yet been worked out. For divalent metals they are of the same order of magnitude. London,[‡] using data due to Wolfsohn,[||] has calculated from formula (2) of this chapter that the interaction energy of two mercury atoms at a distance R (Å. U.) apart is $159R^{-6}$ e.v. Hence (formula (4)) the binding energy of a mercury crystal[††] is 31 kilo-cal./gm. atom which is greater than the experimental value of 18·5.

4. The elastic constants

In the preceding sections we have shown how to calculate the energy of a crystal as a function of the atomic volume; from these calculations the compressibility can be deduced. We shall now show how to calculate the other elastic constants.

For cubic crystals there are, besides the modulus of uniform compression, two independent elastic constants; we may define them by considering the following two types of distortion, *in both of which the volume is unchanged*:

(1) Compression and expansion parallel to two of the cube edges, the volume being kept constant (cf. Fig. 59). If a denotes the lattice constant, ϵa the increase in the length of the cube edge, and W the energy per atom of the crystal, then the elastic constant

$$A = \frac{1}{2}\frac{\partial^2 W}{\partial \epsilon^2}$$

specifies the resistance of the crystal to this distortion.

† Cf. Feinberg, *Phys. Zeits. d. Sowjetunion*, **8** (1935), 407.   ‡ Loc. cit.
|| *Zeits. f. Phys.* **63** (1930), 634.
†† This is calculated for the face-centred cubic structure, but the energy for the actual rhombohedral structure should not differ greatly.

(2) Uniform shear parallel to one surface of the cube. If we denote by γ the angle of distortion shown in Fig. 59, the corresponding elastic constant is

$$B = \frac{1}{2}\frac{\partial^2 W}{\partial \gamma^2}.$$

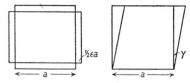

FIG. 59. Two independent distortions of a cubic crystal.

We define further

$$C = \frac{1}{2}\frac{\partial^2 W}{\partial \epsilon_v^2} = \frac{\Omega_0}{2\chi}, \qquad \epsilon_v = \frac{\Delta V}{V_0},$$

where χ is the compressibility and Ω_0 the volume per atom.

The usual elastic constants† c_{rs} are connected with A, B, and C by the following equations:

$$A/\Omega_0 = c_{11}-c_{12}, \qquad 2B/\Omega_0 = c_{44}, \qquad 2C/\Omega_0 = \tfrac{1}{3}(c_{11}+2c_{12}).$$

The change in the energy of a metal crystal on distortion is due partly to the valence electrons and to their electrostatic interaction with the ions, partly to the exchange interaction between the ions themselves. The change in the energy of the electrons has been calculated by Fuchs‡ on the assumption that, in distortions in which the volume of the crystal is unchanged, only their potential energy is altered. This is justified because the wave function of the lowest state is extremely flat (cf. Fig. 34) at the boundaries of the atomic polyhedron, and so the *undistorted* wave function satisfies the boundary conditions approximately at the surface of the distorted polyhedron. The Fermi energy, moreover, is assumed to be a function of the volume only.

Since for Li, Na, and Cu (the metals investigated) the charge density near the edges of the unit polyhedron is about $-e/\Omega_0$, the change in the potential energy of the valence electrons on distortion will be the same as that of a lattice of positive point charges $+e$ embedded in a uniform negative charge distribution $-e/\Omega_0$. This may be calculated by a method due to Ewald;∥ the results are,

---

† For the definition of c_{rs} see, for instance, Geckeler, *Handb. d. Phys.* **6** (1928), 407.
‡ *Proc. Roy. Soc.* A, **153** (1936), 622.   ∥ *Ann. d. Phys.* **64** (1921), 253.

denoting the contributions from this source to A and B by A_{el} and B_{el}, in ergs per atom

Lattice	Face-centred	Body-centred	
A_{el}	0·2115	$\left.\begin{matrix}0{\cdot}1994 \\ 0{\cdot}3711\end{matrix}\right\} \times \dfrac{e^2}{2a}$	(21)
B_{el}	0·4739		

where a is the lattice constant.

We may note that for an isotropic solid $\frac{1}{4}A = B$. The distortion A is thus much easier relative to B than would be the case for an isotropic solid.

If the exchange interaction energy of the ions is known, the calculation of their contributions to A and B is elementary. The force between two ions falls off so rapidly with distance that only ions which are nearest neighbours need be considered. If then $W(r)$ is the energy of a pair of ions at a distance r apart, the following are the contributions to A and B:

	Face-centred	Body-centred	
A_{ion}	$\frac{1}{2}r^2\dfrac{d^2W}{dr^2} + \frac{7}{2}r\dfrac{dW}{dr}$	$\frac{8}{3}r\dfrac{dW}{dr}$	(22)
$2B_{ion}$	$\frac{1}{2}r^2\dfrac{d^2W}{dr^2} + \frac{3}{2}r\dfrac{dW}{dr}$	$\frac{4}{9}r^2\dfrac{d^2W}{dr^2} + \frac{8}{9}r\dfrac{dW}{dr}$	

For copper and the alkali metals the values of the contributions to A and B from the exchange interaction may be calculated from (22) using the forms given in § 2.3 for $W(r)$. A small correction is also introduced by the van der Waals attraction. In the following table we show the contributions to A and B from the various sources. The quantities shown are A/Ω_0, B/Ω_0 in dynes/cm.$^2 \times 10^{11}$; they give the change in the energy per unit volume of the metal.[†]

	Cu		Na	
	A/Ω_0	$2B/\Omega_0$	A/Ω_0	$2B/\Omega_0$
Electrostatic energy of valence electrons and ions, from (21) . .	0·573	2·57	0·143	0·532
Exchange repulsion between ions, from (22)	4·5	6·3	−0·002	0·057
van der Waals attraction between ions, from (22)	−0·012	−0·034	0·000	−0·003
Total	5·1	8·8	0·141	0·580
Observed	4·7	7·5	—	—

† We are grateful to Mr. Fuchs for informing us of certain errors in his published values, which have been corrected in this table.

The observed values for Cu are those of Goens.[†]

It will be noticed that for copper the predominating term in the elastic constants is the exchange repulsion between the ions, whereas this term is relatively unimportant for sodium. Similar results are obtained for lithium and potassium.

For the alkalis no experimental values exist; it is, however, possible to calculate, from the constants obtained, the Debye characteristic temperature for low temperatures, where $c_v \propto T^3$; satisfactory agreement with experiment is obtained.[‡]

4.1. *The Cauchy relations.* If a lattice is considered as a system of point centres of force, certain relations between the elastic constants can be derived. These are known as the 'Cauchy relations'. For a regular lattice the only such relation is $c_{12} = c_{44}$, or in terms of χ, A, and B

$$\Omega_0/\chi = \tfrac{1}{3}A + 2B.$$

The compressibility is therefore given in terms of the elastic constants A and B which define the resistance of the solid to shear.

It is obvious that no such relation can be derived for metals. Several terms in the expression for the energy of the crystal, such as, for example, the Fermi energy, have been explicitly assumed to depend on the atomic volume only, and to remain unchanged in a shear where the volume is unchanged. Such terms will be important in calculating the compressibility, but can have no effect on A and B.

It is, of course, in agreement with experiment that the Cauchy relations are not satisfied by the elastic constants of metals. For copper, for instance, $c_{12} = 11\cdot8$, $c_{44} = 6\cdot1 \times 10^{11}$ dynes/cm.^{2}

5. Work function

The work function represents the work required to move an electron from a point inside to a point outside a metal. Theoretical discussions have been given by Frenkel,[||] Tamm and Blochinzev,[††] Fröhlich,[‡‡] and Wigner and Bardeen.[||||] The work function, being a property of the metal surface, lies, however, outside the scope of this book.

† *Zeits. f. Instrumentenkunde*, **52** (1932), 167.

‡ Cf. Fig. 4, where the observed and calculated values of Θ_D are shown.

|| *Zeits. f. Phys.* **49** (1928), 31. Certain of Frenkel's conclusions have been criticized by Bethe, *Handb. d. Phys.* 24/2 (1933), 427.

†† *Zeits. f. Phys.* **77** (1932), 774.

‡‡ *Phys. Zeits. d. Sowjetunion*, **7** (1935), 509.

|||| *Phys. Rev.* **48** (1935), 84; **49** (1936), 653.

6. Alloys

Very little work has as yet been done on calculating the energy of an alloy as a function of the composition, except for that of Jones relating the electron-atom ratio to the crystal structure. This is discussed in the next chapter. We do not, for instance, know how to calculate the energy of copper dissolved in silver, from which could be calculated the limits of solubility.

THE CRYSTAL STRUCTURE OF METALS AND ALLOYS

1. Introduction

THE chief purpose of this chapter is to describe the form of the 'Brillouin zones'† for the structures of all the known metals and some alloys. The Brillouin zones will be considered in relation to the number of valency electrons per atom of the metal or alloy. For atoms with no incomplete shells we mean by this the number of electrons outside a closed shell, e.g. one for copper, five for bismuth, etc. For alloys we mean the average number, e.g. $\frac{3}{2}$ for CuZn. For metals with incomplete shells the number cannot be specified unless we have independent evidence of the number of 'positive holes' in the d shell (cf. Chap. VI, § 5).

It is of particular interest to consider a zone which could just contain all the available valence electrons. As we have seen in Chap. III, § 1, if such a zone were full the crystal would be an insulator. If the 'overlap' into the next zone is small, the crystal is a poor conductor. Another property associated with a small overlap is, in certain cases, a strong diamagnetism.

We shall discuss in this section, as far as is possible at present, the reasons why the different metals and alloys have their particular crystal structure. We shall find that, other things being equal, the structure is preferred in which the 'overlap' out of the first Brillouin zone is as small as possible. By this principle it has been found possible to account for the well-known Hume-Rothery rules. Other factors are discussed also, such as the relative size of the atoms.

2. The Brillouin zones for various crystal structures

The Brillouin zones for a simple cubic lattice were considered in detail in Chap. II, § 4. In this section we show how to find the zones for any structure.

Let $\mathbf{a}_1$, $\mathbf{a}_2$, $\mathbf{a}_3$ be the three vectors which map out the unit cell, and let $\mathbf{b}_1$, $\mathbf{b}_2$, $\mathbf{b}_3$ be another set of three vectors which satisfy the relations

$$(\mathbf{a}_n, \mathbf{b}_m) = \delta_{nm} = 1 \quad (n = m)$$
$$= 0 \quad (n \neq m).$$

† Cf. Chap. II, § 4.3.

The set of points $n_1\mathbf{b}_1+n_2\mathbf{b}_2+n_3\mathbf{b}_3$, where n_1, n_2, n_3 take all integral values (including zero) form what is known as the 'Reciprocal lattice'. The potential at any point $\mathbf{r}$ in the space lattice

$$\mathbf{r} = x_1\mathbf{a}_1+x_2\mathbf{a}_2+x_3\mathbf{a}_3$$

may be expanded as a Fourier series

$$V(\mathbf{r}) = \sum V_\mathbf{n}e^{2\pi i(\mathbf{nr})},$$

where $\mathbf{n}$ is a vector denoting one of the points of the reciprocal lattice, and is thus defined by $\mathbf{n} = n_1\mathbf{b}_1+n_2\mathbf{b}_2+n_3\mathbf{b}_3$, so that

$$(\mathbf{n}, \mathbf{r}) = n_1x_1+n_2x_2+n_3x_3;$$

the summation is over all points of the reciprocal lattice.

According to the theorem of Bloch† the wave functions for an electron moving in such a crystal have the form

$$\psi_k = u(\mathbf{r})e^{2\pi i(\mathbf{kr})}, \tag{1}$$

where $\mathbf{k}$ is a vector in the reciprocal space.‡ When the components of $\mathbf{k}$ referred to the axes $\mathbf{b}_1$, $\mathbf{b}_2$, $\mathbf{b}_3$ are integers, $\mathbf{k}$ will coincide with a point of the reciprocal lattice.

The planes across which the energy changes discontinuously are most easily found by approaching the problem from the standpoint of nearly free electrons.‖ The perturbation theory is used here in a purely formal manner to obtain the equations of the planes of energy discontinuity; these are, of course, exact. It is not used to obtain approximate energies and wave functions.

Corresponding to (1) we have the free-electron wave function

$$\psi_k = e^{2\pi i(\mathbf{kr})}.$$

Forming the matrix of V with respect to these functions, we find that

$$(\mathbf{k}|V|\mathbf{k}') = 0$$

unless

$$\mathbf{k}' = \mathbf{k}+\mathbf{n}. \tag{2}$$

The discontinuities in the energy occur when two states having initially the same energy combine under the influence of V, i.e. when $|\mathbf{k}'|^2 = |\mathbf{k}|^2$; or, according to (2), when

$$(\mathbf{nk})+\tfrac{1}{2}|\mathbf{n}|^2 = 0. \tag{3}$$

The positions of these planes in the reciprocal lattice are very easily visualized. Imagine the point n_1, n_2, n_3 joined to the origin by a

† Cf. Chap. II, § 4.1.
‡ In this chapter k is equal to $1/\lambda$, instead of $2\pi/\lambda$ as elsewhere in this book.
‖ Cf. Chap. II, § 4.2.

straight line; draw a plane perpendicular to this line through the mid-point. This plane is then represented by equation (3) where the components of **n** are taken to be the numbers n_1, n_2, n_3.

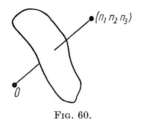

$(n_1\, n_2\, n_3)$

FIG. 60.

If the coefficient $V_\mathbf{n}$ in the development of the potential is zero, the states $\mathbf{k}+\mathbf{n}$ and $\mathbf{k}$ will not combine under a periodic potential and there will therefore be no plane of energy discontinuity corresponding to this value of **n**.

If the structure contains s atoms per unit ɔell, and the positions of these atoms in the unit cell are given by the vectors $u_l\mathbf{a}_1$, $v_l\mathbf{a}_2$, $w_l\mathbf{a}_3$, we call the array of terms (u_l, v_l, w_l) the *Basis* with respect to the axes $\mathbf{a}_1$, $\mathbf{a}_2$, $\mathbf{a}_3$. If, as is frequently the case, one structure is referable to two or more types of axes, there will be two or more different bases. For a structure with s atoms per unit cell, the Fourier coefficients of the potential may be written

$$V_\mathbf{n} = \sum_{t=1}^{s} A_{\mathbf{n}l}\, e^{2\pi i(n_1 u_t + n_2 v_t + n_3 w_t)}.$$

If all the atoms in the unit cell are identical, then all the coefficients $A_{\mathbf{n}s}$ which have the same value of **n** are equal, so that

$$V_\mathbf{n} = A_\mathbf{n} S_\mathbf{n},$$

where
$$S_\mathbf{n} = \sum_{1}^{s} e^{2\pi i(n_1 u_t + n_2 v_t + n_3 w_t)}. \tag{4}$$

$S_\mathbf{n}$ is the well-known structure factor, which is the principal factor determining the intensity of X-ray reflection from a set of planes in the real crystal with the Miller indices (n_1, n_2, n_3). Hence, if the structure factor vanishes, there is no energy discontinuity across the corresponding plane in k-space.

In addition to knowing the form of the Brillouin zones, we must know how many electronic states a zone contains. For this purpose we must know the number of states per cubic centimetre of metal and unit volume of k-space. The number of states per cubic centimetre in an element $d\mathbf{k}$ ($= dk_x\, dk_y\, dk_z$) of k-space depends only on the number of unit cells in one cubic centimetre of the crystal, as can be seen immediately from the way in which **k** is defined in the construction of the Bloch wave function. The number of states does not depend upon the form of the potential V within the unit cell. We can therefore find the density of states in k-space by considering the special case where V is a constant. In this case $\mathbf{k}$ is equal to the momentum of the free electron divided by h. Hence the volume of phase space corresponding to a volume $d\mathbf{k}$ of k-space and a real volume V of the metal is $Vh^3 d\mathbf{k}$; and since there are two states to each cell of volume h^3 in phase-space, it follows that in a volume $d\mathbf{k}$ of k-space and unit volume of metal there are just $2d\mathbf{k}$ electronic states. In other words, the density of states in k-space is equal to 2.

It has already been explained that the quantities $N(E)\,dE$, i.e. the number of states lying between energies E and $E+dE$, and N_{eff}, the effective number of free electrons, are of considerable importance in determining certain physical properties of a metal. Both these quantities are very largely determined by the form of the first few Brillouin zones. To obtain these functions exactly would require a complete determination of the stationary states of the metal by some such method as that of Wigner and Seitz. The mere knowledge of the form of the Brillouin zones and of the number of loosely bound electrons per atom which have to be fitted into these zones can, however, supply interesting information. For example, if a metal forms a well-marked fairly symmetrical zone containing just as many states as there are available valency electrons, then one is safe in deducing that N_{eff} will be small, i.e. that the effective number of free electrons is much less than the actual number of valency electrons.

We give in the following section the geometrical details of the zone structure for all the commonly occurring crystal types in metals and alloys. Only those zones are considered whose boundaries lie close to the surface of the Fermi distribution of the electrons. For example, if we have a metal with n valency electrons per atom, we give the details of the smallest zone which will contain these n electrons. The form and size of a zone depends on the crystal structure alone, and so may easily be found. On the other hand, the exact forms of the energy surfaces within such a zone, and in particular the magnitudes of the energy discontinuities across its boundaries, depend on the field within the crystal, and have not at present been calculated accurately (cf. Chap. II, § 4).

The rules for constructing the Brillouin zones are simple. First, from the known lattice form one writes down according to (3) the equations of all the planes of possible energy discontinuity. Secondly, from the known basis of the unit cell, one determines the structure factor. All planes for which this factor vanishes are excluded, and from the remainder the forms of the various zones are obtained by simple geometry. In the case of some complicated alloy structures, it is often sufficient to pick out the planes of small indices with large values of S_n by noting the strong lines in the X-ray powder photograph of the alloy; this method is available even if the structure is not known.

2.1. *Structures referable to cubic axes.* We denote a set of planes in k-space by means of the same notation which is used to denote sets of planes in a crystal lattice; for example, $\{n_1 n_2 n_3\}$ denotes all those planes for which the components of $\mathbf{n}$ in equation (3) take the values $\pm n_1$, $\pm n_2$, $\pm n_3$. A particular plane is denoted by the corresponding symbol, for example $(\bar{n}_1 n_2 n_3)$. There is one significant difference: in the real lattice the symbol $(n_1 n_2 n_3)$ denotes a whole set of parallel planes, whilst in k-space this symbol refers to one particular plane at a certain fixed distance from the origin. The normal distance from the origin to a plane (n_1, n_2, n_3) will be written $p_{n_1 n_2 n_3}$, so that, if a is the lattice constant, we have, for cubic lattices,

$$p_{n_1 n_2 n_3} = \frac{1}{2a}\sqrt{(n_1^2 + n_2^2 + n_3^2)}.$$

We give below the sets of planes which will be required later in this chapter, the number of planes in each set, and their normal distances $p_{n_1 n_2 n_3}$ from the origin.

Planes	Number of planes in the set	$ap_{n_1 n_2 n_3}$
{100}	6	1/2
{110}	12	$1/\sqrt{2}$
{111}	8	$\sqrt{3}/2$
{200}	6	1
{220}	12	$\sqrt{2}$
{330}	12	$3/\sqrt{2}$
{411}	24	$3/\sqrt{2}$

TYPE A 1.† *Face-centred cubic lattice.* There are 4 atoms in the unit cell, with the basis

$$(000, \ 0\tfrac{1}{2}\tfrac{1}{2}, \ \tfrac{1}{2}0\tfrac{1}{2}, \ \tfrac{1}{2}\tfrac{1}{2}0).$$

† The symbols which denote the structural types are those used by Ewald and Hermann in the *Strukturbericht*. The order in which the different structures are considered in this chapter is not, however, the same.

The structure factor according to (4) is therefore

$$S_{n_1 n_2 n_3} = 1 + \cos \pi(n_2 + n_3) + \cos \pi(n_3 + n_1) + \cos \pi(n_1 + n_2),$$

which gives $\quad S_{100} = S_{110} = 0; \quad S_{111} = S_{200} = 4.$

The zone marked out by the planes {111} and {200} is shown in Fig. 61.

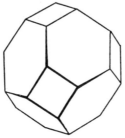

Fig. 61. Zone for face-centred cubic lattice.

Volume of zone	Atomic volume	Number of states per atom	Ratio of inscribed sphere to total volume
$\dfrac{4}{a^3}$	$\dfrac{a^3}{4}$	2	$\dfrac{\pi\sqrt{3}}{8} = 0\cdot681$

Some details of the form of the zone are given in the following table. The point A is the mid-point of a hexagonal face, B the mid-point of a square face, and C is one of the points which are farthest from the centre of the zone. The energy which free electrons would

Points		A	B	C	Surface of sphere with volume half that of zone
ka		0·866	1	1·118	0·782
Energy in volts for	Cu	8·64	11·52	14·40	7·04
	Ag	6·76	9·02	11·27	5·51
	Au	6·79	9·06	11·32	5·53
	Ca	3·64	4·85	6·06	2·97

have at these points in k-space is given, and also the energy of free electrons at the surface of a sphere which contains one half of the volume of the zone, and hence one electron per atom. For monovalent metals with face-centred cubic structure this last energy will be very nearly equal to the actual energy at the surface of the Fermi distribution.

The following metallic elements have the face-centred cubic

structure: Ca, Sr, Ce, Th, Fe, βCo, Rh, Ir, Ni, Pd, Pt, Cu, Ag, Au, βTl, Pb, βLa.

TYPE A 2. *Body-centred cubic lattice.* There are two atoms in the unit cell, with the basis

$$(000, \tfrac{1}{2}\tfrac{1}{2}\tfrac{1}{2}).$$

The structure factor is therefore

$$S_{n_1 n_2 n_3} = 1 + \cos \pi(n_1 + n_2 + n_3),$$
$$S_{100} = 0, \qquad S_{110} = 2.$$

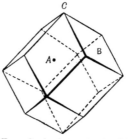

FIG. 62. Zone for body-centred cubic lattice.

The first zone, therefore, is that marked out by the planes {110}. Its form is shown in Fig. 62, and its volume, etc., below.

Volume of zone	Atomic volume	Number of states per atom	Ratio of inscribed sphere to total volume
$\dfrac{2}{a^3}$	$\dfrac{a^3}{2}$	2	$\dfrac{\pi}{3\sqrt{2}} = 0.740$

A is the centre of a face and is one of the points which are nearest to the origin of the zone. Not all the corners of a regular dodecahedron are equivalent. The points C and B lie at different distances from the origin. The points C are at the greatest distance from the origin, as the following table shows.

Points		A	B	C	Surface of sphere with volume half that of zone
ka		0·707	0·866	1	0·620
Energy in volts for	Li	6·12	9·19	12·25	4·71
	Na	4·05	6·08	8·11	3·12
	K	2·77	4·16	5·55	2·13

The inscribed sphere of the first zone for the body-centred cubic structure would just contain 1·480 electrons per atom, whilst the inscribed sphere for the face-centred cubic structure would contain only 1·360 electrons per atom.

The following elements have a body-centred cubic structure: Li, Na, K, Rb, Cs, Ba, βZr, V, Nb, Ta, Cr, Mo, W, αFe.

TYPE A 4. *Diamond structure.* This structure is based upon a cubic lattice, and there are 8 atoms in the unit cell, with the basis

$$(000, 0\tfrac{1}{2}\tfrac{1}{2}, \tfrac{1}{2}0\tfrac{1}{2}, \tfrac{1}{2}\tfrac{1}{2}0; \tfrac{1}{4}\tfrac{1}{4}\tfrac{1}{4}, \tfrac{1}{4}\tfrac{3}{4}\tfrac{3}{4}, \tfrac{3}{4}\tfrac{1}{4}\tfrac{3}{4}, \tfrac{3}{4}\tfrac{3}{4}\tfrac{1}{4}).$$

The structure factor may be written

$$|S_{n_1 n_2 n_3}| = 2|\cos \tfrac{1}{4}\pi(n_1+n_2+n_3)|\{1+\cos \pi(n_2+n_3)+ \\ +\cos \pi(n_3+n_1)+\cos \pi(n_1+n_2)\},$$

giving the values

$\{n_1 n_2 n_3\}$	100	110	111	200	210	211	220	221	222	311
$|S_{n_1 n_2 n_3}|$	0	0	$4\sqrt{2}$	0	0	0	8	0	0	$4\sqrt{2}$

The volume of the zone bounded by the planes {220} is $16/a^3$, and since the atomic volume is $a^3/8$ it follows that this zone contains just 4 states per atom. The elements which take up this structure have 4 electrons external to a closed shell, and are non-conductors. It follows, therefore, that in these structures the four available electrons just completely fill the zone. The form of the zone is identical with that shown in Fig. 62. The only elements which have the diamond structure are the following:

C (diamond),   Si,   Ge,   Sn (grey).

2.2. *Structures based on a hexagonal lattice.* Hexagonal structures are usually referred to 4 axes whose unit vectors may be called $\mathbf{a}_1$, $\mathbf{a}_2$, $\mathbf{a}_3$, $\mathbf{c}$. The first three are coplanar and inclined to each other at angles of 120°, as shown in Fig. 63, and the fourth is perpendicular to the plane which contains the first three. We may write $|\mathbf{a}_1| = |\mathbf{a}_2| = |\mathbf{a}_3| = a, \quad |\mathbf{c}| = c.$

FIG. 63.

Then $(\mathbf{a}_1, \mathbf{a}_2) = (\mathbf{a}_2, \mathbf{a}_3) = (\mathbf{a}_3, \mathbf{a}_1) = -\tfrac{1}{2}a^2, \qquad (\mathbf{a}_1, \mathbf{c}) = 0.$

The volume of the unit cell of the simple hexagonal lattice is $\tfrac{1}{2}\sqrt{3}a^2 c$. A set of reflecting planes is then denoted by a symbol of four letters $\{h j k, l\}$. The advantage of this system, known as the Bravais-Miller method,† is that a set of equivalent planes can be represented by a

† Tutton, *Crystallography and Practical Crystal Measurement*, Chap. XX, London

single symbol, which is not possible if the usual three rectangular axes are employed.

A simple hexagonal lattice may be regarded as two interpenetrating orthorhombic lattices (see Fig. 64), and for the purpose of constructing the Brillouin zones it is convenient to consider it in this way. We shall consider, therefore, three rectangular axes whose unit vectors are A_1 A_2 A_3, where $A_1 = a$, $A_2 = \sqrt{3}a$, $A_3 = c$. Any vector $\mathbf{r}$ in the lattice may be denoted with reference to either system, so that

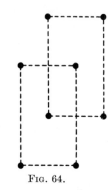

$$\mathbf{r} = \alpha_1\mathbf{a}_1 + \alpha_2\mathbf{a}_2 + \alpha_3\mathbf{c}$$
$$= \beta_1\mathbf{A}_1 + \beta_2\mathbf{A}_2 + \beta_3\mathbf{A}_3,$$

the relations between the coordinates α_1, α_2, α_3 and β_1, β_2, β_3 being

FIG. 64.

$$\beta_1 = \alpha_1 - \tfrac{1}{2}\alpha_2, \qquad \beta_2 = \tfrac{1}{2}\alpha_2, \qquad \beta_3 = \alpha_3. \quad (5)$$

If we denote by x, y, z lengths measured in cm. along the axes A_1, A_2, A_3, the Fourier series which represents the electrostatic potential in a simple hexagonal lattice may be written

$$V = \sum_{\mathbf{n}} V_{\mathbf{n}} \cos^2\tfrac{1}{2}\pi(n_1+n_2)\exp\left\{2\pi i\left(\frac{n_1 x}{A_1} + \frac{n_2 y}{A_2} + \frac{n_3 z}{A_3}\right)\right\}.$$

The equations of the planes of energy discontinuity in k-space are then

$$\frac{n_1 k_x}{a} + \frac{n_2 k_y}{\sqrt{3}a} + \frac{n_3 k_z}{c} = \frac{1}{2}\left\{\left(\frac{n_1}{a}\right)^2 + \frac{1}{3}\left(\frac{n_2}{a}\right)^2 + \left(\frac{n_3}{c}\right)^2\right\}. \quad (6)$$

The perpendicular distance from the origin to a plane denoted by the symbol $(n_1 n_2 n_3)$ will be written $p_{n_1 n_2 n_3}$, where

$$p_{n_1 n_2 n_3} = \frac{1}{2}\sqrt{\left\{\left(\frac{n_1}{a}\right)^2 + \frac{1}{3}\left(\frac{n_2}{a}\right)^2 + \left(\frac{n_3}{c}\right)^2\right\}}.$$

A particular hexagonal structure is usually described by giving the basis referred to the hexagonal axes (as for example in the *Strukturbericht*), i.e. by giving the coordinates $\alpha_1^{(\tau)}$, $\alpha_2^{(\tau)}$, $\alpha_3^{(\tau)}$ for each atom τ in the unit cell. By means of equations (5) the corresponding basis referred to rectangular axes may be obtained at once, and also the structure factor for any set of planes

$$S_{n_1 n_2 n_3} = \cos^2\tfrac{1}{2}\pi(n_1+n_2)\sum_{\tau} \exp\{2\pi i(\beta_1^{(\tau)} n_1 + \beta_2^{(\tau)} n_2 + \beta_3^{(\tau)} n_3)\},$$

where the summation is over all the atoms of the unit cell.

The following table gives the set of planes which are commonly required to find appropriate Brillouin zones for the hexagonal structures:

Planes	Number in set	Distance $p_{n_1 n_2 n_3}$	Notation referred to orthorhombic axes	
$\{1\bar{1}0, 0\}$	6	$1/\sqrt{3}a$	(110)	(020)
$\{\bar{2}11\}$	6	$1/a$	(200)	(130)
$\{1\bar{1}0, 1\}$	12	$\dfrac{1}{\sqrt{3}a}\sqrt{\left\{1+\dfrac{3}{4}\left(\dfrac{a}{c}\right)^2\right\}}$	(021)	(111)
$\{000, 2\}$	2	$1/c$	(002)	
$\{000, 3\}$	2	$3/2c$	(003)	
$\{000, 4\}$	2	$2/c$	(004)	

TYPE A 3. *The hexagonal close-packed structure.* This structure contains 2 atoms in the unit cell, with the basis, referred to hexagonal axes,

$$(000, \tfrac{2}{3}\tfrac{1}{3}\tfrac{1}{2}).$$

According to (5) the basis referred to rectangular axes is

$$(000, \tfrac{1}{2}\tfrac{1}{6}\tfrac{1}{2}),$$

and the structure factor

$$S_{n_1 n_2 n_3} = \cos^2\tfrac{1}{2}\pi(n_1+n_2)\{1+e^{\pi i(n_1+\frac{1}{3}n_2+n_3)}\},$$

giving

$$|S_{\{1\bar{1}0,0\}}| = 1, \quad |S_{\{000,1\}}| = 0, \quad |S_{\{1\bar{1}0,1\}}| = \sqrt{3}, \quad |S_{\{000,2\}}| = 2.$$

The form of the zone bounded by the 20 planes $\{1\bar{1}0, 0\}$, $\{1\bar{1}0, 1\}$, $\{000, 2\}$ is shown in Fig. 65. The number n of states per atom included within this zone is given by

$$n = 2 - \frac{3}{4}\left(\frac{a}{c}\right)^2\left\{1-\frac{1}{4}\left(\frac{a}{c}\right)^2\right\}. \tag{7}$$

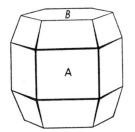

The hexagonal prism bounded by the planes $\{1\bar{1}0, 0\}$ and $\{000, 2\}$ would contain just two electrons per atom.

FIG. 65. Zone for hexagonal close-packed lattice.

Element	c/a	n	E_A volts	E_B volts
Zn	1·861	1·799	7·125	6·171
Cd	1·890	1·805	5·668	4·766
Be	1·585	1·731	9·594	11·478
Mg	1·625	1·743	4·882	5·547

The table shows c/a for certain hexagonal metals, and the energies E_A, E_B of a free electron at the points A, B. $c/a = \sqrt{\tfrac{8}{3}} = 1·63$ corresponds to closest packing. The metals which take the close-packed hexagonal structure have in general two valency electrons per atom, the electrons therefore overlap beyond the first zone. The table shows that, for metals for which $c/a > 1\cdot63$, $E_B < E_A$; whilst the reverse is true for the metals of the second group, for which $c/a < 1\cdot63$. It is probable that for the metals of the first type the electrons will overlap at the points B, but for metals of the second type there will will be no overlap at these points. This difference in the electron distribution in k-space of the two types is without doubt intimately connected with the differences which these groups show in electrical and magnetic properties.

TYPE A 8. *The selenium structure.* The selenium structure is based on a hexagonal lattice with 3 atoms in the unit cell and a basis

$$(000,\ \bar{u}\,u\,\tfrac{1}{3};\ 2\bar{u}\,\bar{u}\,\tfrac{2}{3}).$$

Referred to rectangular axes the basis becomes

$$(000;\ \tfrac{3}{2}\bar{u}\,\tfrac{1}{2}u\,\tfrac{1}{3};\ \tfrac{3}{2}\bar{u}\,\tfrac{1}{2}\bar{u}\,\tfrac{2}{3}),$$

and the structure factor is

$$S_{n_1 n_2 n_3} = \cos^2\tfrac{1}{2}\pi(n_1+n_2)[1+\exp\{2\pi i(-\tfrac{3}{2}un_1+\tfrac{1}{2}un_2+\tfrac{1}{3}n_3)\}+$$
$$+\exp\{2\pi i(-\tfrac{3}{2}un_1-\tfrac{1}{2}un_2+\tfrac{2}{3}n_3)\}].$$

The following table gives the structure factors for various sets of planes:

	$|S_{n_1 n_2 n_3}|$	$u=\tfrac{1}{3}$	Se $u = 0\cdot217$	Te $u = 0\cdot269$
$\{1\bar{1}0,\ 0\}$	$\sqrt{(3+4\cos 2\pi u+2\cos 4\pi u)}$	0	1·41	0·725
$\{\bar{2}11,\ 0\}$	$\sqrt{(1+8\cos^2 3\pi u)}$	3	1·63	2·52
$\{000,\ 2\}$	0	0	0	0
$\{000,\ 3\}$	3	3	3	3

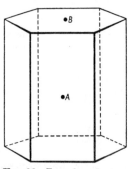

FIG. 66. Zone for selenium structure.

The form of the zone bounded by the planes $\{2\bar{1}1, 0\}$ and $\{000, 3\}$ is shown in Fig. 66. The volume of this zone in k-space is $6\sqrt{3}/a^2 c$, and since the atomic volume in the selenium structure is $a^2 c/2\sqrt{3}$, it follows that this zone contains just 6 states per atom. The elements Se and Te are semi-conductors with high specific resistances, from which it follows that the 6 available electrons external to a closed shell in these atoms just fill this Brillouin zone. Details of the lattice and of the zone structure are given below. k_A, k_B are the wave numbers of the points A, B.

	a	c/a	k_A $\times 10^8$ cm.	k_B $\times 10^8$ cm.	Energy of free electrons at A (electron volts)
Se	4·34	1·14	0·230	0·303	7·94
Te	4·44	1·33	0·225	0·254	7·60

TYPE A 9. *The graphite structure.* This structure is based on a hexagonal lattice with 4 atoms in the unit cell, and with the following basis referred to hexagonal axes

$$(000, \; 00\tfrac{1}{2}, \; \tfrac{1}{3}\tfrac{2}{3}0, \; \tfrac{2}{3}\tfrac{1}{3}\tfrac{1}{2}).$$

Referred to rectangular axes, the basis is

$$(000, \; 00\tfrac{1}{2}, \; 0\tfrac{1}{3}0, \; \tfrac{1}{2}\tfrac{1}{6}\tfrac{1}{2}),$$

and the structure factor

$$S_{n_1 n_2 n_3} = \cos^2 \tfrac{1}{2}\pi(n_1 + n_2)[1 + e^{\pi i n_3} + e^{\frac{2}{3}\pi i n_2} + e^{\pi i(n_1 + \frac{1}{3}n_2 + n_3)}].$$

The value of the structure factor for some planes of low indices is given below:

	$\{1\bar{1}0, 0\}$	$\{\bar{2}11, 0\}$	$\{000, 1\}$	$\{000, 2\}$
S	1	4	0	4

The zone which contains 4 electrons per atom is bounded by the planes $\{2\bar{1}1, 0\}$, $\{000, 2\}$ and its form is shown in Fig. 67. For graphite $a = 2{\cdot}46 \times 10^{-8}$ cm., $c/a = 2{\cdot}76$, and therefore $k_A = 0{\cdot}406, k_B = 0{\cdot}147$ in units of cm.$^{-1} \times 10^8$. The flat nature of this zone is not surprising, for the binding in graphite across the cleavage planes is exceedingly weak. In terms of the Bloch theory (case of

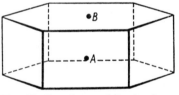

FIG. 67. Zone for graphite structure.

tight binding) this means a very small band width in the direction of the principal axis. The energy discontinuity across the planes B may therefore be expected to be large.

The only other substance known to possess the graphite structure is boron nitride, BN. This compound, like graphite, has 4 valency electrons per atom, which will also just fill the Brillouin zone shown in Fig. 67.

2.3. *Structures based on orthorhombic lattices.* An orthorhombic lattice is one of three unequal rectangular axes. The sides of the unit cell may be called a, b, c. The general equation for the planes of energy discontinuity in k-space is

$$\frac{n_1 k_x}{a} + \frac{n_2 k_y}{b} + \frac{n_3 k_z}{c} = \frac{1}{2}\left\{\left(\frac{n_1}{a}\right)^2 + \left(\frac{n_2}{b}\right)^2 + \left(\frac{n_3}{c}\right)^2\right\} \tag{8}$$

and the structure factor is obtained in exactly the same way as for the cubic structures. The normal distance $p_{n_1 n_2 n_3}$ from the origin of a plane $(n_1 n_2 n_3)$ is given by

$$p_{n_1 n_2 n_3} = \frac{1}{2}\left\{\left(\frac{n_1}{a}\right)^2 + \left(\frac{n_2}{b}\right)^2 + \left(\frac{n_3}{c}\right)^2\right\}^{\frac{1}{2}}.$$

TYPE A 5. *White-tin structure, tetragonal system.* In this structure two sides of the unit cell are equal, $a = b$. There are four atoms in the unit cell, with the basis

$$(000,\ \tfrac{1}{2}\tfrac{1}{2}\tfrac{1}{2},\ \tfrac{1}{2}0\tfrac{1}{4},\ 0\tfrac{1}{2}\tfrac{3}{4}).$$

The magnitude of the structure factor is therefore given by

$$|S_{n_1 n_2 n_3}| = 2\{1 + \cos \pi n_1 \cos \tfrac{1}{2}\pi n_2 + \cos \pi(n_2 + n_3)(\cos \tfrac{1}{2}\pi n_3 + \cos \pi n_1)\}^{\frac{1}{2}}.$$

The only planes with small indices $(\sum n^2 < 9)$ for which the structure factor does not vanish are the following:

	{101}	{200}	{211}	{220}	{112}
$|S|$	$2\sqrt{2}$	4	$2\sqrt{2}$	4	4

The symbol $\{n_1 n_2 n_3\}$, used to denote a set of planes in a tetragonal lattice, is meant to denote only those planes for which the third place in the symbol is either n_3 or $\bar{n}_3$. For example, the plane (110) does not belong to the group $\{101\}$.

Tin has four valency electrons, but the physical properties of white tin, for example its electrical conductivity, indicate that the effective number of free electrons is much less than four. This must be due to the existence of a zone which contains about four states per

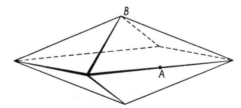

FIG. 68. Zone for white-tin structure.

atom. Such a zone is formed by the group of planes $\{101\}$, its form is shown in Fig. 68. The values of $|\mathbf{k}|$ at the two points A and B are

$$k_A = 0\cdot380 \times 10^8 \text{ cm.}^{-1}, \qquad k_B = 0\cdot205 \times 10^8 \text{ cm.}^{-1}$$

The volume of this zone is

$$\tfrac{1}{3}ca^2\left(\frac{1}{a^2}+\frac{1}{c^2}\right)^3,$$

and, since the atomic volume is $\tfrac{1}{4}a^2c$, it follows that the number of states per atom included is equal to

$$\frac{1}{6}\left(\frac{a}{c}\right)^4\left\{1+\left(\frac{c}{a}\right)^2\right\}^3.$$

For white tin $a = 5\cdot84\times10^{-8}$ cm., $c = 3\cdot15\times10^{-8}$ cm. The number of states per atom within the zone is therefore 4·24. It should be noticed that this zone is not uniquely defined, as are the first zones belonging to the simple structures, such as, for example, the body-centred cubic structure. In the present case there are other planes of energy discontinuity which intersect the zone just described, so that other surfaces may be found which also enclose approximately 4 or 5 states per atom.

2.4. *Structures based on a rhombohedral lattice.* A rhombohedral lattice is one in which there are three equal crystallographic axes inclined to each other at the same angle α which is different from 90°. A face-centred cubic structure forms a simple rhombohedral lattice for which $\alpha = 60°$. If x, y, z represent distances measured along the axes, the Fourier series for the potential may be written

$$V = \sum_{\mathbf{n}} V_{\mathbf{n}}\exp\left\{-\frac{2\pi i}{a}(n_1 x+n_2 y+n_3 z)\right\}$$

and the general equation for the planes of energy discontinuity in k-space may be obtained in the way already explained. The equation is

$$(2n_1-n_2-n_3)k_x+\sqrt{3}(n_2-n_3)k_y+\sqrt{2}(n_1+n_2+n_3)k_z\left/\left(\frac{1-\cos\alpha}{1+2\cos\alpha}\right)\right.+$$

$$+\frac{\sqrt{6}}{2a}\frac{\{(n_1^2+n_2^2+n_3^2)(1+\cos\alpha)-2(n_2 n_3+n_3 n_1+n_1 n_2)\cos\alpha\}}{(1+2\cos\alpha)\sqrt{(1-\cos\alpha)}} = 0. \quad (9)$$

In this equation k_x, k_y, k_z refer to rectangular coordinates in k-space. A plane in k-space specified by the indices $(n_1 n_2 n_3)$ lies parallel to the set of planes in the space lattice with the Miller indices $(n_1 n_2 n_3)$, where the indices n_1, n_2, n_3 refer, of course, to the oblique rhombohedral axes. The volume of the unit cell in a rhombohedral lattice is

$$a^3(1-\cos\alpha)\sqrt{(1+2\cos\alpha)}.$$

a is the length of the side of the unit rhombohedron.

TYPE A 10. *The mercury structure.* This structure is a simple rhombohedral lattice for which $\alpha = 70° 32'$, $\cos\alpha = \frac{1}{3}$, and $a = 3\cdot00$ Å. U. It will appear rather like a distorted face-centred cubic structure. The planes of energy discontinuity are given by the equation

$$(2n_1-n_2-n_3)k_x+\sqrt3(n_2-n_3)k_y+\frac{2}{\sqrt5}(n_1+n_2+n_3)k_z+$$

$$+\frac{3}{5a}\{2(n_1^2+n_2^2+n_3^2)-(n_2n_3+n_3n_1+n_1n_2)\} = 0.$$

The first Brillouin zone associated with this structure is bounded by the set of planes {100}, {111}, and {101} in which all the indices are positive or all negative; for example, (10$\bar1$) does not form a face of the zone. The nature of the zone is shown in Fig. 69. For $\alpha = 60°$, the first two sets of planes would mark out an octahedron and the third set a cube. It is seen that the zone is a distorted form of that shown in Fig. 61. The following details determine the form of the zone. Let O be the origin, then

$$OA = OB = 3/2\sqrt5a, \quad OC = \sqrt6/2\sqrt5a, \quad DF = FH = 2\sqrt3/10a,$$
$$FG = 3/5a, \quad GK = 6\sqrt3/10a.$$

The number of states per atom included within this zone is equal to 2, and the number within the inscribed sphere, which touches at the points C, is equal to $4\pi\sqrt2/15 = 1\cdot185$.

The free electron energies E_A, E_C at the points A and C are as follows (in electron volts):

$$E_A = 7\cdot5, \qquad E_C = 5\cdot0.$$

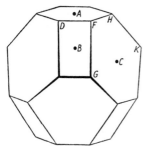

FIG. 69. Zone for mercury structure.

TYPE A 7. *The bismuth structure.* This structure is based on a rhombohedral lattice for which $\alpha = 57° 16'$, and $a = 4\cdot74$ Å.U. The unit cell contains two atoms with the basis

$$(000; \ 2u, 2u, 2u),$$

and the structure factor is

$$S_{n_1 n_2 n_3} = 2\cos\{2\pi u(n_1+n_2+n_3)\}.$$

The structure is almost a simple cubic structure, for the latter would be exactly realized if one were to set $\alpha = 60°$ and $u = \frac{1}{4}$. In the actual bismuth structure the two face-centred lattices of which a simple cubic is composed are displaced slightly relatively to each other.

The metals which take the bismuth structure have 5 valency electrons per atom, and the physical properties of these metals suggest that the *effective* number of free electrons is very small. This implies the presence of an almost completely filled zone. We are therefore interested in a zone which contains 5 states per atom. The form of this zone is shown in Fig. 70. The sides parallel to the principal axes of the zone are formed by the planes $\{1\bar{1}0\}$, whilst the 3 planes forming the top of the zone and the 3 planes forming the bottom of the zone are respectively the sets $\{221\}$ and $\{\bar{2}\bar{2}\bar{1}\}$. The structure factors for the bounding planes are given below:

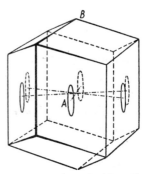

Fig. 70. Zone for bismuth structure. The ellipses show the occupied states, as explained in Chapter VI.

	Bi	Sb	As	*Simple cubic structure*
α	57° 16′	56° 36′	54° 7′	60°
u	0·237	0·233	0·226	0·25
S_{221}	0·794	1·018	1·369	0
$S_{1\bar{1}0}$	2	2	2	2

It is to be observed that if there were no distortion from the simple cubic form, i.e. if u were equal to $\frac{1}{4}$, there would be no energy discontinuity over the planes $\{221\}$, but that even a small relative displacement of the two face-centred lattices can produce a large value of S_{221}. The physical properties of bismuth, particularly the diamagnetism and the effects of very small traces of other metals in solid solution on the electrical conductivity and the diamagnetic susceptibility, show that the valency electrons in bismuth almost completely fill some Brillouin zone. When this is the case, the greater the energy gaps over the zone boundaries the lower will be the total energy, other things being equal.† In this way, then, we can see a simple reason for the relative displacement of the two face-centred lattices; it represents merely the tendency of the total energy to diminish as far as possible.

† This may be seen by reference, for example, to Fig. 22 (a). The energy of electrons just below an energy gap is depressed.

2.5. *Some typical alloy phases.* A solid phase of an alloy is characterized by a particular lattice structure. The different atoms are in general distributed at random through the lattice, and their relative concentrations vary within the range of homogeneity of the phase, but each lattice point is occupied by an atom of some sort, and the lattice type remains the same throughout the range of the phase. The magnitudes of the lattice parameters, in general, vary slightly and continuously with the composition throughout a phase.

A number of commonly occurring phases consist of structures which have already been described in connexion with the crystal structure of the elements. Most metals will form solid solutions with other elements. The structures of these solid solutions are identical with the structures of the solvent metal. An interesting phase which is not a solid solution but has a simple structure is the β-phase characterized by a body-centred cubic lattice. This phase usually occurs when there are approximately 1·5 valency electrons per atom. For example, in the Cu–Zn system it occurs at approximately 50 atomic per cent. Cu.

Two other phases with structures that have already been described are the so-called ϵ- and η-phases with hexagonal approximately close-packed structures. The ϵ-phase, which occurs when the number of valency electrons per atom is approximately 1·75, belongs to the type for which $c/a < \sqrt{\frac{8}{3}}$; the η-phase, which is really a solid solution, belongs to the type for which $c/a > \sqrt{\frac{8}{3}}$.

2.6. *The γ-structure.*[†] This is a rather complicated structure based on a cubic lattice with 52 atoms in the unit cell. The structure factor has been worked out by Bradley and Thewlis,[‡] and we give below their values for the reflecting planes of small indices:

$n_1 n_2 n_3$	{110}	{200}	{211}	{220}	{300}	{222}	{321}	{400}	{330}	{411}
S	0·32	0	0·32	0	0·32	2·68	1·05	0	8·85	5·63

The form of the zone marked out by the planes {330} and {411} is shown in Fig. 71; altogether there are 36 faces. This zone contains exactly 90 states per unit cell of 52 atoms, or 1·731 states per atom. Since this phase invariably occurs when the valency electron to atom ratio is in the neighbourhood of 21/13 (= 1·615), it follows that the zone is almost completely full. The symmetrical nature of the

[†] Cf. Jones, *Proc. Roy. Soc.* A, **144** (1934), 225.
[‡] Ibid. **112** (1926), 678.

zone is seen from the fact that the inscribed sphere contains $9\pi/13\sqrt{2}$ ($= 1\cdot54$) states per atom, i.e. $88\cdot4$ per cent. of the total content of the zone.

The following alloys are known to form a phase having this structure: Cu–Zn, Ag–Zn, Au–Zn, Cu–Cd, Ag–Cd, Au–Cd, Au–Hg, Ag–Hg, Cu–Si, Cu–Al, Cu–Sn, Fe–Zn, Co–Zn, Ni–Zn, Rh–Zn, Pd–Zn, Pt–Zn, Ni–Cd, Na–Pb.

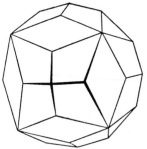

FIG. 71. Zone for γ-structure.

The β-Mn structure. This is a cubic structure with 20 atoms in the unit cell. The structure factor has been given by Preston,† whose values for the planes of low indices are given below:

$n_1 n_2 n_3$	{100}	{110}	{111}	{200}	{210}	{211}	{220}	{300}	{221}	{310}	{311}
S	0	0·63	1·0	0	1·38	0·63	1·26	0	10·1	8·23	6·41

The planes {221} and {310} form a zone with 48 faces whose form is shown in Fig. 72. The volume of this zone is $971/60a^3$, and, since the atomic volume is $a^3/20$, it follows that the zone contains $9\cdot71/6$ ($= 1\cdot62$) states per atom. The number of states per atom contained within the inscribed sphere of the zone is equal to $9\pi/20 = 1\cdot41$. The known alloys which have this structure, e.g. Ag_3Al, have an electron to atom ratio of $3/2$, which is the same as that for the β-body-centred cubic phase.

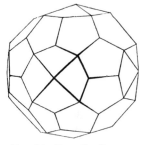

FIG. 72. Zone for β-manganese structure.

The following are the substances so far known which have this structure: β-Mn, $Ag_3Al(\beta')$, Cu–Si(γ), $CoZn_3$, Au_3Al.

2.7. *The fluorspar structure.* There are a number of metallic compounds which possess this structure; they have interesting physical properties. They are the following: Mg_2Sn, Mg_2Pb, Mg_2Si, Mg_2Ge, Li_2S, Na_2S, Cu_2S, Cu_2Se, Be_2C. The valency electron to atom ratio for this set of compounds is $8/3$.

The structure is based on a cubic lattice with 12 atoms in the unit

† *Phil. Mag.* **5** (1928), 1198.

cell. It has the following basis

$$\text{Sn} \quad (000,\ 0\tfrac{1}{2}\tfrac{1}{2},\ \tfrac{1}{2}0\tfrac{1}{2},\ \tfrac{1}{2}\tfrac{1}{2}0),$$

$$\text{Mg} \quad \pm(\tfrac{1}{4}\tfrac{1}{4}\tfrac{1}{4},\ \tfrac{1}{4}\tfrac{3}{4}\tfrac{3}{4},\ \tfrac{3}{4}\tfrac{1}{4}\tfrac{3}{4},\ \tfrac{3}{4}\tfrac{3}{4}\tfrac{1}{4}).$$

The structure amplitude factor may therefore be written in the form

$$S = \{A_n + 2B_n \cos \tfrac{1}{2}\pi(n_1+n_2+n_3)\}\{1 + \cos \pi(n_2+n_3) +$$
$$+ \cos \pi(n_3+n_1) + \cos \pi(n_1+n_2)\},$$

where A_n and B_n are Fourier coefficients referring to the Sn and Mg atoms respectively.

The volume of the zone marked out by the planes $\{220\}$ is $16/a^3$, and therefore, since the atomic volume is $a^3/12$, there are just $8/3$ states per atom enclosed within this zone. Its form is identical with that shown in Fig. 62. As the number of available valency electrons is just equal to the number of states within the zone, we are led to expect a small conductivity. Actually the pure compounds are found to be almost insulators.† That the high resistance is really due to a completed zone, and not due to polar binding, is shown by the fact that the molten alloy, where the zone structure is destroyed, has about the same conductivity as, for example, molten Sn.

3. Factors determining the crystal structure

3.1. *The Hume-Rothery rule.* Hume-Rothery‡ was the first to point out that certain alloy structures with narrow ranges occur at a definite electron-atom ratio; the γ-structure, which occurs at the ratio of 21 valence electrons to 13 atoms, is perhaps the best-known example.

In Chap. I, § 6 we discussed the factors that determine the boundaries of a phase. We plotted the free energy F against c, the concentration, and found (cf. Fig. 9) that, if F increases suddenly as the concentration passes a given value, this concentration will mark the boundary of the phase that includes the pure metal and both boundaries of any narrow intermediate phase. Any such sudden increase of F must be due to the term $E(c)$, the energy at the absolute zero of temperature. Although no precise calculation of $E(c)$ as a function of c has yet been carried out, we shall show below that it is possible to estimate quite simply the value of the concentration for any particular phase beyond which $E(c)$ may be expected to show a

---

† Norbury, *Trans. Far. Soc.* **16** (1921), 570.

‡ *The Metallic State*, p. 328, Oxford (1931).

rapid increase. These critical concentrations are then to be compared with the observed phase boundaries.

We have shown in Chap. II, § 4.6 how to construct a function
$2N(E)$ giving the number of states per unit energy range. The form
of $2N(E)$ for those states lying within the first Brillouin zone is
indicated roughly† in Fig. 73. $N(E)$ is given by the following general
formula

$$2N(E) = \frac{1}{4\pi^3} \int\int \frac{dS}{|\text{grad } E|},$$

where the integration is over the surface in k-space for which the
energy is equal to E. The area lying between the curve and the
E-axis is equal to the total number
of states within the zone. As the
composition of an alloy within a
single phase changes, the form of
the $N(E)$ curve may also change,
but the area which it encloses
remains constant. The maximum
A (Fig. 73) occurs approximately

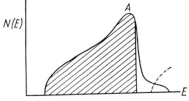

Fig. 73. Density of states in a Brillouin zone.

for that value of the energy for which the energy surfaces just touch the
boundaries of the Brillouin zones,‡ because grad E is then very small
and therefore leads to large values of $N(E)$. If we calculate the total
energy of the electrons as a function of the number per unit volume,
we shall find, therefore, that the energy increases especially rapidly
when the electrons fill up the band to just beyond the point A
(Fig. 73), for example, as shown by the shaded area. The number
of electrons required to do this is given by the number which would
fill the inscribed sphere of the Brillouin zone. This number, which is
most conveniently expressed as a number of electrons per atom and
denoted by n_c, is then to be compared with the electron-atom ratio
of the phase boundary. The values of n_c for the α-, β-, and γ-phases
are respectively 1·362, 1·480, and 1·538.

The following table gives the electron-atom ratio of the boundaries
of α-, β-, and γ-phases. The boundaries of the β-phase invariably
depend in a marked way on the temperature. The width of the phase
becomes smaller as the temperature decreases, and in many cases a

---

† The general form of this curve is given by interpolating between the limiting
cases discussed in Chap. II, § 4.6.

‡ For the body-centred cubic structure investigated on p. 85 the maximum occurs
at exactly this point.

temperature is reached at which the boundaries meet each other. It is the electron concentration corresponding to this composition which is given in Table VI. In some alloys, e.g. Au-Cd, the β-phase appears to have a considerable range of homogeneity even at room temperatures. This may be due partly to the extreme slowness with which thermal equilibrium will be established between different phases at low temperatures.

TABLE VI

Alloy	Electron-atom ratio of maximum solubility in α-phase
Cu–Zn	1·384
Cu–Al	1·408
Cu–Ga	1·406
Cu–Si	1·420
Cu–Ge	1·360
Cu–Sn	1·270
Ag–Cd	1·425
Ag–Zn	1·378
Ag–Hg	1·35
Ag–In	1·40
Ag–Al	1·408
Ag–Ga	1·380
Ag–Sn	1·366
n_c (theoretical)	1·362

Alloy	Electron-atom ratio of the β-phase boundary with smallest electron concentration
Cu–Zn	1·48
Cu–Sn	1·49
Cu–Al	1·48
Au–Zn	1·48
Au–Cd	1·49
Au–Al	1·370
Ag–Cd	1·50
Cu–Si	1·49
n_c (theoretical)	1·480

Alloy	Electron-atom ratio of the γ-phase boundaries	
Cu–Zn	1·58	1·66
Cu–Cd	1·60	1·67
Cu–Sn	1·67	1·67
Cu–Al	1·63	1·77
Ag–Zn	1·58	1·63
Ag–Cd	1·59	1·63
Ni–Zn	1·52	1·76
n_c (theoretical)	1·538	

This table has been constructed by assuming that the various atoms contribute to the total number of valency electrons in the alloy according to the following scheme:

Metal	Cu	Ag	Au	Zn	Cd	Hg	Al	In	Ga	Sn	Si	Ge	Ni
Valency electrons per atom	1	1	1	2	2	2	3	3	3	4	4	4	0

The fact that nickel contributes no valency electrons is particularly to be noted. In Chap. VI, § 4 it is shown that the elements Ni, Pd,

Pt contribute no valency electrons in alloys with Cu, Ag, and Au, so long as the electron-atom ratio is greater than 0·6.

As we have stated, Hume-Rothery first pointed out that particular alloy phases invariably occur at the same valency electron to atom ratios. He proposed the rules that the β-phase is associated with an electron-atom ratio $\frac{3}{2}$, the γ-phase with $\frac{21}{13}$, and the ϵ-phase (hexagonal closest packed) with $\frac{7}{4}$.† The preceding theoretical considerations show the importance of the electron-atom ratio in determining the phase formation. The precise ratios expressed as fractions do not, however, appear to have special significance.

It will be noticed that the explanation of the bismuth structure given on p. 167 is essentially the same as that advanced here for the Hume-Rothery alloys.‡

3.2. *Size of ions.* The elements Cu, Ag, and Au have the close-packed cubic structure, while the alkalis are body-centred. For metals with only one electron per atom, the influence of the Brillouin zones on the energy is small; according to Fuchs‖ the energies of the valency electrons in the two structures will differ, for equal atomic volume, by only about 10^{-3} e.v. per atom. As we have

Metal	Atomic number	Structure	Metal	Atomic number	Structure
Cu	29	cubic close-packed	Ag	47	cubic close-packed
Ni	28	,, ,, ,,	Pd	46	,, ,, ,,
Co	27	hex. ,, ,,	Rh	45	,, ,, ,,
Fe	26	body-centred cubic††	Ru	44	hex. ,, ,,
Mn	25	cubic‡‡	Mo	43	body-centred cubic
Cr	24	body-centred cubic	Nb	42	,, ,, ,,
V	23	,, ,, ,,			

Metal	Atomic number	Structure
Au	79	cubic close-packed
Pt	78	,, ,, ,,
Ir	77	,, ,, ,,
Os	76	hex. ,, ,,
Re	75	,, ,, ,,
W	74	body-centred cubic
Ta	73	,, ,, ,,

† The occurrence of the ϵ-phase appears to coincide with the complete filling of the inner Brillouin zone appropriate to the structure (cf. Chap. V, § 2.2).

‡ Cf. Chap. III, § 8.3 for some remarks on the colour of the brasses Cu–Zn.

‖ *Proc. Roy. Soc.* A, **151** (1935), 585.

†† Cubic close-packed modification (γ-Fe) above 920° C.

‡‡ The structure is complicated, but is nearly body-centred.

seen, however, in the noble metals, the closed d shells of the ions must be regarded as touching, and the intense repulsive forces between them come into play. This repulsion between the ions obviously favours close packing. Fuchs has calculated that the energy per atom of copper in the body-centred structure would be 0·1 e.v. greater.

This conclusion is strengthened by an examination of the crystal structure of the transition metals; when the d shells are nearly full, and may be expected to repel each other, the structure is close-packed; as, however, the number of vacant places in the d shell increases, presumably decreasing the repulsion, other structures occur. This is shown on p. 173.

Note added in proof.

A discussion of the energy of the alloy phases similar to that of this chapter has recently been given by Konobejewski, *Ann. d. Physik*, **26** (1936), 97.

HEAT CAPACITY AND MAGNETIC PROPERTIES OF THE METALLIC ELECTRONS

1. The Fermi distribution law

IN the preceding chapters we have considered the behaviour of the electrons in a metal at the absolute zero of temperature. We have found that there are a series of possible stationary states for an electron; the number of these stationary states, per unit volume of the metal, having energies between the values E, $E+dE$ has been denoted (Chap. II, § 4.6) by $N(E)\,dE$. We have seen that the Pauli exclusion principle forbids more than *two* electrons to be in any one stationary state; it follows that, at the absolute zero of temperature, all states with energy below a certain maximum energy E_{max} are doubly occupied, and the higher states empty. E_{max} is given by the equation

$$2 \int_0^{E_{max}} N(E)\,dE = N, \tag{1}$$

where N is the total number of electrons per unit volume.

At finite temperatures there will be a number of electrons with energies greater than E_{max}; according to the Fermi-Dirac statistics, the probability† that a given quantum state of energy E is occupied is given by the 'Fermi-Dirac distribution function'

$$f(E) = \frac{1}{e^{(E-\zeta)/kT}+1}; \tag{2}$$

the number of electrons having energy between E and $E+dE$ is thus

$$2N(E)f(E)\,dE = \frac{2N(E)\,dE}{e^{(E-\zeta)/kT}+1}. \tag{3}$$

ζ is here the thermodynamic potential per electron, so that

$$\zeta = u - Ts + pv,$$

where u, s, and v are the internal energy, entropy, and volume per electron. It is therefore a function of the temperature and may be

---

† For a proof of the Fermi-Dirac distribution law cf. Sommerfeld and Bethe, *Handb. d. Phys.* **24**/2 (1933), 339, or Fowler, *Statistical Mechanics*, Chap. 21.

determined from the equation

$$2 \int N(E)f(E)\, dE = N, \qquad (4)$$

the integration being over all energies for which $N(E)$ is finite.†

At finite temperatures the constant ζ is approximately, but not exactly, equal to the maximum electronic energy $E_{\max}$ at the absolute zero of temperature. It follows that, for all ordinary metals and all temperatures T up to the melting-point, we may write‡

$$kT \ll \zeta.$$

It follows, therefore, that f is approximately equal to unity for values of E less than ζ, and falls exponentially to zero for E greater than ζ, as illustrated in Fig. 74.

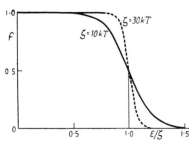

We note that, for $E - \zeta \gg kT$, the number of electrons per energy range dE is, from (2),

$$2N(E)e^{-(E-\zeta)/kT}\, dE.$$

FIG. 74. Fermi distribution function f for the electrons in a metal. The electrons in states for which E/ζ is greater than unity are 'excited'. The functions are calculated for $kT = \frac{1}{10}\zeta$, $kT = \frac{1}{30}\zeta$.

The variation with energy of the number of electrons at a given temperature is, therefore, Maxwellian, and the variation with temperature approximately so, if one takes $E = \zeta$ as the zero of energy.

We note also that the number of electrons in states with energies greater than ζ is, if $N(E)$ may be taken as constant in the range considered,

$$2N(\zeta) \int_{\zeta}^{\infty} \frac{dE}{e^{(E-\zeta)/kT}+1} = 2N(\zeta)kT \log_e 2. \qquad (5)$$

Such states may be termed 'excited'

1.1. *Calculation of ζ.* The constant ζ which occurs in the Fermi distribution function (2) is not exactly equal to $E_{\max}$ except when $T = 0$. At other temperatures it may be calculated from equation

† Compare the classical (Boltzmann) distribution law

$$f_{\text{cl}} = Ce^{-E/kT},$$

where the constant C must be determined by (4). If ζ is the free energy, $C = e^{\zeta/kT}$.

‡ At a temperature of 1,000° C. kT is equal to an energy of $\frac{1}{16}$ of an electron volt; $E_{\max}$ is 1·53 e.v. for Cs, 6·7 for Ag.

(4). In order to solve this equation let us consider the integral

$$I = \int_0^\infty f(E) \frac{d}{dE} F(E) \, dE, \tag{6}$$

where F is any function which vanishes for $E = 0$. By partial integration one obtains

$$I = - \int_0^\infty \frac{df}{dE} F \, dE. \tag{7}$$

Now df/dE is a function which vanishes except in the neighbourhood of $E = \zeta$ (cf. Fig. 74). We therefore expand $F(E)$ by Taylor's theorem,

$$F(E) = F(\zeta) + (E-\zeta)F'(\zeta) + \tfrac{1}{2}(E-\zeta)^2 F''(\zeta) + \ldots,$$

and obtain from (7)

$$-I = L_0 F(\zeta) + L_1 F'(\zeta) + L_2 F''(\zeta) + \ldots, \tag{8}$$

where

$$L_0 = \int_0^\infty \frac{df}{dE} \, dE, \quad L_1 = \int_0^\infty (E-\zeta) \frac{df}{dE} \, dE, \quad L_2 = \tfrac{1}{2} \int_0^\infty (E-\zeta)^2 \frac{df}{dE} \, dE.$$

It is easily seen that $L_0 = -1$, $L_1 = 0$, and, writing $(E-\zeta)/kT = \epsilon$ and remembering that $\zeta/kT \gg 1$,

$$-L_2 \simeq \frac{(kT)^2}{2} \int_{-\infty}^\infty \frac{\epsilon^2 \, d\epsilon}{(e^{\frac{1}{2}\epsilon} + e^{-\frac{1}{2}\epsilon})^2} = \frac{\pi^2}{6}(kT)^2.$$

We see therefore from (6) and (8) that

$$\int_0^\infty f(E) \frac{d}{dE} F(E) \, dE = F(\zeta) + \frac{\pi^2}{6}(kT)^2 F''(\zeta) + \ldots. \tag{9}$$

In our case we set $\quad F(E) = \int_0^E N(E) \, dE,$

and hence obtain from (9) for the total number of electrons per unit volume

$$N = 2 \int_0^\infty f(E) N(E) \, dE = 2 \int_0^\zeta N(E) \, dE + \frac{\pi^2}{3}(kT)^2 \left(\frac{dN}{dE}\right)_{E=\zeta}.$$

If now we subtract equation (1) from the equation above, we obtain

$$2 \int_{E_{\max}}^\zeta N(E) \, dE + \frac{\pi^2}{3}(kT)^2 \left(\frac{dN}{dE}\right)_{E=\zeta} = 0.$$

But we have seen that $E_{max}-\zeta$ is small, as is also the second term in this equation; we may therefore write, to a sufficient order of accuracy,

$$2(\zeta-E_{max})N(E_{max})+\frac{\pi^2}{3}(kT)^2\left(\frac{dN}{dE}\right)_{E=E_{max}} = 0,$$

whence we see that, if T is not too large,

$$\zeta = E_{max}-\frac{\pi^2}{6}(kT)^2\left(\frac{d\log N}{dE}\right)_{E=E_{max}} \tag{10}$$

Since at low temperatures ζ tends to E_{max}, we shall write ζ_0 for E_{max} in our subsequent work.

2. Specific heat

From Fig. 74 we see that, at temperatures much lower than ζ_0/k, the number of electrons in excited states is proportional to kT/ζ_0 and the mean energy of excitation is proportional to kT. The extra energy of the electron gas at a temperature T, due to thermal motion, is therefore proportional to $(kT)^2/\zeta_0$. Differentiating with respect to T, we see that the specific heat, per electron, is of the order

$$k^2T/\zeta_0,$$

which is much less than the classical value $\frac{3}{2}k$, so long as kT is small compared with ζ_0.

We shall now obtain exact formulae for the specific heat.

We require an expression for the total energy of the electrons (the lowest energy of the Fermi distribution being taken as zero). This is equal to

$$\bar{E} = 2\int_0^\infty N(E)f(E)E\,dE.$$

For low temperatures, $kT \ll \zeta_0$, this may be evaluated at once from equation (9), by putting $F(E)$ equal to $\int_0^E N(E)E\,dE$; we obtain

$$2\int_0^\infty N(E)f(E)E\,dE = 2\int_0^\zeta N(E)E\,dE +\frac{\pi^2}{3}(kT)^2\left(\frac{d(EN)}{dE}\right)_{E=\zeta}.$$

Of the two terms on the right, the first may be written

$$2\int_0^{\zeta_0} N(E)E\,dE +2(\zeta-\zeta_0)\zeta_0 N(\zeta_0).$$

The first of these two terms is equal to the energy of the electrons at the absolute zero of temperature; the second, by (10), is equal to

$$-\frac{\pi^2}{3}(kT)^2\zeta_0\left(\frac{dN}{dE}\right)_{E=\zeta_0},$$

whence we see that $\bar{E}$ is equal to

$$2\int_0^{\zeta_0} N(E)E\,dE +\tfrac{1}{3}\pi^2 N(\zeta_0)(kT)^2.$$

To obtain the contribution to the specific heat at constant volume, we must divide by the density ρ and differentiate with respect to T; we obtain

$$c_v = \tfrac{2}{3}\pi^2 k^2 T N(\zeta_0)/\rho, \tag{11}$$

where $N(\zeta_0)/\rho$ is the number of states per unit energy range at the surface of the Fermi distribution for unit mass of metal.

In the particular case of free electrons, or of a band that is nearly empty, we may write

$$N(E) = C\sqrt{E}, \tag{12}$$

where C is a constant.

In this case we have, by integrating (12), $N = \tfrac{4}{3}C\zeta_0^{\frac{3}{2}}$, and hence

$$N(\zeta_0) = \tfrac{3}{4}N/\zeta_0. \tag{13}$$

Writing $kT_0 = \zeta_0$, the specific heat may be written in the form

$$c_v = \tfrac{1}{2}\pi^2 nkT/T_0, \tag{14.1}$$

where $n\ (= N/\rho)$ is the number of electrons per unit mass; for the atomic heat we have[†]

$$C_v = \tfrac{1}{2}\pi^2 n_0\,RT/T_0, \tag{14.2}$$

where n_0 is the number of free electrons per *atom* and R the gas constant.

For high temperatures an asymptotic expansion may be given[‡] in powers of T_0/T, and is

$$C_v = \tfrac{3}{2}n_0 R\left[1-\frac{1}{6(2\pi)^{\frac{3}{2}}}\left(\frac{T_0}{T}\right)^{\frac{3}{2}}...\right]. \tag{15}$$

For intermediate values of T/T_0, C_v may be obtained by numerical integration, and is shown in Fig. 75.

We call T_0 the 'degeneracy temperature' of the electron gas.

---

† Stoner, *Phil. Mag.* **21** (1936), 145, has given the expansion in ascending powers of T/T_0 to a further term; he obtains, subject to (12),

$$C_v = \tfrac{1}{2}\pi^2 n_0\,R[T/T_0 - 2\cdot 96(T/T_0)^3 ...].$$

For arbitrary $N(E)$, see *Proc. Roy. Soc.* A, **154** (1936), 656.

‡ Stoner, loc. cit., or Mott, *Proc. Camb. Phil. Soc.* **32** (1936), 108.

Formulae (14) and (15) are true also for a band that is nearly *full*, so that instead of (12) we have

$$N(E) = C'\sqrt{(E_0 - E)}.$$

N will then refer to the number of unoccupied states (positive holes) per unit volume, and T_0 will be equal to $(E_0 - E_{\max})/k$; we may call kT_0 the degeneracy temperature of the positive holes (i.e. of the unoccupied states).

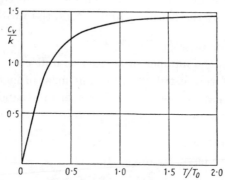

Fig. 75. Heat capacity per electron of a degenerate electron gas.

The specific heat has also been calculated† for a case when $N(E)$ vanishes except in a finite range. If one takes for E

$$E = -\tfrac{1}{3}B(\cos k_x a + \cos k_y a + \cos k_z a),$$

which is the formula for the s band in a simple cubic lattice,‡ then E varies between the limits $\pm B$. If there is one electron per atom, so that the band is half full, then one may obtain, writing $B = kT_B$,

$$C_v = 5{\cdot}68RT/T_B \quad (T \ll T_B)$$
$$= \tfrac{1}{12}R(T_B/T)^2 \quad (T \gg T_B).$$

It will be seen that $C_v \to 0$ as $T \to \infty$.

2.1. *Numerical values.* If the energy is given in terms of the wave number by the formula $E = \alpha\hbar^2 k^2/2m$, then (cf. Chap. II, equation (19))

$$T_0 = \zeta_0/k = \alpha\frac{h^2}{8mk}\left(\frac{3N}{\pi}\right)^{\frac{2}{3}}, \tag{16}$$

where N is the number of electrons per unit volume. α is unity for free electrons. Assuming one electron per atom and setting $\alpha = 1$,

† Unpublished.   ‡ Cf. p. 68.

the following are the theoretical values for certain metals:

	Li	Na	K	Cu	Ag	Au
$T_0 \times 10^{-3}$ (degrees)	55	36·5	24	82	64	64

We see, therefore, from (14) that the contribution to the atomic heat from the electrons is small in normal metals, if the electrons behave as 'free electrons'; for silver at room temperatures it is about 0·04 cal./degree, whereas that due to the lattice vibrations is about 6 cal./degree. We shall see, however, in § 5 that in the *transition metals* the positive holes in the d shells give a much larger contribution.

Explicit formulae for the heat capacity of the electrons in terms of N are, from (14) and (16),

$$C_v = \frac{4\pi^3 m k^2 T}{3\alpha h^2} \left(\frac{3N}{\pi}\right)^{\frac{1}{3}} \frac{A}{\rho},$$

or, with $\alpha = 1$ as for free electrons,

$$C_v = 3\cdot26 \times 10^{-5} n_0^{\frac{1}{3}} (A/\rho)^{\frac{2}{3}} T \text{ cal. per degree per gm. atom} \quad (17)$$

(A = atomic weight, ρ = density, n_0 = number of electrons per atom).

For comparison we give the formula for the heat capacity due to lattice vibrations, which is, at low temperatures,

$$C_v = 468(T/\Theta_D)^3 \text{ cal. degree per gm. atom.} \quad (18)$$

For silver ($\Theta_D = 215°$ K.), formulae (17) and (18) give at $2°$ K.

$$C_v \begin{cases} \text{electrons} & 0\cdot00031 \\ \text{lattice vibrations} & 0\cdot00037. \end{cases}$$

2.2. Comparison with experiment. As we have seen, for normal metals the specific heat due to the free electrons is negligible in comparison with that due to the lattice vibrations except at very low temperatures ($\sim 2°$ K.). In order to test the formula (17), it is natural to choose elements which are not supraconductors; because in supraconductors there is a discontinuity in the specific heat at the transition point.

Keesom[†] and his collaborators have measured the specific heat of the non-supraconductive elements Ag and Zn between $1\cdot7°$ K. and $4°$ K. They have found that the specific heat is greater than that given by extrapolating the Debye T^3 law. It was, however, difficult

† Cf. Keesom and Kok, *Physica*, **1** (1934), 770, where references to earlier work are given.

to calculate the part of the specific heat due to the free electrons, because at somewhat higher temperatures Debye's Θ is not quite constant.[†] However, by extrapolating to lower temperatures the part of the Θ curve belonging to higher temperatures, the part of the atomic heat due to lattice vibrations was estimated, and subtracted from the observed atomic heat. The values obtained for silver are

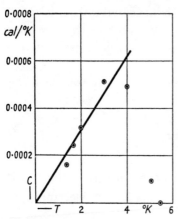

FIG. 76. Linear term in the atomic heat of silver.

⊚ observed (Keesom and Kok); full line calculated from formula (17).

shown in Fig. 76. The full line shows the specific heat of the free electrons calculated from (17) assuming *one* free electron per atom. The apparent falling off at the higher temperatures is not yet established with certainty, since the atomic heat of the free electrons is only a small part of the total heat at these temperatures,[‡] and, as we have stated, the calculation of the energy of vibration depends on an extrapolation from higher temperatures.

Zn shows a similar behaviour; Keesom and Kok (loc. cit.) estimate that the best agreement with experiment would be obtained by assuming rather less than one free electron per atom in formula (17).

Critescu and Simon[||] have observed for Be a maximum in the neighbourhood of 11° K., which is probably to be attributed to the behaviour of the metallic electrons, but is not in agreement with (14).

Transition metals are discussed in § 5.2; they have an electronic specific heat much larger than that of normal metals.

The specific heat of supraconducting elements shows a discontinuity at the transition point. Below the transition point the elements investigated, tin,[††] thallium,[††] lead,[‡‡] and bismuth,[‡‡] follow the T^3 law far more accurately than the non-supraconductors. For thallium, for instance, the value of Θ_D at 1·3 is 78·82° and at 2·36°

---

† Probably for the reason discussed in Chap. I, § 1.2.

‡ About 18 per cent. at 4° K., as compared with 50 per cent. at the lowest temperatures measured, 1·67° K. || *Zeits. f. Phys. Chem.* B, **25** (1934), 378.

†† Cf. Keesom and Kok, *Physica*, **1** (1934), 175, where results for thallium and references to earlier papers are given.

‡‡ Cf. Keesom and van den Ende, *Comm. Leid.* 203 d (1929); 213 c (1931).

(the transition point) 77·23. Directly above the transition point Θ_D is 80·36. No term in the specific heat proportional to T appears therefore to exist. Interesting conclusions have been drawn from this fact by Kok.†

3. Magnetism; introduction

With respect to their magnetic properties, solids may be divided into three classes. These are:

(1) Strongly paramagnetic substances, having a susceptibility dependent upon the temperature. The rare earths and their salts are the best-known example. The strong paramagnetism is due to the presence of a free, or nearly free, magnetic core in the atom. The volume susceptibility is given for cores in doublet S states by the Langevin-Debye formula‡

$$\kappa = \frac{N\mu^2}{kT}. \tag{19}$$

(N = number of magnetic cores per unit volume, μ = magnetic moment of core.) The discussion of this type of magnetism lies outside the scope of this book.∥

(2) Feebly paramagnetic or diamagnetic substances. This class includes the great majority of metals. The atomic susceptibility ($\times 10^6$) varies from -273 (Bi) to 555 (Pd). The dependence on temperature is usually small. The paramagnetic metals Pd, Mn, etc., form an intermediate class between (1) and (2).

(3) Ferromagnetic substances. This class includes only Fe, Co, and Ni and certain alloys and compounds containing these metals or Mn or Cr, and some of the rare earths. They will be discussed in § 7.

In the ferromagnetic metals, we know from the gyromagnetic effect that the magnetic carriers are electrons (Landé g-factor equal to 2). The interaction forces between the cores are therefore sufficiently strong to quench the orbital momentum of the incomplete shells, so that they are not free to turn in a magnetic field. It is highly probable that even in the most paramagnetic metals of class (2), such as Pd, the interaction between the incomplete shells is at least as

---

† *Nature*, **134** (1934), 532.

‡ We omit the factor 3 usually given in the denominator, so that, for the moment of an electron, we must take $e\hbar/2mc$ instead of $\sqrt{3}e\hbar/2mc$.

∥ Cf., for example, Van Vleck, *Electric and Magnetic Susceptibilities*, Oxford (1932).

strong† as in the ferromagnetic elements, though of a different sign, so that the coupling between spin and orbital motion is broken down and the orbital motion quenched. When an external field is applied, therefore, some of the spins will turn over, giving rise to paramagnetism, but there is no question of this happening to any of the orbits. This 'spin paramagnetism' is discussed in the next section. The orbital motion will not, however, be unaffected by the field; currents will be induced, leading to an induced moment. For closed shells these induced currents always give *diamagnetism*. For any electrons which can contribute to an electric current (incomplete shells, valence electrons, etc.) the contribution to the magnetism may be of either sign, but is usually negative. This susceptibility due to orbital motion is discussed in § 6.

4. Weak spin paramagnetism

Pauli‡ was the first to point out that electrons in a metal, obeying the Fermi-Dirac statistics, would yield a small paramagnetism, in general approximately independent of temperature. This was the first application made of the newly discovered Fermi-Dirac statistics to the theory of metals. This paramagnetism will now be discussed.

We denote by μ the Bohr magneton

$$\mu = e\hbar/2mc.$$

Then, in an external magnetic field H, the energy of an electron can take two values: either $-H\mu$ if the spin points parallel to the field, or $+H\mu$ if it points antiparallel. If there is nothing to prevent it, an electron with antiparallel spin will turn its spin parallel to the field, the energy $2H\mu$ being transferred to the lattice vibrations. In a metal, however, this is impossible for most of the electrons, because the states with parallel spin are already occupied. At any finite temperature T, however, a small number of electrons will be in excited states. The number of such electrons is of the order:

$$T/T_0 \times \text{total number of electrons},$$

as formula (5) shows, where T_0 is the Fermi limiting temperature (cf. § 2), equal to about $64{,}000°$ for silver, though much less for the positive holes in the d shells of the transition metals (cf. § 5). Now each of these electrons will behave, qualitatively, as it would in the classical theory, and thus give a contribution to the susceptibility

† Cf. §§ 5, 7.   ‡ *Zeits. f. Phys.* **41** (1926), 81.

of about μ^2/kT (cf. § 3). The volume susceptibility of the metal is thus approximately $N\mu^2/kT_0$, where N is the number of electrons per unit volume. The susceptibility is therefore independent of temperature, so long as T is small compared with T_0.

It is to be emphasized that the electrons in an insulator make no contribution to the spin paramagnetism, because, in any Brillouin zone which contains any electrons, all the states are occupied.

We shall now obtain the exact formula for the susceptibility. At the absolute zero of temperature the calculation is elementary. The occupied electronic states are those which have least energy, so that in the absence of a magnetic field there will be two electrons in each state with energy less than the maximum energy $E_{\max}$, and no electrons in higher states. If now a magnetic field H is imposed, electrons with their spins parallel to the field have their energy lowered by an amount $H\mu$, and electrons with their spins antiparallel raised by the same amount. We denote by[†] $N(E)\,dE$ the number of electronic states, per unit volume, in the energy range E to $E+dE$, in the absence of a field. Then, when the field is applied, the number of states with spin parallel to the field in the energy range E to $E+dE$ is

$$N(E+H\mu)\,dE,$$

and the number with spins antiparallel is

$$N(E-H\mu)\,dE.$$

As in the absence of the field, all states with energy up to a certain maximum ζ'_0 will be full. The magnetic moment σ per unit volume will then be given by

$$\sigma = \mu \int\limits^{\zeta'_0} [N(E+H\mu)-N(E-H\mu)]\,dE, \tag{20}$$

which gives

$$\sigma = \mu \int\limits_{\zeta'_0-H\mu}^{\zeta'_0+H\mu} N(E)\,dE.$$

Even for the strongest fields obtainable, 300,000 gauss, $H\mu$ is only $1{\cdot}7\times10^{-3}$ e.v., and so we may write $H\mu \ll \zeta_0$, $\zeta'_0 \simeq \zeta_0$, and hence

$$\sigma = 2H\mu^2 N(\zeta_0);$$

the volume susceptibility is therefore

$$\kappa = 2\mu^2 N(\zeta_0), \tag{21}$$

[†] $N(E)$ is the number of states, $2N(E)$ the number of electrons. Cf. Chap. II, § 4.6.

where $N(E)$ is the number of quantum states per unit energy interval and per unit volume of the metal.

At finite temperatures the magnetic moment is, by analogy with (20),

$$\sigma = \mu \int_0^\infty [N(E+H\mu)-N(E-H\mu)]f(E)\,dE,$$

where $f(E)$ is the Fermi distribution function (§ 1). Since H may be taken to be small, this gives for the susceptibility

$$\kappa = \frac{\sigma}{H} = 2\mu^2 \int_0^\infty f(E)\frac{dN(E)}{dE}\,dE.$$

For temperatures not too large, a calculation similar to that of § 2.1 gives†

$$\kappa = 2\mu^2 N(\zeta_0)\left[1 + \frac{\pi^2}{6}(kT)^2\left(\frac{d^2\log N}{dE^2}\right)_{E=\zeta_0}\right]. \tag{22}$$

In the particular case of free electrons, or of a nearly empty zone (cf. Chap. II, § 4.6) when $N(E) = C\sqrt{E}$, formula (22) becomes‡ (cf. equation (13))

$$\kappa = \frac{3}{2}\frac{N\mu^2}{kT_0}\left[1 - \frac{\pi^2}{12}\left(\frac{T}{T_0}\right)^2\cdots\right], \tag{23}$$

where $kT_0 = \zeta_0$ and N is the number of electrons per unit volume. In this case, also, we may obtain a formula for the susceptibility valid for large T/T_0, viz.‖

$$\kappa = \frac{N\mu^2}{kT}\left[1 - \frac{2}{3\sqrt{(2\pi)}}\left(\frac{T_0}{T}\right)^{\frac{3}{2}}\cdots\right]. \tag{24}$$

The value of κ for intermediate values may be obtained by numerical integration, and is shown in Fig. 77.

Formulae (23) and (24) are true also for a nearly full zone, where $N(E) = C\sqrt{(E_0-E)}$, if $kT_0 = E_0-\zeta_0$ and N refers to the number of unoccupied states (positive holes) per unit volume (cf. p. 180).

If the energy is given in terms of the wave number $\mathbf{k}$ by $E = \alpha\hbar^2 k^2/2m$ (cf. Chap. II, § 4.6), then T_0 is given by (16), and the volume susceptibility at low temperatures is, from (23),

$$\kappa = \frac{4\pi m\mu^2}{\alpha h^2}\left(\frac{3N}{\pi}\right)^{\frac{1}{3}}. \tag{25}$$

For free electrons $\alpha = 1$.

---

† Cf. Sommerfeld and Bethe, *Handb. d. Phys.* **24**/2 (1933), 476.

‡ Stoner, *Proc. Roy. Soc.* A, **152** (1935), 672, equation (6.1), gives an expression for the variation of κ with H.

‖ Mott, *Proc. Camb. Phil. Soc.* **32** (1936), 108; Stoner, loc. cit.

· Physically the temperature dependence of κ can be described as follows: At low temperatures, as we have seen, the number of

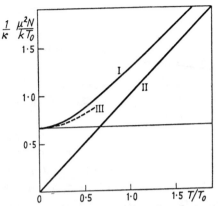

FIG. 77. Theoretical susceptibility of an electron gas.
I. Exact.   II. Classical.   III. Formula (23).

electrons which are free to turn in the field is proportional to T, and by (19) the susceptibility per electron inversely proportional to T. At temperatures comparable with T_0, however, the electron gas becomes non-degenerate and all the electrons (or positive holes) are free to turn; their number cannot therefore increase, and hence the susceptibility decreases as the temperature rises.

4.1. *Numerical values.* For the conduction electrons in such metals as copper or silver or the alkali metals we may assume qualitatively that the electrons behave as though they were free,† i.e. as if the periodic field of the lattice could be neglected; putting $\alpha = 1$ in (25), we thus have for the volume susceptibility

$$\kappa = 1 \cdot 88 \times 10^{-6} (\rho n_0 / A)^{\frac{1}{3}}, \tag{25.1}$$

or for the atomic susceptibility

$$\chi_A = 1 \cdot 88 \times 10^{-6} n_0^{\frac{1}{3}} (A/\rho)^{\frac{2}{3}}. \tag{25.2}$$

Here A is the atomic weight, ρ the density, and n_0 the number of electrons per *atom*. With $n_0 = 1$, this gives $\kappa = 10^{-6}$ for copper, with comparable or smaller values for all the other elements.

For the d electrons in the transition elements, T_0 is much smaller than the value given by (16), and the magnetic susceptibility therefore greater. This is discussed in § 5.

† Cf. Appendix I.

4.2. *Comparison with experiment.* Comparison of formulae (23) or (25) with experiment is difficult, for the following reasons: Firstly, there is often considerable divergence in the experimental values of κ, which may be ascribed to the presence of ferromagnetic impurities. Secondly, in order to obtain the 'susceptibility of the conduction electrons', one must subtract from the observed susceptibility the diamagnetic (and therefore negative) contribution of ions, which is not always easy to estimate accurately. The susceptibility of the ions is often numerically greater than that of the conduction electrons. Thirdly, in order to obtain the spin paramagnetism, one must subtract from the observed values the theoretical diamagnetic susceptibility of the conduction electrons (cf. § 6). For perfectly free electrons this should be exactly one-third of the spin paramagnetism; but this relation is not even approximately true unless the 'effective mass'† of the electrons in the lattice is nearly equal to that of a free electron.

(*a*) *Alkali metals.* Table VII gives the calculated susceptibilities, taken from the article of Sommerfeld and Bethe (p. 475; Sommerfeld and Bethe give volume susceptibilities), and the observed values, taken from Stoner (*Magnetism and Matter*, p. 509, London (1934)).

TABLE VII

Mass susceptibility $\chi \times 10^6$ of the alkalis

		Li	Na	K	Rb	Cs
Density		0·534	0·97	0·86	1·52	1·87
Spin susceptibility calc. from (25), with $\alpha = 1$		1·5	0·68	0·60	0·32	0·24
Diamagnetism of conduction electrons, calc. from (43)		−0·5	−0·23	−0·20	−0·11	−0·08
Diamagnetism of ions (obs.)‡ . .		−0·1	−0·26	−0·34	−0·33	−0·29
Total susceptibility (calc.) . . .		0·9	0·2	0·06	−0·12	−0·15
Observed susceptibility	Honda and Owen‖	0·5	0·51	0·40	0·07	−0·10
	Sucksmith†† . .	..	0·59	0·51	0·07	−0·05
	McLennan, Ruedy, and Cohen‡‡ . .	..	0·59	0·45	0·17	+0·20
	Lane‖‖ . . .	..	0·65	0·54	0·21	+0·22

† The effective mass is defined on p. 95.

‡ Deduced from the measurements of Ikenmeyer, *Ann. d. Physik*, **1** (1929), 169, on the alkali halides, except for Li where it is calculated by wave mechanics.

‖ Ibid. **37** (1912), 657.

†† *Phil. Mag.* **2** (1926), 21.   ‡‡ *Proc. Roy. Soc.* A, **116** (1927), 468.

‖‖ *Phys. Rev.* **35** (1930), 977.

We see that, except for lithium, the observed susceptibility is algebraically greater than the calculated susceptibility by an amount less than, but comparable with, the calculated spin susceptibility. We may conclude from this, either that $N(E)$ is greater and hence the breadth of the Fermi distribution less than would be the case for free electrons, or that 'correlation' forces† are important and increase the paramagnetism (cf. Chap. IV, p. 141). In view of the considerations summarized in Appendix I, we believe that $N(E)$ is *not* much greater than for free electrons, except perhaps for lithium, and that the divergence between theory and experiment is due to the correlation forces.

The dependence of the susceptibility on temperature has been investigated by Sucksmith‡ and found to be small. For Na, for example, χ increased from 0·585 at 0° to about 0·61 at the melting-point (97·5° C.), where it fell to 0·59. The small change in melting shows that the magnetism does not depend much on the crystal structure. It has been suggested by Stoner‖ that the increase with temperature is due to the thermal expansion, which is of course greater for Na than for less compressible metals.

(b) *Metals Cu, Ag, Au, Mg, Ca, Sr, Ba, Al, Pb.* Estimates of the diamagnetism of the ions have been made by Sommerfeld and Bethe, loc. cit. 475, and by Stoner, loc. cit. 511–14. For all these elements except Ba the paramagnetism of the conduction electrons obtained by subtracting this from the observed values lies between 1·5 and $2·5 \times 10^{-6}$ per cm.³, whereas the values calculated from (25) give from 0·6 to 1×10^{-6}; this again we believe to be due to correlation forces, especially since, for the di- and trivalent metals, the formula (20) of Chapter II for $N(E)$ will represent an over-estimate (cf. Fig. 40). Ba has a larger susceptibility ($6·6 \times 10^{-6}$ per cm.³). The susceptibility of some of these metals is discussed further under diamagnetism.

5. The transition metals

5.1. *General discussion.* The free atoms of these metals have an incomplete d shell either in the ground state or (in the case of palladium) in excited states of small energy. In the solid state they are distinguished by ferromagnetism or by strong paramagnetism, and by comparatively low electrical conductivity.

---

† These forces are sometimes referred to as 'exchange' forces, or 'interchange' forces. We prefer to keep this term for discussions which start from the London-Heitler approximation, e.g. § 7.5.

‡ *Phil. Mag.* **2** (1926), 21.   ‖ *Proc. Roy. Soc.* A, **152** (1935), 672.

In this chapter we shall confine our attention for the most part to the triad, nickel, palladium, and platinum, which come before copper, silver, and gold in the periodic table. The other metals have been investigated less fully from a theoretical point of view.

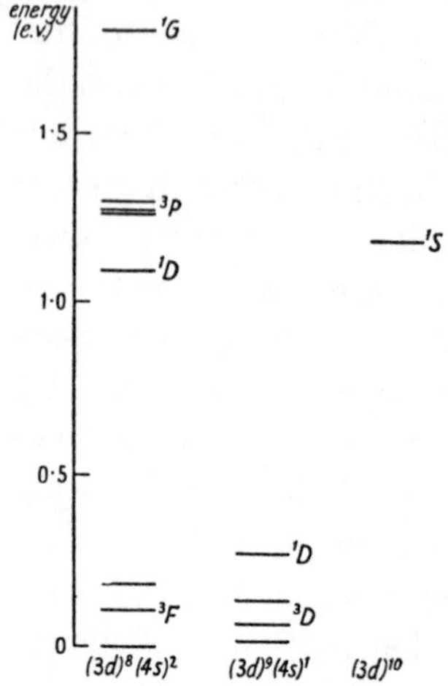

Fig. 78 shows the lowest excited states of nickel; those of platinum and palladium† are similar. It will be noticed that states with the electron configurations $3d^8 4s^2$, $3d^9 4s^1$, and $3d^{10}$ exist within a range of 1·5 e.v. This is less than the binding energy per atom of the metal (4 e.v.). It follows that, in the metal, *the wave function for each atom will be a sum of the atomic wave functions corresponding to several states with different electron configurations*. Thus if ψ_2, ψ_1, ψ_0 represent atomic wave functions formed by the superposition of states with the configurations

$$3d^8 4s^2, \qquad 3d^9 4s^1, \qquad 3d^{10},$$

respectively, the wave function in an atom of the solid will be of the form

$$A_2 \psi_2 + A_1 \psi_1 + A_0 \psi_0.$$

The mean number per atom of electrons in s states is then

$$2|A_2|^2 + |A_1|^2. \tag{26}$$

With such a wave function, at any moment of time some of the ions in the crystal will be closed d shells, others will have one positive hole and others two. The number of ions with two positive holes will be equal to $|A_2|^2$, the number with one to $|A_1|^2$, and so on.

It is characteristic of these metals that the radius of the wave function of the d state is considerably smaller than that of the s state, as shown in Fig. 57, where the wave functions of Cu are shown, which will be similar to those of Ni. Our discussion of their properties will be based on the assumption that, in consequence, nearly all the binding energy is due to electrons with s wave functions, the 'exchange' interaction between the d wave functions being much

FIG. 78. Excited states of nickel, in electron volts, with two, one, or no 'holes' in the d shell.

† For Pd the $4d^{10}$ state is the lowest.

smaller, less than 1 e.v. per atom (cf. Chap. IV, § 3.3). The strong interaction between the s electrons almost quenches their magnetic moment, the s electrons contributing to the metal only a small spin paramagnetism of the type discussed above. On the other hand, the weakness of the interaction between the incomplete d shells makes possible the strong paramagnetism of palladium and of platinum and the ferromagnetism of nickel and of similar elements.

A discussion on the lines of the London-Heitler theory[†] of the cohesive forces of a metal such as nickel, based on a wave function of the type (94), has not yet been given; a discussion of the ferromagnetism has been given by Wolf[‡] and will be referred to later. We shall give here a discussion[||] based on the Bloch theory (Chap. II, § 4), which, though unfitted without important modifications[††] for a quantitative calculation of the energy, yet gives a qualitatively correct account of a variety of phenomena.

The Bloch theory is essentially a *one-electron* theory, the interaction between the electrons being neglected except in so far as it can be represented by a static field.[‡‡] We assume, therefore, the existence in the nickel atom of independent $3d$ and $4s$ states lying close together. When, however, the atoms are brought together into a lattice, each quantized state broadens out into a band of allowed energies (Fig. 79); but, because the $3d$ orbits (wave functions) are small (see above), the broadening of the d band will be considerably less than that of the s band. The density of states[||||] $N(E)$ in the two bands will therefore appear as in Fig. 80; the number of states, $\int N(E)\,dE$, is five per atom in the d band and one in the s band. We see that the density of states is much greater in the d band than in the s band.

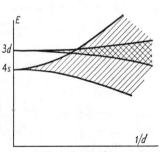

FIG. 79. Allowed energies as a function of interatomic distance (d).

Since in nickel or in palladium there are ten electrons to be shared between the two zones, it is clear that, if in a crystal of N atoms the

† Cf. Chap. IV, § 1.2.   ‡ *Zeits. f. Phys.* **70** (1931), 519.
|| Mott, *Proc. Phys. Soc.* **47** (1935), 571.
†† i.e. the correlation between the positions of the electrons (cf. Chap. IV, § 2.2).
‡‡ Cf. Chap. II, § 2.
|||| $N(E)$ is defined in Chap. II, § 4.6.

number of s electrons is pN, there is also an equal number pN of *positive holes* in the d zone. For copper, on the other hand, with eleven available electrons, the experimental evidence shows that the d band is full (cf. Appendix I), and there is thus *one s* electron per atom. The occupied states in the two metals are shaded in Fig. 80.

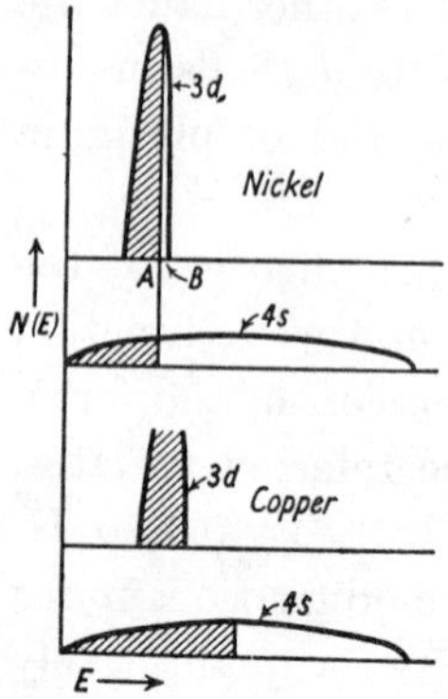

. There is a good deal of evidence that for nickel, and also for the similar metal palladium,† the number p of positive holes per atom is 0·55–0·6. This evidence will be reviewed below. Also for cobalt there are about 0·7 s electrons and 1·7 positive holes in the d band.

We shall see in § 7.3 that for nickel, at any rate, $|A_2|$ in equation (26) is considerably larger than $|A_1|$; in most ions of metallic nickel there will therefore be two positive holes or none.

As Fig. 78 shows, the energy of these positive holes depends markedly on whether their spins are parallel or antiparallel; Fig. 79 is therefore an over-simplification; in reality, any d ion may be excited into a state where the 'holes' have antiparallel spins, the work required for this being of the order of a volt.

FIG. 80. Density of states $N(E)$ in nickel and copper.

X-ray absorption edge

Perhaps the most direct experimental evidence of the truth of the hypothesis that in transition elements there are empty d states with very large $N(E)$ is provided by the work of Veldkamp on the X-ray absorption edge of the metals tantalum and tungsten. This is discussed in Chap. III, § 9.2.

5.2. *Specific heat.* We have seen in § 2 that at ordinary temperatures the conduction electrons in a metal such as copper form a nearly degenerate gas, and make only a small contribution to the heat capacity, of the order per gm. atom

$$\Delta C_v \sim 0\cdot0001T \text{ cal. degree.}$$

The assumption, however, that for the d electrons the density of states is much greater than for the s electrons, leads to the conclusion that, if the d band is not full, it will make a much larger contribution to the specific heat.

For Ni, Pd, and Pt, since the number (0·55–0·6) per atom of d states which are unoccupied is small compared with the total

† For platinum we have no definite evidence.

number (ten), it will probably not involve serious error if we assume for the d band near the surface of the Fermi distribution

$$N(E) = C\sqrt{(E_0 - E)}, \tag{27}$$

where E_0 is the maximum energy of the d band. With this assumption we may apply the formulae of § 2 to the *positive holes* in the d band. We denote by kT_0 the energy interval between E_0 (the head of the band) and the highest occupied state at the absolute zero of temperature. We may then apply formula (14); we obtain for the extra term in the atomic heat due to the electrons†

$$\Delta C_v = \tfrac{1}{2}\pi^2 n_0\, RT/T_0 \text{ cal. per gm. atom}$$

for temperatures T small compared with T_0.

Keesom and Clark‡ have measured the atomic heat of nickel between 1·1 and 19·0° K. and, subtracting the estimated contribution due to the lattice, find an additional term equal to $0·001744T$ cal. per gm. atom. This gives for T_0, assuming n_0 to be 0·6 (cf. p. 197),

$$T_0 = 3{,}470° \quad \text{(nickel)}.$$

This result is discussed further in relation to the ferromagnetism in §§ 7.4, 7.6.

Simon and Pickard‖ have measured the specific heat of palladium in the same range, and find an additional term $0·0031T$ cal. per gm. atom, giving a value of T_0, with $n_0 = 0·55$ (cf. p. 199),

$$T_0 = 1{,}750 \quad \text{(palladium)}.$$

Two other methods of estimating T_0 for this element are discussed in Chap. VII, §§ 7 and 14, which give values of T_0 of the same order of magnitude.

The low values of T_0 will have the further result that the electron gas (or rather the positive hole gas) becomes non-degenerate at comparatively low temperatures, so that each 'positive hole' contributes an amount $\tfrac{3}{2}k$ to the specific heat. As Fig. 75 shows, at a temperature $\tfrac{1}{2}T_0$ the specific heat per particle has risen to $1·25k$, so that we should expect at this temperature that the additional specific heat would be $\Delta C_v = 1·25 \times 0·6 \times R = 1·5$ cal. per gm. atom.

---

† We shall see in § 7 that a further term, proportional to $T^{\frac{3}{2}}$, must be added to the heat capacity of ferromagnetic materials at low temperatures, representing the heat required to demagnetize the body. This has not been observed at present. The existence of the two additional terms was first pointed out by Epstein, *Phys. Rev.* **41** (1932), 91.

‡ *Physica*, **2** (1935), 513; see also Clusius and Goldmann, *Zeits. f. phys. Chem.* B, **31** (1936), 237.   ‖ Pickard, *Nature*, **138** (1936), 123.

We give below the measured atomic heats C_v of palladium and platinum compared with those of copper, silver, and gold:

Observed atomic heats at constant volume (C_v), in cal. degree per gm. atom

$T°\ C.$	Pd†	Pt†	Cu‡	Ag‡	Au‡
500	6·594	6·38	6·2	6·0	6·0
900	7·072	..	..	6·13	..
1,000	7·146	6·65	6·5	..	..
1,300	7·251	..	..	..	6·12
1,500	7·232	..	..	..	..
1,600	..	6·8	..	..	..

The values, especially for palladium, rise considerably above the corresponding values for the noble metals;‖ they are not, however, as great as the low value of T_0 for palladium suggests, since at 1,000° C. the 'positive hole gas' should be practically non-degenerate, giving an excess specific heat $0·6 \times \frac{3}{2}R \simeq 1·8$ cal. per gm. atom. It is not certain, however, what effect the coupling between the positive holes discussed above will have on the specific heat.

Nickel has a very considerable excess specific heat *above* the Curie temperature of about 1·3 cal. per gm. atom. This is discussed further in § 7.4.

5.3. *Paramagnetism.* Palladium, platinum, and some of the other transition metals are strongly paramagnetic. There is no direct proof that this paramagnetism is entirely due to spin; but we know that spin is responsible for the ferromagnetism of nickel both below and above the Curie point (cf. § 7.1), and these paramagnetic metals are probably similar in this respect.

No calculations of the magnetic susceptibility of such substances based on the London-Heitler model have yet been given. If we use the Bloch model, and make the assumption (27) above, then the

---

† Jaeger and Rosenblum, *Proc. Amsterdam Acad.* **33** (1930), 457. Holzmann (*Festschrift J. Siebert,* Hanau (1931)) obtains rather smaller values for Pd, viz. 6·6900 at 800° C., 6·7544 at 900° C.

‡ The mean of several determinations, cf. Eucken, *Handb. d. exp. Phys.* **8** (1929), 1, 221.

‖ For tungsten also the atomic heat C_v rises above the classical value, and according to Magnus and Holzmann (*Ann. d. Physik*, **3** (1929), 585) reaches the value 6·44 at 900° C. Both for tungsten and for platinum it has been suggested that the effect is due to the anharmonic terms in the potential energy of the lattice waves, cf. Chap. I, § 2, but it seems to us probable that it is mainly due to the electrons, as explained here.

susceptibility is given in terms of a single parameter T_0 (cf. (23)) if the number n_0 of elementary magnets per atom is known.

In Fig. 81 we show the observed values of $1/\chi_A$ for palladium and platinum, and also for nickel. The theoretical curves are the same as those shown in Fig. 77, with n_0, here the number of positive holes in the d shell, equal to 0·6 in both cases, and with the following values for T_0:

	Pd	Pt
T_0	500° K.	1,500° K.

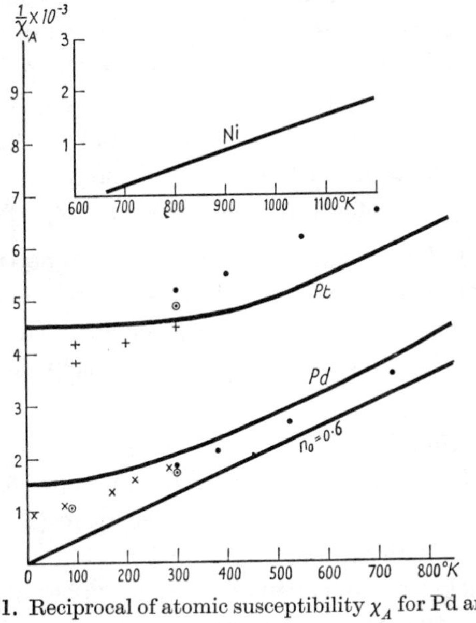

Fig. 81. Reciprocal of atomic susceptibility χ_A for Pd and Pt.

Observer.

• Guthrie and Bourland.[†]	+ Foëx (different specimens).[‡]
⊙ Vogt.[‖]	× Onnes and Oosterhuis.[††]

Full lines are theoretical curves obtained as explained in text.
Inset: reciprocal of atomic susceptibility of nickel ([‡‡]).

The agreement is good considering that the theoretical formula contains only a single unknown parameter T_0, n_0 being determined from other data. On the other hand, we believe that this value of T_0 is too small and that the agreement is mainly accidental, for the following reason.

The parameter T_0 determines also the specific heat; as we may

† *Phys. Rev.* **37** (1931), 308.   ‡ *J. d. Phys. et le Rad.* **4** (1933), 517.
‖ *Ann. d. Physik*, **14** (1931), 1.   †† *Comm. Leiden*, 57 (1914).
‡‡ Honda and Takagi, *Science Reports Tôhoku*, **1** (1911), 229.

see from Fig. 75, at a temperature equal to $\frac{1}{2}T_0$ the electronic heat capacity per particle has already risen to the value $1\cdot25k$; thus at room temperature we should expect an additional contribution to the heat capacity of

$$0\cdot6\times1\cdot2R \simeq 1\cdot4 \text{ cal. per gm. atom,}$$

which is not observed (cf. p. 194).

We believe, then, that the susceptibility of the transition metals is due to electrons in the comparatively narrow d bands, but that, especially for Pd and Pt, the large susceptibility can only be accounted for by the terms in the energy which are neglected in the approximation used here. These terms are discussed in Chap. IV, § 2.2 under the heading of 'correlation forces';[†] it was shown there that, owing to the tendency of electrons with parallel spins to keep away from each other and thus to lower the energy, these terms always *increase* the susceptibility of a metal, if they do not make it ferromagnetic.

5.4. *Alloys of the transition metals.* The magnetic properties of alloys of the transition metals have, of course, been extensively investigated. We can discuss here only certain results which are of especial interest in view of the theoretical considerations of this chapter.[‡]

(a) *Alloys of nickel.* We take for an example the copper-nickel series. These alloys form a face-centred cubic lattice with no super-structure for all compositions. The copper atom ($Z = 29$) contains one electron more than nickel ($Z = 28$). Therefore, to a first approximation, if a nickel atom is replaced by a copper atom in an alloy we may suppose that the lattice is unaltered except that an extra *electron* is added.

Now the addition of extra electrons to the nickel lattice will raise the energy of the surface of the Fermi distribution. In Fig. 80 the point marked A will be moved to the right. As Fig. 80 shows, the density of states in the d band is very much greater than in the s band. Therefore, nearly all of the extra electrons will go into the d band so long as there remain any vacant places in it. We know (p. 222) that the number of vacant places in the d band of nickel is

---

† We prefer not to refer to them as 'exchange forces', since they are quite different from the 'exchange integral' which occurs when we start from the London-Heitler approximation (§ 7.5).

‡ Most of the considerations of this section are taken from a paper by Mott, *Proc. Phys. Soc.* **47** (1935), 571.

equal to the saturation moment in Bohr magnetons, namely 0·6 per atom. Hence, in an alloy of which the atomic composition is 60 per cent. copper, the d band will be just full. In the pure nickel all states below A in Fig. 80 are occupied; in the alloy with 60 per cent. copper all states below B, and in pure copper all states shaded in Fig. 80. Thus in an alloy with x parts of Cu to $1-x$ of Ni the number of holes in the d shell is

$$0\cdot6-x \quad (x < 0\cdot6),$$
$$0 \qquad (x > 0\cdot6).$$

We should expect, therefore, the saturation moment of the copper-

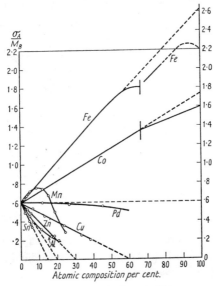

FIG. 82. Atomic moments in nickel alloys in Bohr magnetons per atom. The number of electrons outside an inert gas shell in the atoms shown is:

Mn	Fe	Co	Ni	Pd	Cu	Zn	Al	Sn
7	8	9	10	10	11	12	13	14

nickel alloys to be $0\cdot6-x$ Bohr magnetons per atom. That this is the case has been shown experimentally by Alder,† whose results are shown in Fig. 82. One would expect the alloys with more than 60 per cent. copper to be diamagnetic or weakly paramagnetic; actually they are fairly strongly paramagnetic, probably owing to small traces of undissolved nickel.

Similarly, if an atom of Fe, Co, Pd, Zn, or Al be substituted for

† *Dissertation,* Zürich (1916).

a nickel atom in the alloy, we may suppose the number of electrons
to be increased by −2, −1, 0, 2, 3, respectively, the extra electron
being always added to, or subtracted from, the d band, for the same
reason as before. None of these metals dissolve in nickel to an un-
limited extent without change of structure, but for small concentrations the saturation moment is found,† as for Cu–Ni, to agree with
the assumption that it is equal to the number of holes in the d band,
as shown in Fig. 82.

(b) *Specific heat of Cu–Ni alloys.* Grew‡ has found that the excess
specific heat ΔC_v above the Curie point (cf. § 7.4) tends to zero
approximately at the same composition as the magnetization; this is

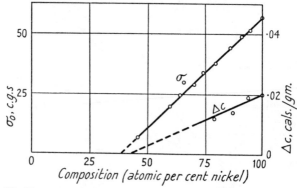

FIG. 83. Excess specific heat of copper-nickel alloys according to Grew.

shown in Fig. 83, where ΔC_v is plotted against atomic composition.
Since, as we shall see below, the excess specific heat is due to the
presence of positive holes in the d band, this gives additional evidence
for the hypothesis that, for alloys with more than 60 per cent.
copper, the d band is full.

The optical properties of Cu–Ni alloys are discussed on p. 125,
and the electrical conductivity on p. 271.

(c) *Alloys of platinum and palladium.* The alloys of these two
metals with copper, silver, and gold have been extensively investigated.‖ The alloys of palladium with silver and gold form a continuous range of solid solutions and no superstructure has been
observed; the alloy with copper can be obtained both with and

† Sadron, *Thèse,* Strasbourg (1932).   ‡ *Proc. Roy. Soc.* A, **145** (1933), 521.
 ‖ The susceptibility of Au–Pd has been measured by Vogt (*Ann. d. Physik,* **14**
(1932), 1), Ag–Pd and Cu–Pd by Svensson (ibid. **14** (1932), 699), Au–Pt by Johansson
and Linde (ibid. **5** (1930), 762).

without a superstructure (cf. Chap. I, § 7). In the platinum alloys there are 'Mischungslücken'; for this reason we shall discuss the palladium alloys.

Fig. 84 shows the susceptibility of the palladium-gold series of alloys plotted against the atomic composition. It will be seen that

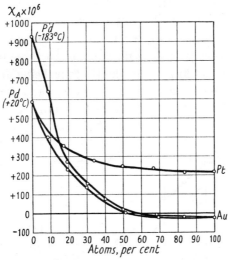

FIG. 84. Atomic susceptibility of gold-palladium and platinum-palladium alloys according to Vogt.

the susceptibility drops to the small negative value for gold at about the composition 55 per cent. gold. Pd–Pt is also shown. The curves for Pd–Cu and Pd–Ag are similar. For the Pt–Au alloys[†] the transition from the paramagnetic to the diamagnetic alloys is less sharp.

The theoretical explanation is similar to that given for the copper-nickel alloys. Each atom of gold contributes an extra electron which goes into the d band, as long as there is any room there; when the d band is full the alloy is diamagnetic. We deduce that there are in palladium 0·55 positive holes in the d band per atom, and thus about the same as for nickel (0·6).

According to equation (21) the paramagnetic susceptibility should be proportional to $N(E_{max})$, where E_{max} is written for ζ_0. The curve of Fig. 84 may thus be taken to show the variation of $N(E_{max})$ with atomic composition x. The experimental curve has roughly the form

$$N(E_{max}) \sim (0·55 - x)^2. \tag{28}$$

† Vogt, loc. cit. 23.

The thermal expansion coefficients of Ag–Pd and Cu–Pd alloys, when plotted against atomic composition,† show a kink at about 50 per cent. Pd, indicating that there is some change in the rate of change of the binding forces at this composition. This is just what we should expect from our theory, since the number of s electrons is constant up to about 55 per cent. Cu or Ag, and the s electrons are mainly responsible for the binding forces.

(d) *Palladium and hydrogen.* ‚ Palladium absorbs hydrogen very strongly; the hydrogen atoms do not replace those of palladium in the lattice, the exact location of the H atoms being unknown.‡

The paramagnetic susceptibility decreases and finally disappears when the metal absorbs hydrogen, as shown in Fig. 85, which is due

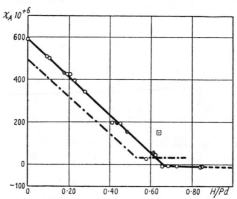

Fig. 85. Atomic susceptibility of the system Pd–H, plotted against the ratio of the number of H atoms to the number of Pd atoms.

⊕ Biggs.‖   —·—• Aharoni and Simon.††   ⊡ Oxley.‡‡   ○ Svensson.‖‖

to Svensson (ref. below). We may deduce that the hydrogen electron goes into the d shell, thereby decreasing the number of positive holes and hence the paramagnetism.

We know that the palladium atom has 0·55 holes per atom in the d shell, and since the susceptibility drops to zero when there are about 0·55 hydrogen atoms to each palladium atom, we may deduce that the dissolved *hydrogen atom is completely ionized*, unlike the silver atoms discussed in the last section, which keep part of their s electrons.

† Johansson, *Ann. d. Physik*, **76** (1925), 445.
‡ Cf. Hanawalt, *Phys. Rev.* **33** (1929), 444, for the effect on the lattice constant, etc.
‖ *Phil. Mag.* **32** (1916), 131.   †† *Zeits. f. phys. Chem.* B, **4** (1929), 175.
‡‡ *Proc. Roy. Soc.* A, **101** (1922), 264.   ‖‖ *Ann. d. Physik*, **18** (1933), 294.

6. Diamagnetism of the conduction electrons

According to the laws of classical mechanics the diamagnetic susceptibility of a free-electron gas is zero. This appears to have been shown first by Bohr.[†] The proof depends on the fact that, whether a magnetic field is present or not, the energy of each electron is proportional simply to the square of its velocity, and the velocity does not change in a magnetic field; the expression for the energy (more generally for the free energy) does not therefore contain the field, and the magnetic moment vanishes. A theorem due to Van Leeuwen[‡] shows that the moment vanishes, according to classical mechanics, even if the electrons move in a field which varies from point to point.

These results are not valid if we apply quantum mechanics to the motion of the electrons, as was first shown by Landau.[||] According to quantum mechanics, the state of motion of a free electron can be specified by the three components of its velocity only if no magnetic field is present. In the presence of a field an electron moves in circular orbits round the direction of the field, and the energy of motion is quantized, according to familiar laws. Any attempt to determine the three components of the velocity of an electron at a given instant would disturb it, and might cause it to jump from one quantized state to another. It follows that we cannot know the energy and velocity of an electron at the same time. We cannot therefore argue, as in the classical mechanics, that the energy is a quadratic function of the velocity, and must investigate the magnetic moment in another way.

Throughout this section we shall obtain the magnetic moment σ from the free energy F by means of the formula

$$\sigma = -\left(\frac{\partial F}{\partial H}\right)_{v,T} \tag{29}$$

Our aim is therefore to calculate the free energy.

6.1. *Diamagnetism of free electrons.* We consider first a gas of free electrons, the lattice field being neglected (approximation of Chap. II, § 3).

Let the magnetic field H be parallel to the z-axis. An electron can

---

[†] Bohr, *Dissertation*, Copenhagen (1911). Cf. the discussion by Van Vleck, loc. cit. 100.

[‡] *Dissertation*, Leiden (1919), Cf. also Van Vleck, loc. cit. 94.

[||] *Zeits. f. Phys.* **64** (1930), 629. Cf. also Darwin, *Proc. Camb. Phil. Soc.* **27** (1932), 86, Teller, *Zeits. f. Phys.* **67** (1930), 311, and Stoner, *Proc. Roy. Soc.* A, **152** (1935), 672.

describe a circular orbit of any radius perpendicular to this field with frequency

$$eH/2\pi mc.$$

This circular motion may be resolved into two simple harmonic oscillations at right angles. According to the quantum theory, the only values which the energy of a particle vibrating with this frequency can have are given by

$$E_l = \frac{eHh}{2\pi mc}(l+\tfrac{1}{2}),$$

where l is a positive integer. The state of motion of an electron is therefore completely determined when we know l, the components of the velocity parallel to H, and the position of the centre of the orbit. Clearly the energy is independent of the position of the orbit, so that, if p_z is the component of the momentum parallel to H, we may write for the total energy E,

$$E = 2\mu H(l+\tfrac{1}{2})+p_z^2/2m, \tag{30}$$

where μ is the Bohr magneton defined by

$$\mu = e\hbar/2mc.$$

The quantum state l is degenerate, even if p_z is known. We require an expression for the number of quantum states with given quantum number l and for which p_z lies in the range p_z' to $p_z'+dp_z'$. If we assign two states to each cell h^3 of phase-space, the number required will be equal to the number of states for which, in the absence of a field

$$2\mu H(l-\tfrac{1}{2}) \leqslant \frac{1}{2m}(p_y^2+p_x^2) \leqslant 2\mu H(l+\tfrac{1}{2}),$$

$$p_z' \leqslant p_z \leqslant p_z'+dp_z'.$$

If V is the volume of space occupied, and if we write $p_x^2+p_y^2 = p^2$, the corresponding volume of phase-space is clearly

$$2\pi V p\,dp dp_z = 4\pi V\mu mH\,dp_z.$$

Allowing two states for each cell h^3 of phase-space, this gives for the required number of states, substituting for μ,

$$2VeH\,dp_z/ch^2. \tag{31}$$

(a) *A non-degenerate electron gas.* We consider first a non-degenerate electron gas, as in this case the calculation is particularly simple. We have to calculate the magnetic moment σ at a temperature T of an electron gas containing N electrons per unit volume, neglecting, of course, the moment due to the electron spin. σ is obtained from the free energy F by equation (29). Since the gas is non-degenerate, we use classical statistics; the number of particles n_i which have energy E_i at temperature T is thus proportional to $e^{-E_i/kT}$, and the entropy of the state with total energy $U = \sum_i n_i E_i$ is given by

$$S = k \log \frac{N!}{n_1!\, n_2! \ldots n_i! \ldots}, \tag{32}$$

where k is Boltzmann's constant.† Making use of Stirling's formula in the approximate form

$$\log N! = N \log(N/e), \tag{33}$$

and of the equation $N = \sum_i n_i$, we obtain easily the well-known formula

$$F = U - TS = -kTN \log Z,$$

where

$$Z = \sum_i e^{-E_i/kT} \tag{34}$$

Thus, from (29), we have for the magnetic moment

$$\sigma = kTN \frac{\partial}{\partial H} \log Z. \tag{35}$$

According to (30), (31), and (34) we obtain

$$Z = \sum_{l=0}^{\infty} \int_{-\infty}^{\infty} \frac{2eVH}{ch^2} e^{-[\mu H(2l+1)+p_z^2/2m]/kT}\, dp_z.$$

$$= \frac{eVH}{ch^2} \frac{\sqrt{(2\pi mkT)}}{\sinh(\mu H/kT)}.$$

Hence from (35)

$$\sigma = -N\mu \left[\coth\!\left(\frac{\mu H}{kT}\right) - \frac{kT}{\mu H} \right].$$

Therefore, when $H\mu \ll kT$, the susceptibility κ per unit volume may be written

$$\kappa = -\frac{1}{3} \frac{N\mu^2}{kT}.$$

Since $\mu = eh/4\pi mc$, this susceptibility vanishes when we make the transition to the classical theory by putting h equal to zero.

The spin paramagnetic moment of a non-degenerate electron gas under the same conditions, viz. $\mu H \ll kT$, is given by (cf. § 4)

$$\kappa = N\mu^2/kT.$$

For small fields, therefore, the diamagnetic susceptibility is numerically just one-third of the spin paramagnetic susceptibility.

---

† The Boltzmann distribution law is obtained by making the expression (32) for S a maximum for a given total energy and fixed N.

(b) *A degenerate electron gas; Landau's formula.* Since electrons obey the Fermi-Dirac statistics, the entropy of the electron gas, according to the quantum theory, is not given by (32) but by the following formula:

$$S = k \log \prod_i \frac{z_i!}{n_i!(z_i-n_i)!},$$ (36)

where z_i denotes the number of states with energy E_i, and n_i the number of electrons in these states. When S is a maximum for a given total energy $U = \sum n_i E_i$ and for a given number of electrons N, we have the result

$$\frac{n_i}{z_i} = \frac{1}{e^{(E-\zeta)/kT}+1},$$ (37)

which is the well-known Fermi-Dirac distribution law (cf. § 1). From (36), (37), and (33) we obtain for the free energy

$$F = N\zeta - kT \sum \log(1+e^{(\zeta-E_i)/kT}).$$ (38)

The parameter ζ may be determined in terms of N, the number of electrons per unit volume, by the equation†

$$\partial F/\partial \zeta = 0.$$ (39)

The free energy of a degenerate electron gas in the presence of an external magnetic field is therefore, according to (30), (31), and (38),

$$F = N\zeta - \frac{8\pi m}{h^3}\mu HkT \sum_{l=0}^{\infty} \int_{-\infty}^{\infty} \log\left[1+e^{[\zeta-2\mu H(l+\frac{1}{2})-p_z^2/2m]/kT}\right] dp_z.$$ (40)

The summation can be evaluated by means of Euler's formula‡ which, to the degree of approximation required here, states that

$$\sum_{l=a}^{b-1} f(l+\tfrac{1}{2}) = \int_a^b f(x)\,dx - \tfrac{1}{24}|f'(x)|_a^b.$$ (41)

This formula is applicable only when the function $f(x)$ is approximately linear in the range (ab) between any two values of x which differ by unity; in other words, when

$$f(x+\tfrac{1}{2})-f(x-\tfrac{1}{2})-f'(x) \ll f(x).$$

The integrand in (40) varies appreciably in a range of l equal to $kT/\mu H$ at some points in the interval $0 < l < \infty$. Euler's formula in the form (41) may therefore only be applied to evaluate F when

$$\mu H \ll kT.$$

† Cf. § 1.   ‡ Bromwich, *Theory of Infinite Series*, p. 238.

Under these conditions we obtain, according to (40) and (41),

$$F = N\zeta - \frac{8\pi m}{h^3}\mu H kT \int\limits_{-\infty}^{\infty} dp_z \int\limits_{0}^{\infty} \log[1 + e^{(\zeta - 2\mu Hx - p_z^2/2m)/kT}]\, dx +$$

$$+ \frac{2\pi m}{3h^3}(\mu H)^2 \int\limits_{-\infty}^{\infty} \frac{dp_z}{1 + e^{(p_z^2/2m - \zeta)/kT}}.$$

By a suitable change of variables the integrals may be put into the following more convenient forms

$$F = N\zeta - \frac{4\pi(2m)^{\frac{3}{2}}}{h^3}kT \int\limits_{0}^{\infty} \log(e^{(\zeta - x)/kT} + 1)x^{\frac{1}{2}}\, dx +$$

$$+ \frac{\pi(2m)^{\frac{3}{2}}}{3h^3}(\mu H)^2 \int\limits_{0}^{\infty} \frac{x^{-\frac{1}{2}}\, dx}{1 + e^{(x - \zeta)/kT}}.$$

The integrations may be carried out[†] and the results expressed as series of ascending powers of kT. We only require the first terms, which are readily obtained, giving

$$F = N\zeta - \frac{16\pi(2m)^{\frac{3}{2}}\zeta^{\frac{5}{2}}}{15h^3} + \frac{2\pi(2m)^{\frac{3}{2}}\zeta^{\frac{1}{2}}}{3h^3}(\mu H)^2. \tag{42}$$

By (29) we have for the magnetic moment

$$\sigma = -\left(\frac{\partial F}{\partial H}\right)_\zeta - \left(\frac{\partial F}{\partial \zeta}\right)_H \frac{d\zeta}{dH}.$$

The second term vanishes, by (39); substituting from (42) we have therefore

$$\sigma = -\frac{4\pi}{3h^3}(2m)^{\frac{3}{2}}\zeta^{\frac{1}{2}}\mu^2 H.$$

Since we only need σ to the first power of H, we may replace ζ by its value when $H = 0$, which we shall write ζ_0. According to (39) and (42),

$$\zeta_0 = \frac{h^2}{2m}\left(\frac{3N}{8\pi}\right)^{\frac{2}{3}},$$

a formula already obtained in Chapter II.[‡] Hence

$$\kappa = \frac{\partial \sigma}{\partial H} = -\frac{4\pi m}{3h^2}\left(\frac{3N}{\pi}\right)^{\frac{1}{3}}\mu^2. \tag{43}$$

This formula was first obtained by Landau (loc. cit.). Stoner[||] has given a formula for the rate of variation of κ with T and H.

[†] Cf. Fowler, *Statistical Mechanics*, p. 544.
[‡] Equation (19); ζ_0 and E_{max} are identical.
[||] *Proc. Roy. Soc.* A, **152** (1935), 672.

We see, as for a non-degenerate gas, that the value of κ, for small fields, is numerically equal to one-third of the spin paramagnetic susceptibility (cf. § 4). Note, however, that this result is only valid for 'free' electrons, i.e. when the effective mass is approximately equal to the mass m of a free electron. It must not be applied, for instance, to the positive holes in the d shells of transition metals (cf. equation (56) below).

(c) *Application of Landau's formula to real metals.* It would appear from the preceding section that the diamagnetism of free electrons, degenerate or otherwise, depends essentially on the existence of discrete energy values, for if l is regarded as a continuous variable and the summations are replaced by integrations, the free energy becomes independent of H. However this conclusion is not really correct. Indeed, if the diamagnetism depended upon the existence of a set of sharply defined energy levels such as (30), the theory could have little practical significance, because such a set of levels could not exist in a real metal, where the electrons are continually being disturbed by the heat motion of the lattice. The average time between two collisions of an electron with the vibrating lattice is of the order of magnitude 10^{-14} sec., a result which is easily obtained from the observed electrical conductivity.† On the other hand, the period of an electron moving in a field of one kilogauss is approximately 3×10^{-10} sec. Thus, even in a moderately high magnetic field, an electron can perform only a very small part of its orbit in the time between successive collisions caused by the heat motion.

If τ is the average time between two collisions, the coupling between the atoms of the lattice and the electron is such that one can only define the individual energies of the electron and the vibrating atoms with an uncertainty h/τ; only the total energy of the electron *and* the atoms is exactly determinable. Hence, in place of an energy spectrum consisting of a series of sharp lines $E_l = 2\mu H(l+\frac{1}{2})$, there will in reality be a series of lines of approximate width h/τ. The form of these lines we may assume to be given by a function $\phi(x)$ which has a **maximum** at $x = 0$ and for which

$$\int_{-\infty}^{\infty} \phi(x)\,dx = 1. \tag{44}$$

$\phi(x)$ should fall rapidly to zero for $x > h/\tau$. In the former calcula-

† Cf. Chap. VII, § 6.3.

tions we have tacitly assumed that $\tau = \infty$, thereby giving $\phi(x)$ the properties of a δ function.

We shall now show that this does not affect the diamagnetism provided that $kT > h/\tau$. If we write for the energy

$$E = E_1 + p_z^2/2m, \tag{45}$$

we may put for the number of states which lie in the interval E_1 to $E_1 + dE_1$, p_z to $p_z + dp_z$, instead of (31),

$$N(E_1)\,dE_1\,dp_z = \frac{8\pi m}{h^3}\mu H \sum_{l=0}^{\infty} \phi\{E_1 - 2\mu H(l+\tfrac{1}{2})\}\,dE_1\,dp_z. \tag{46}$$

We have already pointed out that for all values of H such as are ordinarily employed in susceptibility measurements, e.g. up to 30 kilogauss, $h/\tau \gg \mu H$. Hence it is legitimate to apply Euler's summation formula (41) directly to (46); we obtain

$$N(E_1) = \frac{4\pi m}{h^3}\left\{\int_{-\infty}^{E_1} \phi(x)\,dx + \frac{(\mu H)^2}{6}\left|\frac{\partial \phi}{\partial x}\right|_{E_1}^{\infty}\right\} \tag{47}$$

Substituting from (45) for E, (38) may be written

$$F = n\zeta - kT \int_{-\infty}^{\infty} dp_z \int_{-\infty}^{\infty} \log[1 + e^{(\zeta - E_1 - p_z^2/2m)/kT}]N(E_1)\,dE_1, \tag{48}$$

and, substituting for $N(E_1)$ from (47), we obtain by partial integration with respect to E_1 (making use of (44)) an expression for F, which is essentially identical with (42), and leads therefore at once to Landau's formula for the diamagnetic susceptibility. There is one necessary assumption involved in passing from (48) to (42). The assumption is that $\log[1 + \exp\{((\zeta - E_1 - p_z^2/2m)/kT\}]$ does not vary appreciably with E_1 in a range h/τ. In other words, it is necessary that

$$kT \gg h/\tau. \tag{49}$$

The condition (49) for the applicability of Landau's formula to metals was first deduced by Peierls.[†]

6.2. *Effect of the lattice field on the diamagnetism.* It is not possible in the case of electrons in a periodic lattice to proceed directly by finding the stationary states in an external magnetic field, as was done for free electrons. The appearance of the lattice potential in the wave equation makes such a procedure wholly impracticable.

Peierls[‡] has obtained a formula for the diamagnetic susceptibility

---

[†] *Zeits. f. Phys.* **80** (1933), 763.　　　　[‡] Loc. cit. 786.

by a method which does not involve a detailed knowledge of the stationary states. Using wave functions obtained for the limiting case of tight binding,† he has deduced the following formula for the volume susceptibility:

$$\kappa = -\frac{2}{3}\left(\frac{e}{4\pi ch}\right)^2 \int\int \frac{1}{|\text{grad } E|}\left\{\frac{\partial^2 E}{\partial k_x^2}\frac{\partial^2 E}{\partial k_y^2} - \left(\frac{\partial^2 E}{\partial k_x\,\partial k_y}\right)^2\right\} dS, \quad (50)$$

where the integration is over the surface of the Fermi distribution. This formula reduces to Landau's expression (43) when $E = \hbar^2 k^2/2m$ (free electrons).

We shall not give a general proof of Peierls's formula; we shall, however, obtain *ab initio* a formula for the susceptibility in a certain special case which is of considerable interest. This is the case (cf. p. 83) in which the surfaces of constant energy in k-space form a family of similar ellipsoids. We shall suppose that the magnetic field is in the direction of one of the principal axes and choose the coordinates so that this direction is that of the k_z-axis. We may write, therefore, in the absence of a field

$$E = \frac{\hbar^2}{2m}\{\alpha_1 k_x^2 + \alpha_2 k_y^2 + \alpha_3 k_z^2\}. \quad (51)$$

In Chap. III, § 2 we have shown that the velocity of an electron is given by the equation

$$\mathbf{v} = \frac{1}{\hbar}\text{grad } E \quad (52)$$

and that an external force $\mathbf{F}$ produces an acceleration according to the formula
$$\dot{\mathbf{k}} = \mathbf{F}/\hbar. \quad (53)$$

The x and y components of the force on an electron moving in the magnetic field H are, by (52), $e\hbar H\alpha_2 k_y/mc$ and $-e\hbar H\alpha_1 k_x/mc$, respectively. Hence the equations of motion are, by (53),

$$\dot{k}_x = \frac{eH}{mc}\alpha_2 k_y, \qquad \dot{k}_y = -\frac{eH}{mc}\alpha_1 k_x;$$

differentiating with respect to the time we obtain

$$\ddot{k}_x + \left(\frac{2\mu' H}{\hbar}\right)^2 k_x = 0, \quad (54)$$

where $\mu' = \sqrt{(\alpha_1\alpha_2)}\mu, \qquad \mu = e\hbar/2mc,$

with a similar equation for $\dot{k}_y$. The equations (54) are formally the same as the classical equations of motion in a magnetic field, and

† Cf. Chap. II, § 4.4.

differ from those for a free electron only in having μ' in place of μ for the magnetic moment.

The formulae for the diamagnetism of free electrons (§ 6.1) may therefore be applied, and we may write for the energy levels

$$E_l = 2\mu'H(l+\tfrac{1}{2}). \tag{55}$$

The number of states which have a given value of l and lie in a range k_z to k_z+dk_z may be shown, in the same way as for free electrons, to be

$$\frac{8\pi m}{h^2}\, \frac{\mu'H}{\sqrt{(\alpha_1\alpha_2)}}\, dk_z.$$

The expression for the free energy of the electrons is therefore identical with (40) if μ is replaced by μ' and the integral is multiplied by $(\alpha_1\alpha_2\alpha_3)^{-\frac{1}{2}}$. The magnetic moment, therefore, is given by

$$\sigma = -\frac{4\pi(2m)^{\frac{3}{2}}}{3h^3}\, \frac{\zeta^{\frac{3}{2}}\mu'^2 H}{(\alpha_1\alpha_2\alpha_3)^{\frac{1}{2}}},$$

and, since in this case according to (39) we have

$$\zeta_0 = \frac{h^2}{2m}\left(\frac{3N}{8\pi}\right)^{\frac{2}{3}}(\alpha_1\alpha_2\alpha_3)^{\frac{1}{3}},$$

the diamagnetic susceptibility per unit volume is

$$\kappa = -\frac{4\pi m\mu^2}{3h^2}\left(\frac{3N}{\pi}\right)^{\frac{1}{3}}\left\{\frac{(\alpha_1\alpha_2)^2}{\alpha_3}\right\}^{\frac{1}{3}}. \tag{56}$$

Note that the suffix 3 refers to the direction of the magnetic field. This result could have been obtained by substituting (51) into Peierls's formula (50).

Note that if $\alpha_1 = \alpha_2 = \alpha_3 = \alpha$, one obtains

$$\kappa = \alpha\kappa_1, \tag{57}$$

where κ_1 is the value (43) for free electrons. Thus small α (large effective mass) gives small diamagnetism. Contrast the behaviour of the paramagnetic susceptibility (equation (25)), which becomes large when α is small.

The formula (56) is of interest because it may be applied with fair approximation to certain actual metals. It has already been shown that in most metals, with the exception of the monovalent group, the surface of the Fermi distribution overlaps the first Brillouin zone. A simple idealized case is shown in Fig. 86, where the curved lines represent in k-space the boundary of the Fermi distribution of electrons, and the straight lines planes of energy discontinuity. In the neighbourhood of the points A and B the surfaces of constant energy approximate closely to a family of ellipsoids. The susceptibility of the electrons in the regions beyond A and B taken together may be calculated by means of formula (56). N in (56) would then refer to the number of electrons per unit volume of the metal which lie in the regions beyond A and B. As for all structures planes of energy discontinuity occur in pairs symmetrically placed with respect to the origin, *when the overlap is small* it is always possible to regard the 'overlapping' electrons as forming a number of ellipsoidal distributions in k-space. (56) may then be used to give a good idea of the magnitude of the diamagnetism in such cases. The value which

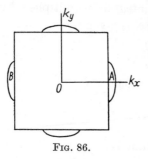

FIG. 86.

N takes may be estimated from other physical properties of the metal.

The values of the parameters α_1, α_2, α_3 were discussed in Chap. II, § 4.5. We found that, while α_2 and α_3 are comparable with unity, α_1 is given approximately by

$$\alpha_1 = 1 + 4E_0/\Delta E, \qquad (58)$$

where E_0 is the energy of a free electron at the point A in Fig. 86 (6–10 e.v.) and ΔE the energy gap. If ΔE is of the order of 1 e.v., we see that, in the region near the energy discontinuity where formula (58) holds, α_1 may be as great as 30 times α_2 or α_3. It follows, therefore, from (56) that electrons in states near A, B will make large contributions to the diamagnetism in the direction of the y-axis, but small ones only in the direction of the x-axis.

6.3. *Comparison with experiment.* For comparison with experiment we shall use the susceptibility per unit mass, χ. Landau's formula (43) gives

$$\chi_L = -0.623 n_0^{\frac{1}{3}} \rho^{-\frac{2}{3}} A^{-\frac{1}{3}} \times 10^{-6} \text{ c.g.s. units}, \qquad (59)$$

where n_0 is the number of free electrons per atom in the metal, ρ is the density, and A the atomic weight. We have already seen that free electrons give a paramagnetic effect due to electron spin, which is just three times as great as the Landau diamagnetism. The contribution of the free electrons to the total susceptibility is thus $2|\chi_L|$. In Table VIII we give the observed susceptibilities for a number of diamagnetic metals in the liquid and solid phases, and for comparison a calculated susceptibility obtained by adding together the contributions from the ions and the conduction electrons, assuming the latter to be perfectly free. The table shows that the susceptibility calculated in this way gives approximately the observed value in the liquid phase. In certain metals, notably Bi, Sb, and the γ-alloys, the susceptibility in the solid state is far greater than the calculated value assuming free electrons. This we believe to be due to the fact that in these metals the conduction electrons slightly overlap an almost full Brillouin zone (cf. Chap. V). We shall consider the cases of the γ-phase alloy Cu–Zn and the pure metal Bi.

TABLE VIII

Mass susceptibility of diamagnetic metals in solid and liquid phases, compared with Landau's formula

Metal	$\chi_s \times 10^6$ (observed)	$\chi_l \times 10^6$ (observed)	n_0	$\chi_L \times 10^6$	$\chi^{(n)} \times 10^6$	$\{\chi^{(n)} + 2\|\chi_L\|\} \times 10^6$
Ag	$-0\cdot26$	$-0\cdot28$	1	$-0\cdot027$	$-0\cdot391$	$-0\cdot336$
Au	$-0\cdot15$	$-0\cdot17$	1	$-0\cdot015$	$-0\cdot296$	$-0\cdot266$
Zn	$-0\cdot157$	$-0\cdot09$	2	$-0\cdot053$	$-0\cdot236$	$-0\cdot131$
Hg	$-0\cdot15$	$-0\cdot18$	2	$-0\cdot024$	$-0\cdot237$	$-0\cdot189$
Ga	$-0\cdot23$	$-0\cdot04$	3	$-0\cdot067$	$-0\cdot183$	$-0\cdot049$
Ge	$-0\cdot10$	$-0\cdot30$	4	$-0\cdot077$	$-0\cdot147$	$+0\cdot007$
Pb	$-0\cdot12$	$-0\cdot08$	4	$-0\cdot033$	$-0\cdot163$	$-0\cdot097$
Sb	$-0\cdot55$	$-0\cdot04$	5	$-0\cdot061$	$-0\cdot167$	$-0\cdot045$
Bi	$-1\cdot02$	$-0\cdot08$	5	$-0\cdot039$	$-0\cdot140$	$-0\cdot062$
Cu–Zn (γ-brass)	$-0\cdot77$	$-0\cdot10$	$1\cdot66$	$-0\cdot046$	$-0\cdot259$	$-0\cdot351$

χ_s = mass susceptibility of solid

χ_l = mass susceptibility of liquid

n_0 = number of valency electrons per atom

χ_L = mass susceptibility according to Landau's formula (59)

$\chi^{(n)}$ = mass susceptibility of ion with n_0 positive charges.

The values for the γ-phase alloy refer to the limit of the phase with high electron concentration. The value given for $\chi^{(n)}$ in this case is an average value.

The γ-phase alloys obey the Hume-Rothery rule; a discussion of their electronic structure is given in Chapter V. According to some recent measurements of C. S. Smith,† the mass susceptibility of the γ-phase of the alloy Cu–Zn increases linearly from $-0\cdot23 \times 10^{-6}$ c.g.s. units at the copper-rich end of the phase, where the electron-atom ratio is $1\cdot58$, to $-0\cdot77 \times 10^{-6}$ c.g.s. units at the zinc-rich end, where the electron ratio is $1\cdot66$. The form of the Brillouin zone (Fig. 71) shows that we may expect overlapping of electrons to occur when the electron ratio is somewhat greater than $1\cdot54$, which suggests that

† *Physics*, **6** (1934), 47.

the increasing diamagnetism is mainly due to the electrons lying beyond this zone. A rough quantitative test can be made of this hypothesis. As there are 36 faces on the zone, all equidistant from the origin, we have to consider 18 ellipsoidal distributions of the type considered in § 6.2. A short calculation, taking into account the proper orientation of the axes of the ellipsoids, gives for the susceptibility†

$$\chi = -\frac{0 \cdot 623 \times 12 \cdot 5}{\rho^{\frac{2}{3}} A^{\frac{1}{3}}} \left(\frac{n_0'}{18}\right)^{\frac{1}{3}} \frac{(\alpha_1 \alpha_2)^{\frac{2}{3}}}{\alpha_3^{\frac{1}{3}}} \times 10^{-6},$$

where α_1, α_2, α_3 are defined above, α_1 referring to the direction normal to a plane of energy discontinuity; n_0' is the total number of electrons per atom overlapping the Brillouin zone. Assuming a value of 1 e.v. for the energy discontinuity and calculating the α's by formula (58), we obtain $\chi = -2 \cdot 01 \times 10^{-6} (n_0')^{\frac{1}{3}}$, and if at the zinc-rich end of the phase the value of n_0' is of order 0·05 (in the whole range of the phase the electron-atom ratio varies by 0·08), we obtain for the susceptibility approximately $-0 \cdot 74 \times 10^{-6}$. The strongest evidence in favour of this view of the origin of diamagnetism of these alloys is that the molten alloy, in which the zone structure is lost, has a comparatively small susceptibility.

Bismuth may be considered from a similar point of view.‡ The Brillouin zone containing five electrons is shown in Fig. 70. Electrons will overlap at the points A, and positive holes will exist at the points B. We require first to know the number n_0' of overlapping electrons per atom.

Bismuth contains 5 valence electrons, lead or tin 4, and tellurium 6. Therefore, in a bismuth alloy containing x per cent. of tin or lead atoms (x small), the number of valence electrons is decreased by $\frac{1}{5}x$ per cent. Similarly, in an alloy with tellurium, the number of electrons will be increased. Now Goetz and Focke,‖ and also Shoenberg and Uddin,†† have measured the magnetic anisotropy of these alloys, i.e. the ratio $\chi_\perp/\chi_\parallel$ between the susceptibilities perpendicular and parallel to the principal axis. They find that the admixture of 0·1 per cent. of Sn or Te atoms makes a large difference to the magnetic anisotropy, as the following table shows.

† Cf. Jones, *Proc. Roy. Soc.* A, **144** (1934), 225.

‡ Ibid. **147** (1934), 396.

‖ *Phys. Rev.* **45** (1934), 170.

†† *Proc. Roy. Soc.* A (in press).

Magnetic anisotropy of alloys of bismuth with tin and tellurium

Atomic per cent.		Atomic per cent.	
Sn	$\chi_\perp/\chi_\parallel$	Te	$\chi_\perp/\chi_\parallel$
0·01	1·470	0·01	1·329
0·03	1·570	0·03	1·197
0·09	1·803	0·09	0·903
0·81	2·965	0·27	0·697
2·43	4·655	··	··

One must deduce that n_0', the number of overlapping electrons per atom, is of the order[†] 10^{-4}. Similar results have been obtained for the electrical conductivities of these alloys.[‡] The fact that $\chi_\perp/\chi_\parallel$ is increased by the addition of tin, but decreased by the addition of tellurium, suggests that the positive holes are responsible for $\chi_\perp$ and the overlapping electrons for $\chi_\parallel$.

For the overlapping electrons we take the energy surfaces of the form (51), with k_3 the wave vector parallel to the principal axis. Then $\chi_\parallel$ is given by (56), with α_3 referring to the direction of the principal axis. With n_0' as obtained above, formula (56) gives the right order of magnitude for the susceptibility when $(\alpha_1\alpha_2)^{\frac{3}{2}}/\alpha_3^{\frac{1}{2}} \sim 10^3$.

Assuming $\alpha_3 \sim 1$, this gives $\alpha_1\alpha_2 \sim 3\times10^4$. Such a highly eccentric form for the surfaces of constant energy is surprising, but, as we shall see below, the magnetic behaviour of bismuth at low temperatures leads to the same conclusion.

With these values of α we have, for the energy interval between the bottom of the second zone and the surface of the Fermi distribution,

$$E = n_0^{\frac{2}{3}}(\alpha_1\alpha_2\alpha_3)^{\frac{1}{3}}\frac{h^2}{8m}\left(\frac{3}{\pi\Omega_0}\right)^{\frac{2}{3}} \sim 0.23 \text{ e.v.},$$

where Ω_0 is the volume per atom and[∥] $n_0 = \frac{1}{3}n_0'$. For the velocities $\hbar\alpha_1 k_x/m$, etc., we have

$$v_x = n_0^{\frac{1}{3}}\alpha_1^{\frac{1}{2}}(\alpha_1\alpha_2\alpha_3)^{\frac{1}{3}}(3/\pi\Omega_0)^{\frac{1}{3}}h/2m. \tag{60}$$

It would be tempting to attribute the large value of $\alpha_1\alpha_2$ to a very small energy-gap, and to assume it to be caused only by that component of α referring to the direction perpendicular to the plane of energy discontinuity (cf. equation (58)). Formula (60), however, shows this hypothesis to be untenable, since it would give the

---

[†] We believe that the estimate 10^{-3} given by Jones, loc. cit., was too large.
[‡] Cf. Chap. VII, § 14.
[∥] Cf. Fig. 70; n_0 is the number of electrons in each overlapping ellipse.

component of v in the same direction much greater than the corresponding value for free electrons. We must therefore assume

$$\alpha_1 \sim \alpha_2 \sim 2 \times 10^2, \qquad \alpha_3 \sim 1.$$

The surfaces of constant energy are thus ellipsoids with two short axes, and the long axis pointing in the direction of the crystal axis (cf. Fig. 71).

Formula (56) shows that, with these values of α_1, α_2, the overlapping electrons will make a negligible contribution to the component $\chi_\perp$ perpendicular to the direction of the principal axis. $\chi_\perp$ must therefore be due to the positive holes in the first zone. We have not been able to make any estimate of the energy surfaces for the positive holes; since $\chi_\perp/\chi_\parallel$ is about 1·6, they must, if they are ellipsoids, be even more eccentric.

The magnetostriction† of single crystals of bismuth may also be discussed with the same model; for details the reader is referred to the original paper.‡

6.4. *Diamagnetism at low temperatures and high fields.* Peierls‖ has discussed the diamagnetism of electrons in metals under conditions such that

$$\mu H > kT.$$

As a preliminary study he considers the idealized case of a two-dimensional electron gas at the absolute zero of temperature. Following him we may imagine a magnetic field perpendicular to the plane of the gas. The stationary states have the energies

$$E_l = \mu H(2l+1).$$

As for an actual three-dimensional gas (cf. p. 202), we may suppose that each energy level E_l is degenerate and has a statistical weight proportional to H, which we may write βH, where β is a constant of the same type as the coefficient of H in (31). Let there be N electrons, which, according to Fermi statistics, will completely fill the r lowest states and partly fill the $(r+1)$th state, where r is given by

$$r\beta H < N < (r+1)\beta H,$$

except of course when $\beta H/N > 1$.

The energy of the rth state is $\mu H(2r-1)$, since the energy of the first state is μH. The total energy U is equal to the energy of the

† Kapitza, *Proc. Roy. Soc.* A, **135** (1932), 537.
‡ Jones, loc. cit. 411.   ‖ *Zeits. f. Phys.* **81** (1933), 186.

electrons in the fully occupied states, which is

$$\beta H \sum_{1}^{r} \mu H(2r-1) = \beta \mu H^2 r^2,$$

and the energy of the $N-r\beta H$ electrons in the $(r+1)$th state, which is $\mu H(N-r\beta H)(2r+1)$, giving a total of

$$U = \mu H[(2r+1)N - \beta Hr(r+1)].$$

Hence the total moment σ is given by

$$-\sigma = \frac{\partial U}{\partial H} = \mu[(2r+1)N - 2r(r+1)\beta H]. \tag{61}$$

Fig. 87 shows the average moment σ/N plotted as a function of $\beta H/N$ according to (61). It is a discontinuous function, each branch

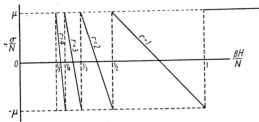

FIG. 87. Magnetic moment of a two-dimensional metal plotted against field.

corresponding, by (61), to one value of r. When $\beta H/N > 1$, the first state ($r = 0$), whose energy is μH, is not completely full, and the total energy, therefore, is $N\mu H$, and the average moment $-\mu$. The discontinuities in σ/N arise in the following way: when H is so great that $\beta H > N$, all the electrons lie in the state of lowest energy. The total energy is therefore μHN, and the moment is independent of the field. As H decreases a field strength is reached for which $\beta H = N$; for any further decrease in the field some electrons move up into the next highest level. At this point, therefore, the derivative of the energy suddenly becomes negative.

For an actual three-dimensional electron gas, for different values of the temperature, Peierls has calculated the magnetic moment σ as a function of $\mu H/\zeta_0$, where ζ_0 is the energy at the top of the Fermi distribution. He obtains the result

$$\sigma = \frac{-(2e\zeta_0)^{\frac{3}{2}}}{\pi^{\frac{1}{2}}h^{\frac{3}{2}}c^{\frac{3}{2}}\mu^{\frac{1}{2}}} f\left(\frac{\mu H}{kT}, \frac{\zeta_0}{kT}\right).$$

Fig. 88 shows the function f for $T = 0$ plotted against $\mu H/\zeta_0$.

The difference in form between the two curves of Figs. 87 and 88 is due to the fact that, in the three-dimensional gas, the

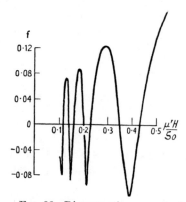

energy of an electron can vary continuously by changing its momentum in the field direction, as equation (30) shows.

If the field H is very great, only the state $l = 0$ (equation (30)) will be occupied by electrons. This corresponds to the part of the curve in Fig. 88 lying to the right of $\mu H/\zeta_0 = 0\cdot4$. For a given value of p_z the difference of energy between the states $l = 0$ and $l = 1$ is $2\mu H$. Hence the condition that the first state only should be occupied is that

FIG. 88. Diamagnetic moment of a degenerate electron gas at the absolute zero of temperature.

p_z should lie in the range given by

$$0 < p_z^2/2m < 2\mu H;$$

in other words, that p_z should be numerically less than the quantity

$$p_0 = \sqrt{(4m\mu H)}.$$

But from (31), under the same conditions, we have for the total number N of electrons per unit volume

$$N = 4eHp_0/ch^2.$$

Hence, eliminating p_0 from the last two equations, we obtain the following equation for the field at which the second level begins to fill:

$$N = \frac{4}{\sqrt{\pi}} \left(\frac{eH}{ch}\right)^{\frac{3}{2}}. \tag{62}$$

For values of H greater than this critical value, the moment at a given field strength is easily calculated. The total energy according to (30) with $l = 0$ and $p_z = p_0$ is

$$W = \mu NH + \frac{N^3}{24m} \left(\frac{ch^2}{e}\right)^2 \frac{1}{H^2},$$

and therefore the average moment per electron is

$$-\mu + \frac{N^2}{12m} \left(\frac{ch^2}{e}\right)^2 \frac{1}{H^3}.$$

The negative moment decreases, therefore, as the field decreases, as is shown in Fig. 88.

To obtain the magnetic moment as a function of H for low but finite temperatures requires a laborious calculation. Fig. 89 shows

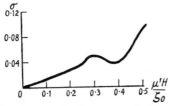

FIG. 89. Magnetic moment of a degenerate electron gas at a finite temperature ($kT/\zeta_0 = 0.15$).

the result of such a calculation according to Peierls. This calculation may be compared with the experimental results of de Haas and van Alphen,[†] who have measured the specific magnetic moment of very pure single crystals of bismuth at different field strengths. For low fields and high temperatures the moment varies linearly with H, giving a definite diamagnetic susceptibility independent of the field. At high fields and low temperatures, however, the moment varies with the field in the manner shown in Fig. 90. In these experiments the magnetic field was always at right angles to the direction of the

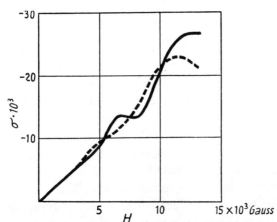

FIG. 90. Magnetic moment of bismuth as a function of the external field at 14·2° K. (observed by de Haas and van Alphen).

- - - - - Field perpendicular to binary axis.
——— Field parallel to binary axis.

principal axis of the bismuth crystal, and Fig. 90 corresponds to two definite settings of the binary axes relative to the field direction.

† *Comm. Leiden*, **212** a (1930); *Proc. Amsterdam Ac.* **33** (1930), 1106.

When the magnetic field was parallel to the principal axis the moment was found to be directly proportional to the field strength; in other words, for this direction bismuth was found to behave normally.

The similarity of the curves in Figs. 89 and 90 suggests that Peierls's theory is essentially correct. A more critical test is provided by the calculation of the position of the first bend in the moment-field curve, which occurs at $H \simeq 20$ kilogauss. We have seen that this bend occurs when the electrons completely fill the first level ($l = 0$) and just begin to fill the second level for which $l = 1$. Assuming free electrons, therefore, the required critical value of H is given by equation (62). Applied to bismuth this equation gives $H = 6 \cdot 5 \times 10^5$ kilogauss, a value of entirely the wrong order of magnitude.

The model for bismuth discussed in § 6.3 gives, however, a much lower value of the critical field. We believe the susceptibility parallel to the principal axis to be due to 'overlapping' electrons; for such electrons the critical field may be calculated in a way similar to that by which (62) was obtained. We find

$$ N' = \frac{4}{\sqrt{\pi}} \left(\frac{eH}{hc}\right)^{\frac{3}{2}} \frac{(\alpha_1 \alpha_2)^{\frac{1}{4}}}{\alpha_3^{\frac{1}{2}}}; $$

setting $N' = 2 \cdot 84 \times 10^{18}$, which corresponds to 10^{-4} electrons per atom, and giving α_1, α_2, α_3 the values assumed in § 6.3, this gives

$$ H \simeq 85 \text{ kilogauss.} $$

This is not so much larger than the 20 kilogauss observed for the perpendicular field. Actually no effect has been observed in the parallel direction, the field 85 kilogauss not having been reached.

We note that the critical field is determined by the quantity $(\alpha_1 \alpha_2 / \alpha_3^2)^{\frac{1}{4}} / N'$, while the paramagnetism itself depends on $(\alpha_1 \alpha_2 / \alpha_3) N'$. It is not therefore possible to determine N' uniquely, unless we make some assumption about one of the coefficients α.

It has not yet been possible to make a theory applicable to the case when the field is perpendicular to the axis. It should be noted, however, that $\chi_\perp > \chi_\parallel$ and so we should expect the critical field to be even less for this case.

6.5. *Effect of cold work in the diamagnetism of metals.* Honda and Shimizu† reported in 1930 that they had observed large changes in

† *Nature*, **126** (1930), 990.

the diamagnetic susceptibilities of metals Cu and Ag when subjected to cold work, which was interpreted as being the result of the consequent internal strains in the lattice. Later, however, Kussmann and Seemann[†] showed that it was in fact due to the presence of minute quantities of iron, which were brought out of solid solution in the metals by the cold work and which therefore acted as a ferromagnetic impurity. Recently Shimizu[‡] has reported a very small residual effect which must presumably be due to lattice distortion.

7. Ferromagnetism

Ferromagnetic materials are believed to be 'spontaneously magnetized'; that is to say, any small block of magnetic material of volume $d\tau$ is believed to have a magnetic moment, $I\,d\tau$, where I depends on the temperature. This hypothesis has to be reconciled with the fact that an ordinary piece of iron, for example, can exist in an apparently unmagnetized state. It is therefore generally accepted that any apparently unmagnetized block of ferromagnetic material is divided into 'domains', each domain being spontaneously magnetized in a direction uncorrelated with that of all the other domains. The evidence for this view, and the size of the 'domains', will not be discussed here;[||] in this chapter we discuss mainly the magnitude of the spontaneous magnetization and its dependence on temperature.

7.1. *Nature of the elementary magnets.* Below the Curie point, we know from the experimental work on the gyromagnetic effect[††] that in iron, cobalt, and nickel the Landé g-factor is 2; above the Curie point, the investigations of Sucksmith[‡‡] on an alloy of copper and nickel lead, within an accuracy of 10 per cent., to the same result. We may assume, then, that in most ferromagnetic materials the orbital motion makes no contribution to the magnetic moment, but that the magnetism is entirely due to the spins of electrons. The electron has an angular momentum $\frac{1}{2}\hbar$; but, as we shall see below, there is evidence that pairs of electrons may be coupled together in the same atom with their spins in the same direction, as are, for instance, the electrons in an atom in a triplet state. We shall therefore assume for our elementary magnets an angular momentum $j\hbar$, where

$$j = \tfrac{1}{2},\ 1,\ 1\tfrac{1}{2},\ldots,$$

† *Zeits. f. Phys.* **77** (1932), 567.   ‡ *Science Reports Tôhoku,* **22** (1933), 915.
|| Cf., for example, Stoner, *Magnetism and Matter*, p. 120, London (1934).
†† Cf. Stoner, loc. cit.   ‡‡ *Helv. Phys. Acta,* **8** (1935), 205.

according as one, two, or three electrons are coupled together, and a magnetic moment†

$$jg\mu \quad (\mu = e\hbar/2mc; \; g = 2).$$

In a magnetic field H the energy of the elementary magnets can take the values

$$-m_j Hg\mu \quad (m_j = j, j-1, ..., -j),$$

the energy of a single electron having thus *two* values, that of a pair of electrons, three, and so on.

7.2. *The Weiss molecular field.* The important phenomenological theory of Weiss‡ has proved of great value in correlating experimental observations. We discuss here only the form the theory takes when modified‖ by the quantum theory, since this form is obviously appropriate when the magnets are electrons.

We consider a ferromagnet containing N elementary magnets per gramme atom. Then, according to Weiss, in an external field H the mean magnetic field acting on any elementary magnet is not H but a field H_i in the same direction as H and as the magnetization, given by

$$H_i = H + \lambda\sigma,$$

where σ is the intensity of magnetization per gramme atom and λ is a constant characteristic of the material.†† The 'field' $\lambda\sigma$ is called the molecular field and λ the molecular field constant. Its origin was first explained by Heisenberg on the basis of quantum mechanics (see below). To obtain agreement with experiment one must assume $\lambda\sigma \sim 10^7$ gauss, and hence $\lambda\sigma \gg H$ for all fields up to the value of 300,000 gauss which is the greatest yet obtained.

At the absolute zero of temperature all the elementary magnets point along the direction of the field $(m_j = j)$, however weak H may be. The intensity of magnetization is therefore

$$\sigma_0 = N\mu gj = \frac{Ne\hbar}{mc}j \tag{63}$$

independently of the field.

At the absolute zero the work required to change the quantum number m_j to $m_j - 1$ is

$$\mu g(H + \lambda\sigma_0),$$

---

† Cf. the footnote on p. 183.
‡ *J. de Phys.* **6** (1907), 661.
‖ Stoner, *Phil. Mag.* **10** (1930), 27; **12** (1931), 737.
†† The usual notation for the 'field' is NI; hence $\lambda = N/$(volume per gm. atom).

and hence at low temperatures, where $\sigma_0 - \sigma$ is small, we have by Boltzmann's law,

$$\frac{\sigma_0 - \sigma}{\sigma} \simeq e^{-\mu g(H + \lambda\sigma_0)/kT}. \tag{64}$$

At low temperatures the right-hand side is small, and hence the magnetization has nearly the saturation value, again almost independently of the field. The hypothesis of Weiss thus accounts for the spontaneous magnetization.

At higher temperatures an elementary calculation gives†

$$\frac{\sigma}{\sigma_0} = \frac{2j+1}{2j} \coth \frac{(2j+1)a}{2j} - \frac{1}{2j} \coth \frac{a}{2j}, \tag{65}$$

where
$$a = \mu g j(H + \lambda\sigma)/kT. \tag{66}$$

In the particular case $j = \frac{1}{2}$, this reduces to

$$\sigma/\sigma_0 = \tanh a. \tag{67}$$

These equations have to be solved for σ. When $a \ll 1$, (65) gives

$$\frac{\sigma}{\sigma_0} = \frac{j+1}{3j} a. \tag{68}$$

In the absence of a field the spontaneous magnetism therefore disappears at the Curie temperature Θ given by

$$\Theta = \frac{j+1}{3j} \frac{\lambda\sigma_0 g j\mu}{k}, \tag{69}$$

or, writing $g j\mu = \mu_0 =$ moment of each carrier,

$$\Theta = \frac{j+1}{3j} \frac{\lambda\sigma_0 \mu_0}{k}. \tag{70}$$

Below the Curie point the spontaneous magnetization deduced from (65) in the absence of a field is given in Table IX.

TABLE IX

	σ/σ_0		
T/Θ	$j = 1$	$j = \frac{1}{2}$	Observed (nickel)
0·0	1·00	1·00	1·00
0·1	1·00	1·0000	0·997
0·2	0·99	0·9999	0·991
0·3	0·99	0·9974	0·979
0·4	0·97	0·9856	0·961
0·5	0·93	0·9575	0·937
0·6	0·87	0·9073	0·895
0·7	0·80	0·8287	0·837
0·8	0·67	0·7104	0·742
0·9	0·49	0·5256	0·595
1·0	0·0	0·00	0·0

† Cf. Stoner, *Magnetism and Matter*, p. 354.

Above the Curie point the atomic susceptibility is, by (66) and (68),

$$\chi_A = \frac{C}{T-\Theta},\tag{71}$$

where

$$C = \frac{j+1}{3j}\frac{\sigma_0\mu gj}{k}.\tag{72}$$

The energy per gm. atom due to magnetization is

$$E = -\tfrac{1}{2}\lambda\sigma^2,\tag{73}$$

and hence the contribution to the atomic heat due to demagnetization

$$C_v = \frac{dE}{dT} = -\lambda\sigma\frac{d\sigma}{dT}\ \text{erg degree.}\tag{74}$$

7.3. *Review of the experimental material and comparison with the Weiss theory.*

(a) *Numerical values of the saturation moment σ_0 at low temperatures.* According to the values given by Stoner,[†] these are as follows:

Element	σ_0 in erg gauss^{-1} per gm. atom	Bohr magnetons per atom $p = \sigma_0/N_L$
Fe	12,230	2·22
Co	9,500	1·71
Ni	3,370	0·606

The non-integral number of Bohr magnetons per atom is to be noted. The numbers in the last column give the actual number of spins $e\hbar/2mc$ which are responsible for the magnetization. In accordance with the considerations of § 5.1, we assume that this number is equal[‡] to the *actual number of positive holes* in the d shell. The number of electrons in s states will be

Fe $2\cdot22-2 = 0\cdot22$

Co $1\cdot71-1 = 0\cdot71$

Ni $0\cdot606 = 0\cdot606.$

In the notation of equation (26) we shall have for nickel

$$2|A_2|^2+|A_1|^2 = 0\cdot606.\tag{75}$$

(b) *Dependence of the spontaneous magnetization on temperature.* The ratios σ/σ_0 plotted against T/Θ for pure iron, cobalt, or nickel lie approximately[‖] on the same curve. This curve agrees fairly well

---

† Loc cit. 366.   ‡ See note on p. 239.

‖ Cf., for example, Potter, *Proc. Roy. Soc.* A, **146** (1934), 379.

with the theoretical curve for $j = \frac{1}{2}$, but lies slightly above it, as the figures of Table IX show. (Cf. Van Vleck, loc. cit. 334, where a figure is given showing the curves for $j = \frac{1}{2}, j = 1$.) The experimental values of σ/σ_0 thus favour $j = \frac{1}{2}$, i.e. *uncoupled* electrons, if the Weiss theory is assumed to be correct.

(c) *Calculation of j (the angular momentum of the elementary magnets) from the susceptibility above the Curie point.* The susceptibility of nickel above the Curie point is shown in Fig. 81. For nickel in the range 500° C. to 1,050° C. it will be seen that $1/\chi$ is a linear function of T; if we write for the susceptibility per gm. atom.

$$\chi_A = \frac{C}{T - \Theta_p} \tag{76}$$

then C has the value $C = 0.325$ (nickel).

For cobalt and iron it is doubtful whether the $1/\chi$, T curve ever becomes linear, and the C values seem too uncertain to be worth comparing with experiment.

For nickel the 'paramagnetic Curie temperature' Θ_p is equal to about 650° K., and differs appreciably from the ferromagnetic Curie point 631° K. Between the two Curie temperatures $1/\chi$ is not linear in T. The existence of the two Curie temperatures has been discussed by Ludloff[†] in a paper based on quantum mechanics, which is referred to further below.

The saturation moment σ_0 is given in terms of the Curie constant C by formula (72); if we assume that the elementary magnets have an angular momentum given by $j = \frac{1}{2}$, this gives for σ_0 for nickel

$$\sigma_0 = Ck/\mu = 4,850 \text{ erg gauss per gm. atom,}$$

which is different from the value 3,370 obtained by direct measurements at low temperatures and corresponds to a number of Bohr magnetons per atom, $p_{\text{par}} = 0.87$. On account of this result it is usually stated that the paramagnetic and ferromagnetic 'magneton numbers' are not the same. In view of the considerations of § 5, it is improbable that the number of elementary magnets actually changes, and we think it more likely that the assumption $j = \frac{1}{2}$ is at fault. As we saw in that section, there will be in the crystal some ions (3d shells) with two positive holes, some with one, and some with none. The ions with no positive holes have, of course, no angular momentum ($j = 0$); those with one positive hole will have $j = \frac{1}{2}$, and, as the energy levels (Fig. 78) of atomic nickel show, those with

† *Zeits. f. Phys.* **91** (1934), 742.

two positive holes will have angular momentum $j\hbar$ with $j = 1$. In other words, the spins in the two 'holes' are parallel; the state with antiparallel spins is higher by an electron volt than that with parallel spins, and an energy of one electron volt is much greater than the exchange interaction ($\sim k\Theta$) between the ions.

In § 5 we denoted by $|A_{2j}|^2$ ($j = 0, \frac{1}{2}, 1$) the number of ions with 0, 1, and 2 'holes'. If $|A_2|^2$ is appreciable, some of the spins within each atom will be coupled and the assumption that $j = \frac{1}{2}$ will be wrong.

We shall now use the ferromagnetic and paramagnetic magneton numbers to obtain $|A_{2j}|^2$. From (72) we have for the paramagnetic magneton number for nickel

$$g^2 \sum \frac{j(j+1)}{3} |A_{2j}|^2 = \frac{kC}{N\mu^2} = 0.87;$$

for the ferromagnetic magneton number

$$g \sum j |A_{2j}|^2 = 0.606$$

and finally

$$\sum |A_{2j}|^2 = 1.$$

It is not actually possible† to solve these equations with positive $|A_1|^2$, but if we take

$$|A_2|^2 = 0.303 \qquad |A_1|^2 = 0 \qquad |A_0|^2 = 0.597,$$

we obtain for the paramagnetic magneton number 0.81, which is near the observed value. We must thus conclude that few if any of the ions have only one positive hole; the elementary magnets (positive holes) are coupled together, there being either two or none in each atom.

It is not difficult to understand qualitatively the reason for this coupling. We have seen in Chapter IV that electrons tend to keep away from each other; the same will be true of the positive holes which are the elementary magnets in nickel. Now we know from the spectrum of nickel (Fig. 78) that two holes can be in the same atom with quite low energy, so long as their spins are parallel; therefore the state of lowest energy will be reached when two 'holes' or none are in each atom, for then each pair of holes will be surrounded mainly by neutral ions.

It follows that any positive ion in nickel can strongly attract the conduction electrons outside its own atomic polyhedron, for, although of course each polyhedron is neutral on the average, nevertheless

† See note on p. 239.

when a positive hole is in a given polyhedron, the polyhedron is positive and the surrounding ones negative. This is the probable explanation of the fact that the transition metals have higher binding energy than the noble metals.

(*d*) *Calculation of the molecular field constant* λ *from the Curie temperature* Θ_f. The molecular field constant λ may be calculated for any assumed value of j from formula (69) and from the experimental values of Θ_f and σ_0 given above; the following table gives the results (c.g.s. units).

	Curie temperature Θ_f, in degrees K.	Molecular field constant λ	
Element		$j = \frac{1}{2}$	$j = 1$
Fe	1,043	1,280	960
Co	1,393	2,200	1,650
Ni	631	2,950	2,210

(*e*) *Calculation of the molecular field constant from the susceptibility above the Curie point and the Curie temperature.* According to formulae (69), (72), we have for the molecular field constant

$$\lambda = \Theta/C \qquad (77)$$

independently of j. The value obtained, however, depends upon whether we take for Θ the ferromagnetic or paramagnetic Curie points. For nickel we obtain

$$\lambda = \Theta_f/C = 1,950,$$
$$\lambda = \Theta_p/C = 2,000.$$

The values obtained agree approximately with that obtained above with $j = 1$, and thus lend support to the conclusion that the electron spins are coupled.

7.4. *Internal energy of a ferromagnetic.* The energy E at constant pressure of a ferromagnetic substance is a function of the temperature T and of the magnetic field H,

$$E = E(T, H).$$

We may divide this up into the energy E_L of the lattice, given approximately by a Debye function, and the energy E_e of the electrons. According to the Weiss theory, that part of E_e which is due to the magnetization is

$$-\tfrac{1}{2}\lambda\sigma^2 - H\sigma. \qquad (78)$$

We shall see on p. 229 that a further term must be added, giving the kinetic energy of translation of the magnetic electrons.

(a) *Magneto-caloric effect.* If a magnetic substance is placed in a field H and the field is suddenly changed by an amount ΔH, the internal energy changes by

$$\Delta E = \text{work done}$$
$$= -\sigma \Delta H. \tag{79}$$

The temperature therefore changes: for a substance having the magnetic energy given by the Weiss theory,

$$\Delta E = \frac{dE_L}{dT}\Delta T - \lambda\sigma\,\Delta\sigma - H\,\Delta\sigma - \sigma\,\Delta H, \tag{80}$$

and hence in an adiabatic change the change of temperature ΔT is given by

$$\frac{dE_L}{dT}\Delta T = (\lambda\sigma + H)\,\Delta\sigma. \tag{81}$$

For dE_L/dT we may write $3RJ$ (J = mechanical equivalent of heat). Hence

$$\Delta T = \frac{(\lambda\sigma + H)\,\Delta\sigma}{3RJ}. \tag{82}$$

The change in T for given change in H and hence in σ can thus be used to determine λ.

Measurements can only be made in the neighbourhood of the Curie temperature, where σ is sensitive to H. In this region $H \ll \lambda\sigma$ and can be neglected. Fig. 91 shows the results of measurements of Weiss and Forrer[†] for nickel; similar results have been obtained by Potter[‡] for iron. Considerably above the Curie point (i.e. some $60°$ above Θ_f), we obtain for λ the following values:

$$\text{Ni} \quad . \quad\quad . \quad\quad . \quad\quad . \quad\quad . \quad 1{,}770$$
$$\text{Fe} \quad . \quad\quad . \quad\quad . \quad\quad . \quad\quad . \quad 700$$

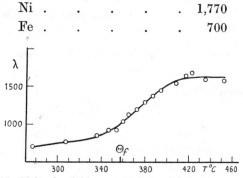

FIG. 91. Molecular field constant λ for nickel, from magneto-caloric effect (Weiss and Forrer).

† *Ann. d. Physique*, **12** (1929), 304.
‡ *Proc. Roy. Soc.* A, **146** (1934), 362.

which agree approximately with those obtained above. On the other hand, near and below the Curie temperature λ appears to fall to much lower values. This has been interpreted by Stoner† as meaning that the observed change $\Delta\sigma/\Delta H$ in the intensity of magnetization is not equal to the change in the intrinsic magnetization, even for fields above 5,000 gauss.

It is possible that for ferromagnetics we must in any complete theory make a distinction between 'long-distance order' and 'order of neighbours' (cf. Chap. I, § 7). The Curie point will then be the temperature at which a spontaneous magnetization extending over domains comprising millions of atoms disappears; but a small distance above the Curie point groups of four or five electron spins may still be coupled together. If this is the case, one would not expect the Weiss theory to be even approximately valid in this region.

The molecular field constant λ may thus be obtained from three independent sets of experimental data; for nickel the summarized results are

λ

Magneto-caloric effect 1,770

From Curie constant and Curie temperature . . 1,950–2,000

From Curie temperature and saturation intensity $\begin{cases} j = 1 & 2,100 \\ j = \frac{1}{2} & 2,800 \end{cases}$

The agreement between the values is best if we assume $j \sim 1$, which is in agreement with our deductions from the susceptibility at high temperatures. On the other hand, the magnetization-temperature curve deduced from the Weiss theory for $j = 1$ is not in such good agreement with experiment as that for $j = \frac{1}{2}$.

(b) *Specific heat.* According to the Weiss theory the atomic heat C_v of a ferromagnetic should be given by

$$C_v = \frac{1}{J}\left[\frac{dE_L}{dT} - \tfrac{1}{2}\lambda\frac{d(\sigma^2)}{dT}\right]. \tag{83}$$

The Weiss theory gives (with $j = \frac{1}{2}$)

$$\Theta\frac{d}{dT}\left(\frac{\sigma^2}{\sigma_0^2}\right) = -3\cdot 0 \quad (T = \Theta - 0),$$
$$= 0 \qquad (T = \Theta + 0).$$

We should therefore expect a discontinuity in the atomic heat of

$$\Delta C_v = \frac{3}{2}\frac{\lambda\sigma_0^2}{J\Theta} = \frac{3}{2}\frac{k\sigma_0}{J\mu} = 1\cdot 8 \text{ cal. per gm. atom for nickel.} \tag{84}$$

† *Phil. Trans. Roy. Soc.* **235** (1936), 165.

The specific heat of nickel near the Curie point has been measured by Sucksmith and Potter,† Lapp,‡ Klinkhardt,‖ Ahrens,†† and Grew.‡‡ The form of the curve depends markedly on the purity of the specimen. Fig. 92 shows some of the results obtained.

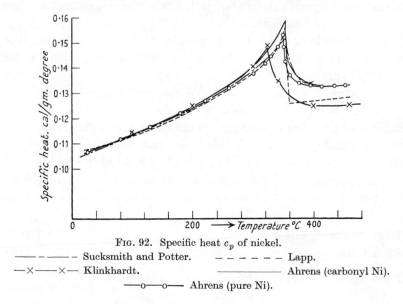

FIG. 92. Specific heat c_p of nickel.

——‒——‒— Sucksmith and Potter.   — — — — — Lapp.
—×——×— Klinkhardt.   ——————— Ahrens (carbonyl Ni).
                  ——o——o—— Ahrens (pure Ni).

We give below the maximum and minimum values of the atomic heat at constant pressure.

	C_{max}	C_{min}	$C_{max} - C_{min}$	Distance in degrees from max. to min.
Sucksmith and Potter .	8·9	7·9	1·0	25
Lapp	9·3	7·4	1·9	7
Klinkhardt . . .	8·8	7·3	1·5	55
Ahrens	9·0	7·8	1·2	30
Grew	8·5	7·5	1·0	50
Weiss theory ($j = \frac{1}{2}$)			1·8	0

It will be seen that, except for Lapp's result, the jump in the atomic heat is rather less than the calculated value and is not sharp. A further result of the theory is that the entropy in the specific

† *Proc. Roy. Soc.* A, **112** (1926), 157.
‡ *Ann. d. Physique*, **12** (1929), 442.
‖ *Ann. d. Physik*, **84** (1927), 167.
†† Ibid. **21** (1934), 169.
‡‡ *Proc. Roy. Soc.* A, **145** (1933), 509.

heat 'bump' is, with $j = \frac{1}{2}$, per gm. atom

$$\int \frac{C_e \, dT}{T} = n_0 \, R \log_e 2, \qquad (85)$$

where n_0 is the number of magnets per atom and R the gas constant.

It will be seen that the *minimum* values of C_p above the Curie point exceed the classical value 5·98 considerably. The dilatation correction[†] for nickel at 650° K. is

$$C_p - C_v = 0 \cdot 3.$$

Thus above the Curie temperature there is an excess specific heat of about 1·5 cal. per gm. atom. This is referred to in the literature as the 'term of unknown origin'. We believe that it is due to the kinetic energy of the 'positive holes in the d band' to which the magnetism is due.[‡] We saw in § 5.2 that the degeneracy temperature T_0 for these holes is 3,470° for ferromagnetic nickel; for paramagnetic nickel, where two holes may be in each state instead of one, we shall have

$$T_0 = 3{,}470/2^{\frac{2}{3}} = 2{,}180° \text{K}.$$

At the relevant temperature of 700° the specific heat per particle is (cf. Fig. 75) about $1 \cdot 1k$, giving for the atomic heat

$$C_e = 1 \cdot 1 \times 0 \cdot 6R = 1 \cdot 3 \text{ cal. per gm. atom,}$$

in fair agreement with the observations.

When copper is alloyed with nickel, the excess specific heat disappears at about the same composition as the ferromagnetism, cf. § 5.4.

7.5. *Application of quantum mechanics; the Heisenberg theory.* Until the discovery of quantum mechanics no satisfactory explanation of the physical nature of the Weiss molecular field could be given, the order of magnitude of all purely magnetic forces being much too small. Heisenberg‖ and Frenkel,[††] independently, were the first to point out that the explanation was provided by quantum mechanics, according to which, in a many-electron system, a strong coupling exists between the spin directions of the electrons and their orbital motion. In the hydrogen molecule H_2, for instance, if the electrons are in their ground states the spins must be antiparallel ($s = 0$), and if the spins are to be set parallel to each other the molecule must be excited to the first triplet state ($s = 1$), which requires the

---

† Cf. Chap. I, § 2.   ‡ Cf. § 5.
‖ *Zeits. f. Phys.* **49** (1928), 619.   †† Ibid. **49** (1928), 31.

expenditure of work equal to about 9 e.v. The authors quoted suggested that *in ferromagnetic substances this effect is reversed,* and that in the state of lowest energy the magnetic moment is different from zero. (We refer here, of course, to the state of lowest energy of the whole crystal, considered as a single giant molecule.) The work required to raise a gramme atom of the crystal to the *excited* state with zero moment is of the order of $R\Theta$ (Θ = Curie temperature), and thus of the order of magnitude of 0·1 e.v. per atom. Thus the effect considered certainly provides forces sufficiently great to account for ferromagnetism.

On the basis of these ideas, a more or less quantitative quantum-mechanical theory has been built up by Heisenberg, Bloch, Slater, and others. We may say that the main result of their researches is to show that quantum mechanics is capable of yielding results qualitatively in agreement with experiment and thus with those deduced from the Weiss hypothesis of the molecular field. The actual formulae that have been obtained, for instance, for the dependence of the saturation moment on temperature, depend for the most part on the drastic simplifying assumptions that have had to be made to solve the equations involved. We would therefore deprecate any serious attempt to compare these formulae with experiment at this stage in the development of the theory. There is, however, one important exception, namely the theorem of Bloch that, *at low temperatures,* $T \ll \Theta$, the saturation moment σ is given by

$$\sigma = \sigma_0[1 - AT^{\frac{3}{2}}], \qquad (86)$$

where σ_0, A are constants. This appears to be a rigorous deduction from the theory, and not to depend upon any simplifying assumptions.

An account of the quantum-mechanical theory of ferromagnetism has appeared in several text-books;[†] we therefore give here an outline only.

We discuss first a very idealized ferromagnetic consisting of a three-dimensional lattice of atoms, each with one valence electron in an s state. The atoms are supposed to be so far apart that the interaction between them may be considered small. We therefore use, instead of, as elsewhere in this book, the Bloch approximation (wave

---

[†] Cf., for instance, the full account by Bloch, 'Molekulartheorie des Magnetismus', *Handb. d. Radiologie*, 6/2, 2te aufl. (1934). Also Van Vleck, loc. cit., and Sommerfeld and Bethe, loc. cit. 585.

functions for each electron extending throughout the crystal), the 'London-Heitler-Heisenberg approximation', in which the wave function for each electron vanishes except in the neighbourhood of one atom.

Let $V(r)$ be the potential energy function in the neighbourhood of an atom with its centre at the origin, and $\phi(r)$ the orbital wave function. Let $\mathbf{r}_f$ denote any lattice point; then for an atom with its centre at that lattice point, we write

$$V_f(\mathbf{r}) = V(|\mathbf{r}-\mathbf{r}_f|),$$
$$\phi_f(\mathbf{r}) = \phi(|\mathbf{r}-\mathbf{r}_f|).$$

The wave function of the atom depends also on the Pauli spin variable[†] ζ, defined so that $-H\mu\zeta$ is the energy of the electron in a magnetic field H; then ζ can take two values only, $+1$ if the electron spin points parallel to the field, and -1 if it points antiparallel. Let $u(\zeta)$ be the spin wave function; for the two stationary states of the spin, u will be equal to u_α, u_β defined by

$$u_\alpha(+1) = 1 \qquad u_\beta(+1) = 0,$$
$$u_\alpha(-1) = 0 \qquad u_\beta(-1) = 1.$$

u_α is clearly the wave function of an atom in which the spin is known to point *parallel* to a given magnetic field, and u_β the wave function when the spin is known to point antiparallel.

The complete wave function of an atom with its centre at the lattice point $\mathbf{r}_f$ is thus:

$$\psi_f(\mathbf{r}, \zeta) = \phi_f(\mathbf{r})u(\zeta). \tag{87}$$

Let us now consider a crystal consisting of N such atoms, which are so far apart that the perturbation of one atom by its neighbours may be neglected. We denote the spatial coordinates of the electrons by $\mathbf{r}_1, ..., \mathbf{r}_s, ..., \mathbf{r}_N$; and the spin coordinates similarly by $\zeta_1, ..., \zeta_N$. For a discussion of ferromagnetism we must investigate the energy of the crystal when the spins all point in the same direction; if the energy is then a minimum, the crystal will behave as a ferromagnet.[‡] We therefore first calculate the energy of the crystal when the spins are all pointing in the direction of some external field H, and then investigate the *change* in the energy when a few of the spins are turned in the opposite direction.

In the former case the spins are all in the same state, namely that

---

[†] *Zeits. f. Phys.* **43** (1927), 601.

[‡] Bloch (loc. cit.) has shown that this is not true except for a three-dimensional lattice.

with wave function $u_\alpha(\zeta)$. The wave function for the system as a whole, even neglecting the interaction between the atoms, is therefore non-degenerate. By the Pauli exclusion principle in its wave-mechanical form, the wave function must be antisymmetrical in the coordinates of the electrons; in other words, if we exchange any pair of coordinates such as $\mathbf{r}_1, \zeta_1$ and $\mathbf{r}_2, \zeta_2$, the wave function must change sign.

As is well known, the only antisymmetrical wave function which can be formed from the functions (87) is the determinant

$$\Psi = \begin{vmatrix} \phi_1(\mathbf{r}_1)u_\alpha(\zeta_1) & \phi_1(\mathbf{r}_2)u_\alpha(\zeta_2) & \cdots \\ \phi_2(\mathbf{r}_1)u_\alpha(\zeta_1) & \phi_2(\mathbf{r}_2)u_\alpha(\zeta_2) & \cdots \\ \cdot \quad \cdot \quad \cdot \quad \cdot \quad \cdot \quad \cdot \quad \cdot \quad \cdot \end{vmatrix}, \tag{88}$$

which we therefore take to be the wave function, in the approximation of zero order. It may also be written

$$\begin{vmatrix} \phi_1(\mathbf{r}_1) & \phi_1(\mathbf{r}_2) & \cdots \\ \phi_2(\mathbf{r}_1) & \phi_2(\mathbf{r}_2) & \cdots \\ \cdot \quad \cdot \quad \cdot \quad \cdot \quad \cdot \end{vmatrix} u_\alpha(\zeta_1)u_\alpha(\zeta_2).$$

We calculate the energy W_0 of the whole crystal from the formula

$$W_0 = \sum_\zeta \int \Psi^* H \Psi \, d\tau \Big/ \sum_\zeta \int \Psi^* \Psi \, d\tau, \tag{89}$$

where the summation is over all spin variables and the integration over the spatial coordinates of all the electrons of the lattice. H is here the Hamiltonian function, and is equal to

$$H = -\frac{\hbar^2}{2m}\nabla_s^2 + \sum_f \sum_s V_f(\mathbf{r}_s) + \sum_s \sum_{s'} \frac{e^2}{|\mathbf{r}_s - \mathbf{r}_{s'}|}.$$

In evaluating (89) it is easily seen that $\sum_\zeta u^* u = 1$. In the discussions usually given of this problem it is assumed that the $\phi(\mathbf{r})$ are orthogonal,† i.e. that

$$\int \phi_f^*(\mathbf{r})\phi_{f'}(\mathbf{r}) \, d\tau = 0 \quad (f \neq f'). \tag{90}$$

The denominator in (89) is thus equal to N. In the numerator there occur N equal terms of the type

$$\int \phi_1^*(\mathbf{r}_1)\phi_2^*(\mathbf{r}_2)\ldots H\phi_1(\mathbf{r}_1)\phi_2(\mathbf{r}_2)\ldots d\tau_1 \, d\tau_2\ldots .$$

This term we set equal to $N\epsilon_0$; ϵ_0 is approximately equal to the energy of an unperturbed atom. It is actually equal to the energy of the atoms together with their Coulomb interaction.

† Cf. a recent discussion by Van Vleck, *Phys. Rev.* **49** (1936), 232.

We also obtain $N(N-1)$ 'cross terms' of the type

$$\int \phi_f^*(\mathbf{r}_1)...H\phi_{f'}(\mathbf{r}_1)... \, d\tau_1...; \tag{91}$$

we shall assume these to be zero unless $f = f'$ for all pairs of wave functions occurring except two, these two being wave functions for atoms f, f' which are *nearest neighbours* in the crystal. Then, by virtue of (90), (91) reduces to

$$I = \tfrac{1}{2} \int\int \phi_f^*(\mathbf{r}_1)\phi_{f'}^*(\mathbf{r}_2)\phi_f(\mathbf{r}_2)\phi_{f'}(\mathbf{r}_1) \times -$$
$$\times \left[\frac{2e^2}{|\mathbf{r}_1-\mathbf{r}_2|} + V_f(\mathbf{r}_1) + V_f(\mathbf{r}_2) + V_{f'}(\mathbf{r}_1) + V_{f'}(\mathbf{r}_2) \right] d\tau_1 \, d\tau_2. \tag{92}$$

This is the famous 'exchange integral'.

The energy of the whole crystal is thus

$$W_0 = N(\epsilon_0 - zI), \tag{93}$$

where z is the number of nearest neighbours of each atom.

We now suppose *one* electron to be pointing antiparallel to the field, so that the total magnetic moment of the crystal is $(N-1)\mu$. If the interaction between the atoms be neglected, we can arrange that this electron shall be in any one of the N atoms; thus for the whole crystal there are N independent states with given magnetic moment, all having the same energy. If the spin in the atom f is pointing antiparallel to the field, the wave function of the system as a whole is

$$\Psi_f = \begin{vmatrix} \phi_1(\mathbf{r}_1)u_\alpha(\zeta_1) & ... & \phi_f(\mathbf{r}_1)u_\beta(\zeta_1) & ... & \phi_N(\mathbf{r}_1)u_\alpha(\zeta_1) \\ \phi_1(\mathbf{r}_2)u_\alpha(\zeta_2) & ... & \phi_f(\mathbf{r}_2)u_\beta(\zeta_2) & ... & \phi_N(\mathbf{r}_2)u_\alpha(\zeta_2) \\ \cdot & \cdot & \cdot & \cdot & \cdot \end{vmatrix}. \tag{94}$$

Owing to the interaction between the atoms, however, these N states represented by the wave functions Ψ_f are not *stationary* states of the crystal as a whole. This is because the antiparallel spin will not stay localized in one atom, but will travel from atom to atom. The N *stationary* states with moment $(N-1)\mu$ have wave functions of the form

$$\sum_f a(f)\Psi_f, \tag{95}$$

where the coefficients $a(f)$ must now be determined.

For the moment we limit ourselves to the case of a linear lattice, so that the atoms form an ordered sequence

$$1, 2, ..., f-1, f, f+1, ..., N.$$

If, then, we substitute (94) into the Schrödinger equation

$$(H - W)\Psi = 0,$$

multiply by Ψ_f^* for each f in turn, integrate over all $\mathbf{r}$, and sum over ζ, we obtain, as may easily be seen, the set of equations

$$(W-W_0)a(f) = I[a(f+1)+a(f-1)-2a(f)]. \tag{96}$$

These are the 'secular equations' of the problem. The energy values $W-W_0$ can only be determined uniquely if we assign appropriate boundary conditions to $a(f)$. We assign the usual boundary condition, that $a(f)$, after a large but finite number N of atoms, must repeat itself periodically. If we then set

$$a(f) = a_\xi(f) = e^{i\xi f}, \tag{97}$$

equations (96) are solved with

$$W-W_0 = 2I(1-\cos\xi), \tag{98}$$

where ξ may take any of the values

$$\xi = 2\pi n/N, \quad n \text{ integral}.$$

We obtain thus a *band* of energy levels having the same magnetic moment. The work required to decrease the total magnetic moment from $N\mu$ to $(N-1)\mu$ may thus have any value from practically zero to $4I$.

The wave function (95), namely

$$\Psi = \sum e^{i\xi f}\Psi_f,$$

represents what we may call a 'spin wave', i.e. a state of affairs in which the single reversed spin is travelling through the crystal with wave number ξ/a, where a is the interatomic distance.

This result may easily be generalized to the physically interesting case of three dimensions; we obtain for the three cubic structures, with $\xi = 2\pi n_1/G$, $\eta = 2\pi n_2/G$, $\zeta = 2\pi n_3/G$, n_1, n_2, n_3 integral, and G a large number:

Simple cubic: $G^3 = N$

$$W-W_0 = 2I(3-\cos\xi-\cos\eta-\cos\zeta). \tag{99.1}$$

Body-centred cubic: $G^3 = 2N$

$$W-W_0 = 8I(1-\cos\tfrac{1}{2}\xi\cos\tfrac{1}{2}\eta\cos\tfrac{1}{2}\zeta). \tag{99.2}$$

Face-centred cubic: $G^3 = 4N$

$$W-W_0 = 4I(3-\cos\tfrac{1}{2}\eta\cos\tfrac{1}{2}\zeta-\cos\tfrac{1}{2}\zeta\cos\tfrac{1}{2}\xi-\cos\tfrac{1}{2}\xi\cos\tfrac{1}{2}\eta). \tag{99.3}$$

In all three structures, for small ξ, η, ζ,

$$W-W_0 \simeq I(\xi^2+\eta^2+\zeta^2).$$

If I, the exchange integral, is positive, we see therefore that $W-W_0$

is positive, and the energy of the state with the spins all in the same direction is a minimum.† The crystal is therefore ferromagnetic.

The theory outlined above has been worked out for s electrons, whereas we have seen (cf. § 5) that d wave functions are responsible for ferromagnetism in real metals. The conclusion reached, that I must be positive to give ferromagnetism, will still hold. The exchange integral has been worked out in several cases for s functions (cf. Chap. IV) but not for d functions, for which the algebraical difficulties are considerable. For the cases investigated it has been found to be negative, which agrees with the observed fact that most metals are not ferromagnetic. We can, however, see from general considerations under what conditions it will be positive.‡

In the exchange integral (92) the interaction between the nuclei and the interaction between the electrons give positive contributions; on the other hand, the interaction of the electrons with the nuclei gives a negative contribution. We denote by $\rho(\mathbf{r})$ the charge density defined by
$$\rho(\mathbf{r}) = \phi_1(\mathbf{r})\phi_2(\mathbf{r}).$$

Now the electron interaction
$$e^2 \int\int \rho(\mathbf{r}_1)\rho(\mathbf{r}_2)\frac{1}{|\mathbf{r}_1-\mathbf{r}_2|}\,d\tau_1\,d\tau_2$$

is clearly large, and the interaction of the electrons with the nuclei small, if ρ is small near the nuclei and concentrated in between the atoms. The first condition is fulfilled for high azimuthal quantum number (d and f functions); the second for large interatomic distance, so that, in the region where the atoms overlap, ϕ is decreasing exponentially.

Our conditions for ferromagnetism are therefore:

The atoms must have incomplete shells of high azimuthal quantum number rather far apart. If, however, the shells are too far apart, as in salts of the rare earths, the interaction is too small to give ferromagnetism except perhaps at very low temperatures.‖

To show how these conditions are fulfilled best by iron, nickel, cobalt, we reproduce a table due to Slater.††

---

† Teller (*Zeits. f. Phys.* **62** (1930), 102) has shown that no minimum with lower energy exists.

‡ Sommerfeld and Bethe, loc. cit. 595.

‖ Urbain, Weiss, and Trombe have recently shown that metallic gadolinium is ferromagnetic with a Curie temperature of 16° C. (*Comptes rendus*, **200** (1935), 2132, and **201** (1935), 652).   †† *Phys. Rev.* **36** (1930), 57.

Ratio of interatomic distance to radius of incomplete shell

Metal	Ti	Cr	Mn	Fe	Co	Ni	Pd	Pt	Ce	Yb
Ratio	2·24	2·36	2·94	3·26	3·64	3·96	2·82	2·46	3·2	5·28

7.6. Dependence of the saturation moment on temperature; the $T^{\frac{3}{2}}$ law.
According to the phenomenological theory of Weiss, the work required to decrease the saturation moment of a crystal at the absolute zero of temperature by two Bohr magnetons is a definite amount $2\lambda\mu\sigma_0$ ($\lambda =$ molecular field constant), and hence at low temperatures the magnetization is given by

$$\sigma_0 - \sigma = \sigma_0 e^{-2\lambda\mu\sigma_0/kT}. \tag{100}$$

According to the quantum theory, on the other hand, as formula (99) shows, the work required may have any one of a whole set of values ranging from practically zero to $12I$. This leads to a dependence on temperature at low temperatures quite different from (100), as may be seen qualitatively as follows:

So long as only a small number of electrons have their spins antiparallel to the field, we may assume that for each such electron there exists a spin wave with wave number (ξ_i, η_i, ζ_i) and energy given by (99). At low temperatures only the states with low energies will be occupied, and for these we may write, from (99), for a simple cubic lattice
$$E = W - W_0 = I\rho^2 \quad (\rho^2 = \xi^2 + \eta^2 + \zeta^2).$$
The number of states with energy less than E is, per unit volume,[†]

$$\left(\frac{G}{2\pi}\right)^3 \frac{4\pi}{3} \rho^3 \propto E^{\frac{3}{2}}. \tag{101}$$

Now the spin waves obey the Einstein-Bose statistics, because an exchange of any two of the electrons with spins antiparallel to the field does not result in a new state; therefore the probable number of spin waves in any state ξ, η, ζ with energy E is given by the Einstein-Bose distribution function

$$\frac{1}{e^{E/kT} - 1}, \tag{102}$$

a function which tends rapidly to zero for $E > kT$. Therefore only states with energy below kT are likely to be occupied. Thus by (101) the number of electrons with antiparallel spins is proportional to $T^{\frac{3}{2}}$.

In the same way it follows that the internal energy is proportional

[†] Cf. (99) for G^3 in terms of N.

to $T \times T^{\frac{3}{2}}$, and hence the specific heat due to demagnetization is proportional to $T^{\frac{3}{2}}$.

To obtain exact formulae we proceed as follows:

For the magnetization we have, from (101) and (102),

$$\frac{\sigma_0 - \sigma}{\sigma_0} = \frac{2}{(2\pi)^3} \int_0^\infty \frac{4\pi\rho^2 \, d\rho}{e^{I\rho^2/kT} - 1}$$

$$= \frac{1}{2\pi^2}\left(\frac{kT}{I}\right)^{\frac{3}{2}}$$

$$= 0 \cdot 1323(kT/I)^{\frac{3}{2}} \tag{103}$$

for the simple cubic structure; for the other cubic structures we have

$$\frac{\sigma_0 - \sigma}{\sigma_0} = 0 \cdot 0661(kT/I)^{\frac{3}{2}} \qquad \text{b. c.}$$

$$= 0 \cdot 0331(kT/I)^{\frac{3}{2}} \qquad \text{f. c.}$$

Similarly for the internal energy, we have

$$U = \frac{G^3}{(2\pi)^3} \int_0^\infty \frac{I\rho^2 4\pi\rho^2 \, d\rho}{e^{I\rho^2/kT} - 1}$$

$$= 0 \cdot 045 G^3 kT(kT/I)^{\frac{3}{2}},$$

where G^3 is given in terms of the number of atoms by (99). Hence for the heat capacity due to demagnetization we have

$$0 \cdot 113 G^3 k (kT/I)^{\frac{3}{2}},$$

or, in terms of the magnetization,

$$C_v = 1 \cdot 18 R \frac{\sigma_0 - \sigma}{\sigma} \text{ cal. degree per gm. atom,} \tag{104}$$

the latter formula being valid for all structures at low temperatures.

The formula (103), viz.

$$\sigma_0 - \sigma = \text{const. } T^{\frac{3}{2}} \quad (T \ll \Theta),$$

is the only quantitatively exact result which has been deduced from quantum mechanics in the field of ferromagnetism. It is in much better agreement with experiment than the Weiss law (100), and for iron between 20° and 90° K., according to Weiss, Forrer, and Fallot,[†] it fits the observed values better than a T^2 law.

As regards the numerical magnitudes, Weiss and Forrer[‡] find at 288° K.

† *Bulletin d. Soc. Française de Physique* (1934), 122.
‡ *Ann. d. Physique*, **12** (1929), 279.

Element	Structure	$\dfrac{\sigma_0 - \sigma}{\sigma_0}$	I (electron volts)
Ni	f.c.c.	0·054	0·018
Fe	b.c.c.	0·018	0·059

The values of I, the exchange integral, deduced are also shown.

The term in the specific heat proportional to $T^{\frac{3}{2}}$ has not yet been observed, being masked by the term proportional to T discussed in § 5.2. The latter term does not arise with the simple model used here, in which there is only one electron per atom, in an s state. The reason for this is easy to see if we consider the metal from the point of view of the Bloch model; at the absolute zero of temperature, *all* the states in the s zone with one spin direction are full, and all the states with the other spin are empty. A term in the specific heat proportional to T only arises when a zone is partly full.

When we come to extend the theory to the case where the numbers of spins pointing in the two directions are comparable, i.e. to phenomena near and above the Curie temperature, the mathematical difficulties are very great and have not yet been overcome.[†] In this connexion also it would be of interest to investigate the case when I is negative (paramagnetism) and to see whether the results obtained by the Bloch model ($c_v \propto T$, $\chi = A + BT^2$...) remain valid.

It may be shown that there are ${}^N C_{\frac{1}{2}N+n} - {}^N C_{\frac{1}{2}N+n+1}$ states with given total magnetic moment $2n\mu$, and these form a band; further that the mean energy of the band is

$$\overline{W - W_0} = \tfrac{1}{2} I z (\tfrac{1}{4} N^2 - n^2)/N,$$

but it has not yet proved possible to determine the lowest state or density of states within the bands.

Heisenberg in his original paper[‡] assumed for the density a Gaussian distribution about the mean value, which gives a behaviour very similar to that of the Weiss theory, but does not give the $T^{\frac{3}{2}}$ law at low temperatures. Ludloff[||] has more recently given a treatment which approximates to that of Bloch at low temperatures and to that of Heisenberg at high temperatures, and which moreover gives a paramagnetic Curie point differing slightly from the ferromagnetic as is observed.

† A solution for a one-dimensional lattice has been given by Bethe, *Zeits. f. Phys.* **71** (1931), 205, but it has not been possible to generalize his solution for three dimensions.

‡ Ibid. **49** (1928), 619. || Ibid. **91** (1934), 742.

Bloch[†] has also investigated the magnetic behaviour of a metal, assuming the wave function of each electron to be a plane wave extending throughout the crystal; he finds that under certain circumstances such a model gives ferromagnetism. This conclusion has, however, been criticized by Wigner[‡] (cf. Chap. IV, § 2.2).

Møller[||] has extended the Heisenberg theory to the case when each atom has more than one electron. Wolf[††] has given a quantum-mechanical discussion of the magneton numbers above and below the Curie point; this paper has already been referred to. Fay[‡‡] has discussed a more accurate distribution function than that of Heisenberg.

Bloch and Gentile[||||] have given a theoretical discussion of the dependence of magnetization on direction in a single metal crystal; they find that the effect can be accounted for by the interaction between an electron's spin and its orbital motion.

[†] *Zeits. f. Phys.* **57** (1929), 545.   [‡] *Phys. Rev.* **46** (1934), 1002.

[||] *Zeits. f. Phys.* **82** (1933), 559.   [††] Ibid. **70** (1931), 519.

[‡‡] *Proc. Nat. Ac. Sc.* **21** (1935), 537.   [||||] *Zeits. f. Phys.* **70** (1931), 395.

Note of recent developments

Slater (*Phys. Rev.* **49** (1936), 537) has given a discussion of ferromagnetism on the basis of the Bloch theory, and has shown that it is most likely to occur in the iron group of metals. In a further paper (ibid. 931), he discusses the dependence of magnetization on temperature. He finds that, at the absolute zero of temperature, the saturation moment expressed in Bohr magnetons per atom (0·6 for nickel), and the number of positive holes in the d band are not identical. For nickel, however, the difference is small ($\sim 6\%$), but may be larger for iron (cf. p. 222).

As we saw on p. 224, the behaviour of nickel above the Curie point suggests that the number of spins is slightly greater than 0·6.

THE ELECTRICAL RESISTANCE OF METALS AND ALLOYS

A THEORY of metallic conduction has to explain, among others, the following experimental results:

(1) The Wiedemann-Franz law, which states that the ratio of the thermal to the electrical conductivity is equal to LT, where T is the absolute temperature and L is a constant which is the same for all metals.

(2) The absolute magnitude of the electrical conductivity of a pure metal, and its dependence on the place of the metal in the periodic table; e.g. the large conductivities of the monovalent metals and the small conductivities of the transition metals.

(3) The relatively large increase in the resistance due to small amounts of impurities in solid solution, and the Matthiessen rule, which states that the change in the resistance due to a small quantity of foreign metal in solid solution is independent of the temperature.

(4) The dependence of the resistance on temperature and on pressure.

(5) The appearance of supraconductivity.

With the exception of (5), the theory of conductivity based on quantum mechanics has given at least a qualitative understanding of all these results.

1. Former theories

Shortly after the discovery of the electron Riecke,[†] Drude,[‡] Lorentz,[||] and others[††] recognized that a current in a metal is carried by electrons, and developed a theory of metallic conductivity on this basis, an outline of which we shall now give. We consider a metal containing N electrons per unit volume, and suppose that each electron can move quite freely for a mean time 2τ, after which it suffers a collision and its momentum is destroyed. τ is called the 'time of relaxation'. During the time 2τ the equation of motion in an external field F is

$$m\ddot{x} = eF.$$

[†] *Ann. d. Phys. u. Chem.* **66** (1898), 353 and 545.
[‡] *Ann. d. Physik,* **7** (1902), 687, where earlier refs. are given.
[||] *Theory of Electrons,* Leipzig (1909).
[††] Cf., for instance, Grüneisen, *Handb. d. Phys.* **13** (1928), 65; or Richardson, *Electron Theory of Matter,* p. 406, Cambridge (1916).

Our hypothesis is that immediately after a collision the mean value of the component of the velocity of the electron in the direction of the field is zero; after a time 2τ it is therefore $2eF\tau/m$.

Taking a time average, we obtain for the mean velocity of drift in the direction of the field

$$eF\tau/m.$$

The current j is obtained by multiplying by Ne, and hence for the conductivity we have

$$\sigma = j/F = Ne^2\tau/m. \tag{1}$$

The authors quoted assumed further that the electrons in a metal behaved like a perfect gas, so that, if u^2 is the mean of the square of the velocity of an electron,

$$\tfrac{1}{2}mu^2 = \tfrac{3}{2}kT. \tag{2}$$

With the value of u so obtained, they defined the *mean free path l* by

$$l = 2\tau u.$$

In terms of the mean free path the conductivity is

$$\sigma = Ne^2l/2mu.$$

This formula gives the right order of magnitude for the conductivity of, say, silver *at room temperature*, if one assumes that N is of the order of magnitude of the number of atoms per unit volume and l of the interatomic distance. On the other hand, to account for the fact that the conductivity is inversely proportional to the temperature, whereas the mean square velocity u^2, according to (2), is proportional to T, it was necessary to assume that l increased to very much larger values at low temperatures, which was difficult to understand from the classical point of view. Moreover, under pressure the conductivity of most metals increases, whereas one would expect that, as the atoms are pressed closer together, the mean free path and hence the conductivity would decrease.

A further difficulty of the theory was that, according to (2), the electrons should contribute an amount $\tfrac{3}{2}Nk$ to the heat capacity per unit volume, or, to the atomic heat, $\tfrac{3}{2}R \times$ number of free electrons per atom. The measurements of the specific heat of good conductors (cf. Chap. I, § 1) made it certain that, at room temperature, any contribution to the atomic heat due to the free electrons was very much less than R. It was therefore necessary to assume that only a very small proportion of the atoms were 'ionized'.

The most important success of the gas-kinetic theory was the explanation of the law of Wiedemann and Franz. Drude assumed that the conduction of heat in good conductors was due to the motion of the free electrons and obtained, on the same assumptions, the following formula for the thermal conductivity κ:

$$\kappa = \tfrac{1}{2}Nluk. \tag{3}$$

From (1) and (3) one obtains, by (2),

$$\frac{\kappa}{\sigma}\frac{1}{T} = 3\left(\frac{k}{e}\right)^2,$$

showing that the quantity on the left is independent of temperature and the same for all metals. This result is in fair agreement with experiment for a number of metals.[†]

Prior to the introduction of wave mechanics the most important advance in the theory was that of Wien.[‡] Wien assumed that the mean velocity u of the electrons in a metal was independent of temperature, and that the mean free path was inversely proportional to the mean square of the amplitude, X, of the atomic oscillations, so that

$$l \propto 1/X^2.$$

These assumptions, later supported by quantum mechanics, enabled an account to be given of the change of resistance with temperature both above and below the Debye characteristic temperature. On the same basis Grüneisen[||] was able to explain the change of resistance under pressure.

Sommerfeld[††] was the first to apply the ideas of quantum mechanics to problems of metallic conduction. The advance made by Sommerfeld was the application of the Fermi-Dirac statistics to the electrons in a metal. As we saw in Chap. VI, § 2, he was thus able to explain why the *specific heat* of the conduction electrons is negligibly small.

In the theory of Sommerfeld the field acting on an electron due to the ions and to the other electrons is neglected.[‡‡] Each electron, therefore, is accelerated by the applied field, just as in the classical theory, and the formula for the conductivity is given as before by equation (1). It should be noticed that if one introduces the conception of the 'mean free path' l, defined as before by

$$l = 2\tau u,$$

then u, the mean velocity, is much greater than in the gas-kinetic

† Cf. § 15.　　　　　　　　　‡ *Preuss. Akad. Wiss. Berlin, Sitz. Ber.*, 1913, p. 184.
|| *Verh. d. Deuts. Phys. Ges.* **15** (1913), 186.
†† *Zeits. f. Phys.* **47** (1928), 1.　　　　　　　　‡‡ Cf. Chap. II, §§ 2 and 3.

theory of Drude, because the mean kinetic energy, according to the Fermi-Dirac statistics, is much greater than $\frac{3}{2}kT$.

In Sommerfeld's papers no theory was given by which τ and its dependence on temperature could be calculated.

Bloch[†] made two important advances in the theory. Firstly, he introduced the conception of electrons moving in an electrostatic field having the periodicity of the lattice. As we have seen in Chap. III, § 1, this model makes it clear why some solids are insulators and others conductors and how metals can have a non-integral effective number of 'free' electrons per atom. Secondly, he gave a quantum-mechanical justification of Wien's hypothesis that the cause of electrical resistance lies in the heat motion of the metal atoms. Bloch's theory and its extension by other workers will be explained in the next sections.

2. Dependence of resistance on temperature; qualitative discussion

Since the detailed theory discussed in the following sections is rather complicated we shall give first an elementary discussion of the dependence of resistance on temperature.

As we have seen in Chapter III, in a perfectly periodic lattice a beam of electrons moving in a given direction will continue to move in that direction indefinitely. *A perfect lattice has therefore no resistance whatever.* If, however, the lattice is not perfectly periodic, the electrons will eventually be scattered. It is from this scattering that resistance arises. To calculate the resistance, therefore, our problem, just as in the classical theory, is to obtain the probability of scattering and hence the time between collisions 2τ and the mean free path l.

The departures from periodicity in the lattice, which give rise to the resistance, may be due to

(1) The displacement of the atoms from their mean positions due to their thermal motion.

(2) The presence of foreign atoms in solid solution.

(3) The break-down of the lattice in the liquid and amorphous states.

(2) and (3) are dealt with in §§ 12 and 10; we shall deal here only with the thermal motion.

We shall take an Einstein model for the metal crystal, supposing each atom to be able to vibrate about its mean position with frequency ν; then the Einstein characteristic temperature is defined (cf. Chap. I, § 1) by

$$h\nu = k\Theta. \tag{4}$$

We denote the restoring force when an atom is displaced a distance X from its mean position by $-bX$, so that the equation of motion of an atom is

$$M\ddot{X} + bX = 0,$$

where M is the mass of the atom. Therefore

$$b/M = 4\pi^2\nu^2. \tag{5}$$

If $\overline{X^2}$ denotes the mean square of the displacement in a given direction, we have

$$\tfrac{1}{2}b\overline{X^2} = \text{mean potential energy} = \tfrac{1}{2}kT \tag{6}$$

for temperatures above the characteristic temperature; for lower temperatures†

$$\tfrac{1}{2}b\overline{X^2} = \frac{\tfrac{1}{2}h\nu}{e^{h\nu/kT} - 1}. \tag{7}$$

Now, as we have stated above, an electron may be scattered by a displaced atom, and it will be shown below that the probability for scattering is proportional to X^2, i.e. to the square of the displacement. Since the resistance R of a metal is proportional to the scattering probability, we may write

$$R \propto \overline{X^2}. \tag{8}$$

But, from (4), (5), and (6), for $T \gg \Theta$

$$\overline{X^2} = kT/b = h^2T/4\pi^2Mk\Theta^2, \tag{9}$$

whence, from (8), $R \propto T/M\Theta^2$.

We thus obtain the result that at high temperatures the *resistance is proportional to the absolute temperature*. The dependence of R on Θ will be used in § 3 to compare the resistivities of different metals and in § 8 to obtain the pressure coefficient of the resistance.

At lower temperatures $(T < \Theta)$ the mean square displacement drops below the value (6). According to (7), we should expect the resistance to be proportional to the total energy. This prediction is only in rough agreement with experiment; we must remember that at low temperatures, according to the Debye theory of specific heats,

† It is not actually possible to separate potential and kinetic energy; the expression on the right is half the total energy. The zero-point energy is omitted as it has no influence on the resistance (cf. § 5.4).

the atoms are not displaced at random from their mean positions as assumed above (cf. Chap. I, § 1). The resistance at very low temperatures is discussed further in § 9.

Fig. 93 shows, for the two metals copper and lead, ρ/T and C_p plotted against T, where ρ is the resistivity. It will be seen that, at

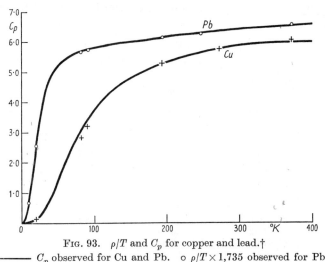

Fig. 93. ρ/T and C_p for copper and lead.†
——— C_p observed for Cu and Pb. o $\rho/T \times 1{,}735$ observed for Pb.
+ $\rho/T \times 1{,}570$ observed for Cu.

temperatures $T \sim \Theta$, there is in fact a close correspondence between the energy and the resistance. It is even possible to deduce the characteristic temperature Θ from the resistance curves, as the following values† show (cf. also p. 14):

		Pb	Au	Pt	Ag	Cu
Θ_D	from specific heat	88	180–90	230	215	315–25
	from resistance	92	190–200	230	230	346–75

3. Dependence of the resistance on the position of the metal in the periodic table

We have seen that the resistance of a pure metal is proportional to the mean square of the amplitude of the thermal oscillations of its atoms, i.e. to $T/M\Theta^2$. The quantity $\sigma T/M\Theta^2$, therefore, represents the conductivity for given amplitude of thermal oscillation. The

† From Grüneisen, *Handb. d. Phys.* **13** (1928), 22.

other factors which determine the resistance do not depend on the elastic properties of the metal, and hence we should compare the values of $\sigma T/M\Theta^2$ rather than the values of the conductivity σ itself if we wish to discuss the influence on the conductivity of the crystal structure, atomic volume, etc.

Table X shows the values at $0°$ C. of $\sigma/M\Theta^2$ for all the metals for which the resistance has been measured and is a normal function of temperature.

TABLE X

Metal	Conductivity at $0°$ C. $\sigma \times 10^{-4}$ ohm cm.	M	Θ	$\sigma/M\Theta^2 \times 10^2$
Li	11·8	6·940	363	12·9
Be	18	9·02	1,000	2·0
Na	23	22·997	202	24
Mg	25	24·32	357	8·1
Al	40	26·97	395	9·5
K	15·9	39·096	163	15·3
Ca	23·5	40·08	230	11·1
Ti	1·2	47·90	342	0·21
Cr	6·5	52·01	495	0·51
Fe	11·2	55·84	420	1·14
Co	16	58·94	401	1·7
Ni	16	58·69	375	1·9
Cu	64·5	63·57	333	9·1
Zn	18·1	65·38	213	6·1
Ga	2·45	69·72	125	2·25
As	2·85	74·91	291	0·45
Rb	8·6	85·44	85	14
Sr	3·3	87·63	171	1·3
Zr	2·4	91·22	288	0·32
Mo	23	96·0	380	1·7
Ru	8·5	101·7	426	0·46
Rh	22	102·91	370	1·6
Pd	10	106·7	270	1·3
Ag	66·7	107·880	223	12·4
Cd	15	112·41	172	4·5
In	12	114·76	198	2·7
Sn	10	118·70	260	1·2
Sb	2·8	121·76	140	1·2
Cs	5·6	132·91	54	14
Ba	1·7	137·36	113	1·0
La	1·7	138·92	152	0·53
Ce	1·4	140·13	..	..
Pr	1·6	140·92	..	..
Hf	3·4	178·6	213	0·42
Ta	7·2	180·88	228	0·77
W	20	184·0	333	1·0

TABLE X (cont.)

Metal	Conductivity at $0°$ C. $\sigma \times 10^{-4}$ ohm cm.	M	Θ	$\sigma/M\Theta^2 \times 10^2$
Os	11	191·5	256	0·9
Ir	20	193·1	**316**	1·0
Pt	10·2	195·23	**240**	0·91
Au	49	197·2	**175**	8·1
Hg	4·4†	200·61	80	3·4
Tl	7·1	204·39	**140**	1·8
Pb	5·2	207·22	**86**	3·4
Bi	1·0	209·00	100	0·5

The experimental values of σ are taken from Grüneisen, *Handb. d. Phys.* **13** (1928), 11, and from Borelius, *Handb. d. Metallphysik*, **1** (1935), 321. The values of Θ shown in bold-face type are values deduced from the electrical resistance in the range $T \sim \Theta$ by Grüneisen, *Ann. d. Physik*, **16** (1933), 530 (cf. § 2). The others are taken from Table III.

The *periodic* change with atomic number of the series of values obtained is striking; in particular:

I. The large values of $\sigma/M\Theta^2$ for the noble metals and alkalis, which have one electron outside a closed shell.‡ $\sigma/M\Theta^2$ drops in most cases by a factor between 2 and 4 on passing to the divalent metals next to them in the periodic table. We shall see in § 6.3 that this is due to the small 'effective number of free electrons' in the divalent metals.

II. The low values of $\sigma/M\Theta^2$ for the transition metals. We shall see in § 6.3 that the incomplete d shells in these metals give rise to a much larger scattering probability and hence a shorter mean free path than for the metals with closed cores.

4. Detailed theory of conductivity; general outline

We shall now proceed to a more detailed calculation of the conductivity.

We have seen (§ 1) that according to the classical theory the conductivity σ is given by

$$\sigma = Ne^2\tau/m, \qquad (10)$$

where N is the number of electrons per unit volume and 2τ the time

† Solid mercury at $0°$ C. under a pressure of $7,640$ kg./cm.² (Bridgman, *Proc. Amer. Acad. Arts Sci.* **56** (1921), 99).

‡ The elements copper and gold are not always monovalent, but there is strong evidence that in the metallic state there is only one electron outside a closed d shell (cf. Appendix I).

between collisions. In the quantum theory the formula (10) is still qualitatively valid, but with the following modifications:

(1) As was shown in Chapter III, between collisions the acceleration of an electron in a lattice due to an external field F is in general less than for a free electron; we therefore introduced the 'effective' number N_{eff} of free electrons per unit volume, defined so that $N_{\text{eff}} e^2 F \delta t / m$ is the current produced by a field F in time δt. We must therefore replace N in formula (10) by N_{eff}.

(2) τ must be calculated by a quantum-mechanical method, on the basis of the ideas of § 2.

To obtain τ we shall calculate the probability $\bar{P}$ per unit time that an electron suffers a collision; we shall then have

$$\tau \sim 1/\bar{P}.$$

The calculation of $\bar{P}$ is thus the main purpose of this chapter.

Two methods are possible for the calculation of $\bar{P}$; we may use for the lattice vibrations either an Einstein model or a Debye model.† With the former method we treat each atom as vibrating independently of all the others, and calculate the scattering by each atom separately. With the Debye model we analyse the heat motion of the crystal into sound waves, and calculate the probability that an electron is deflected through its interaction with each sound wave. The Einstein model has the advantage of simplicity, and yields a formula which is very convenient for comparison with experiment; it will, however, give incorrect results at low temperatures ($T \ll \Theta$), when the Debye model must be used.

When an electron moving in a metal is deflected, there is a transfer of momentum and of energy between the electron and the lattice. Since the atoms of the lattice are much heavier than the electron, the energy lost or gained by the electron is much less than its total energy. We shall see (§ 5.3) that the energy lost or gained is less then $k\Theta$.

We assume, of course, that the electrons obey the Fermi-Dirac statistics. An important consequence of this assumption is that *only electrons in states with nearly the maximum energy can actually be scattered*; the other electrons cannot be scattered at all, because all states with nearly equal energy are already occupied. Thus we need

† The Debye model was used by Bloch in his original papers on this subject.

only calculate the scattering probability for electrons of nearly the maximum energy $E_{\max}$.

The calculation which follows consists of two parts; we first find the probability $P(\mathbf{kk'})$ that the electron makes a transition from a state $\mathbf{k}$ to a state $\mathbf{k'}$, and then use these transition probabilities to calculate the current.

5. Calculation of the probability of scattering

As we have already seen, an electron in the lattice in the state $\mathbf{k}$ may, under the influence of any irregularity in the crystalline field, make a transition to any other state $\mathbf{k'}$. We shall first of all treat the irregularity as *static*, so that the electron can only make transitions to states of the same energy. This treatment is obviously appropriate when the irregularity is due to foreign atoms in solid solution. When it is due to the thermal vibration of the atoms of a pure crystalline metal, the field in the crystal is continually changing; but since the velocity with which the atoms vibrate is small compared with that of the electrons, we shall obtain the correct transition probabilities if we treat the atoms as momentarily at rest in their displaced positions. This is proved in § 5.3.

Let then $U(x, y, z)$ denote the *difference* between the actual potential in the lattice and the potential $V(x, y, z)$ that would exist in the perfectly regular lattice. In this section we shall treat U as small, and shall obtain the transition probabilities by the perturbation method of Dirac. The use of a perturbation will certainly not lead to serious error for the thermal part of the resistance, since at ordinary temperatures the displacement of the atoms is small compared with the interatomic distance; for foreign atoms in solid solution, however, the perturbation method leads in certain cases to incorrect conclusions; for this case an alternative method is given in § 12.2.

The Schrödinger equation of an electron is

$$i\hbar \frac{\partial \Psi}{\partial t} = H\Psi + U\Psi, \tag{11}$$

where H is the Hamiltonian for an electron in the unperturbed lattice. Let E_k and $\psi_k(\mathbf{r})$ be the energies and wave functions for an electron in the unperturbed lattice; we shall assume that ψ_k is normalized so that

$$\int \psi_k^* \psi_k \, d\tau = 1,$$

where the integration is over unit volume; if, therefore, we are

3595.17

considering electrons in a volume Ω, we must take for our normalized wave function $\Omega^{-\frac{1}{2}}\psi_k$. We write for the wave function at time t

$$\Psi = \Omega^{-\frac{1}{2}} \sum_{k'} a_{k'}(t)\psi_{k'}(\mathbf{r})e^{-iE_{k'}t/\hbar}. \tag{12}$$

$|a_{k'}|^2$ then denotes the probability that the electron is in the state $\mathbf{k}'$ at time t. Initially, at time $t = 0$, we shall assume that the electron is in the state $\mathbf{k}$, so that

$$\left.\begin{aligned} a_k(0) &= 1, \\ a_{k'}(0) &= 0 \quad (\mathbf{k}' \neq \mathbf{k}). \end{aligned}\right\} \tag{13}$$

We substitute (12) into (11), multiply by $\psi_k^* e^{iE_k t/\hbar}$, and integrate over the volume Ω; neglecting terms of the second order (the product $Ua_{k'}$), we obtain

$$i\hbar \frac{da_{k'}}{dt} = \frac{1}{\Omega} U_{kk'} e^{i(E_{k'}-E_k)t/\hbar},$$

where

$$U_{kk'} = \int U\psi_k^* \psi_k \, d\tau.$$

Integrating with respect to t and using the initial conditions (13), we obtain

$$a_{k'}(t) = \frac{1}{i\hbar} \frac{e^{ixt}-1}{ix} \frac{U_{kk'}}{\Omega},$$

where x is written for $(E_{k'}-E_k)/\hbar$; the probability that after a time t the electron is in the state $\mathbf{k}'$ is thus

$$|a_{k'}|^2 = \frac{1}{\hbar^2} \frac{1}{\Omega^2} |U_{kk'}|^2 \frac{2(1-\cos xt)}{x^2}.$$

The function on the right has for large t a strong maximum at $x = 0$, i.e. for transitions in which energy is conserved.

The energies E_k lie so close together as to form virtually a continuum. As is usual in quantum-mechanical problems of this type, we must integrate the transition probability over a large number of final states, for all of which $\mathbf{k}'$ has nearly the same value, in order to obtain a result of physical significance.

Fig. 94 shows the 'k-space' of the electrons; the points A and B represent respectively the initial and final states $\mathbf{k}$ and $\mathbf{k}'$, and the curved line represents the section made with the plane of the paper by the surface for which $E_{k'} = E_k$.

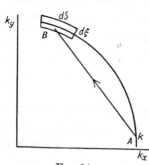

FIG. 94.

Since our initial state is assumed to be near the surface of the Fermi distribution, the curved line represents this surface. Then the physical quantity which we require is the probability per unit time, $P(\mathbf{kk'})\,dS$, that the electron makes a transition to a state lying in an area dS of the Fermi distribution. To calculate this probability we take a small volume $dS\,d\xi$ as shown in the figure. The number of states in this volume is $\dfrac{\Omega}{8\pi^3}\,dS\,d\xi$, and the probability that after a time t the electron is in a state within it is obtained by multiplying this number by $|a_{k'}|^2$. We then integrate the probability across the surface of the Fermi distribution, and obtain

$$P\,dS = \frac{dS}{8\pi^3\hbar^2\Omega}\,\frac{\partial}{\partial t}\int |U_{kk'}|^2\,\frac{2(1-\cos xt)}{x^2}\,d\xi.$$

For points in the neighbourhood of the surface of the Fermi distribution we may write

$$x = \frac{1}{\hbar}\,\frac{\partial E}{\partial k_n}\,\xi.$$

We may, moreover, take t to be large compared with $\hbar/E$. The integrand then has a strong maximum in the neighbourhood of $\xi = 0$, and we obtain for the integral

$$2\pi t\hbar|U_{kk'}|^2\Big/\frac{\partial E}{\partial k_n},$$

and hence $$P(\mathbf{kk'})\,dS = \frac{1}{4\pi^2\hbar}\,\frac{dS}{\Omega}\,|U_{kk'}|^2\Big/\frac{\partial E}{\partial k_n}. \tag{14}$$

This is the required transition probability, giving the probability per unit time that an electron makes a transition from a state $\mathbf{k}$ to an element dS of the surface in k-space having the same energy.

From (14) we note that, apart from the matrix element $U_{kk'}$, which depends on the nature of the perturbation, the transition probability is proportional to

$$dS\Big/\frac{\partial E}{\partial k_n},$$

and hence (cf. Chap. II, § 4.6) to the density of states at the surface of the Fermi distribution.

It is of interest to discuss the result (14) for the case of free electrons, i.e. for the case when the influence of the lattice field is negligible. In this case we may think of the electron as having a velocity $\mathbf{v}$ in the same direction as the wave vector $\mathbf{k}$, and ask for the probability per unit time that it is deflected through an angle θ into a solid angle $d\omega$. With $dS = k^2 d\omega$ and $\mathbf{k} = m\mathbf{v}/\hbar$, (14) gives

$$P \, dS = \frac{v}{\Omega} \left| \frac{2\pi m}{h^2} \int e^{i(\mathbf{k}-\mathbf{k}' \cdot \mathbf{r})} U(x,y,z) \, d\tau \right|^2 d\omega. \quad (15)$$

In the ordinary collision theory it is usual to express the collision probabilities as *areas*; since (15) refers to one electron per volume Ω moving with velocity v, we see that the effective area that an electron must hit if it is to be scattered into a solid angle $d\omega$ is

$$I(\theta) \, d\omega = \frac{P\Omega}{v} \, dS = \left| \frac{2\pi m}{h^2} \int e^{i(\mathbf{k}-\mathbf{k}' \cdot \mathbf{r})} U(x,y,z) \, d\tau \right|^2 d\omega. \quad (16)$$

This is, of course, just the formula given by the usual Born collision theory.[†]

5.1. *Calculation of matrix element for temperature resistance.* We have now to calculate the term $U_{kk'}$ in (14). The resistance due to foreign atoms in solid solution will be treated in § 12; we shall discuss here the resistance due to the thermal motion of the atoms.

We shall first calculate $U_{kk'}$ using an Einstein model for the heat motion; that is to say, we shall consider that a single atom is displaced from its mean position by a vector (X, Y, Z), all the other atoms being undisplaced. We shall then assume that, within the atomic polyhedron[‡] of the displaced atom, the potential at a given distance from the nucleus is the same as it would be if the atom were not displaced.[||] Outside the polyhedron of the displaced atom we shall assume the potential to be unchanged. This will certainly be a good approximation near the displaced nucleus within the inner shells of the atom, where the field is strongest. Near the boundaries of the polyhedron it may not be so satisfactory. The field outside the polyhedron of the displaced atom could in principle be calculated by a method similar to that of Chap. II, § 5, and will fall off exponentially with increasing distance. Its neglect will lead to too small values of the resistance (cf. p. 265).

It follows from these assumptions, if $V(x, y, z)$ is the original potential within the polyhedron, that the change U in the potential in the lattice when the atom is displaced is given by

$$U = V(x-X, y-Y, z-Z) - V(x, y, z)$$

[†] Cf. Mott and Massey, *The Theory of Atomic Collisions*, pp. 87 and 88, equations (1) and (5).   [‡] Cf. Chap. II, § 4.5.

[||] This is the assumption of 'Starre Ionen', introduced by Nordheim, *Ann. d. Physik*, **9** (1931), 607.

within the polyhedron considered, and zero outside. For small displacements (X, Y, Z) this may be written

$$-U = X\frac{\partial V}{\partial x} + Y\frac{\partial V}{\partial y} + Z\frac{\partial V}{\partial z}. \tag{17}$$

Since V has a singularity at the nucleus, the expansion is not valid at regions near the nucleus, but such regions may easily be seen to make a negligible contribution to the integral in (18) below. Using (17), therefore, we have

$$-U_{kk'} = X\int \psi_{k'}^{*}\frac{\partial V}{\partial x}\psi_{k}\, d\tau + \dots . \tag{18}$$

Now for the mean value of X^2 we obtained (equation (9))

$$\overline{X^2} = \frac{kT}{M\Theta^2}\frac{\hbar^2}{k^2} \tag{19}$$

and, further, $\overline{XY} = 0$. Hence the mean value of $|U_{kk'}|^2$ is

$$\frac{kT}{M\Theta^2}\left|\frac{\hbar}{k}\int \psi_{k'}^{*}\,\mathrm{grad}\,V\psi_{k}\, d\tau\right|^2. \tag{20}$$

To obtain the transition probability for the whole metal we substitute (20) into (14) and multiply by the number of atoms in the volume Ω. We obtain

$$P\, dS = \frac{\hbar}{4\pi^2 k}\frac{dS}{\Omega_0}\frac{T}{M\Theta^2}\left|\int \psi_{k'}^{*}\,\mathrm{grad}\,V\psi_{k}\, d\tau\right|^2 \Big/ \left(\frac{\partial E}{\partial k_n}\right)_{k=k'},$$

where $\Omega_0 = \Omega/N$ is the volume per atom.

The integral $\qquad \int \psi_{k'}^{*}\,\mathrm{grad}\,V\psi_{k}\, d\tau, \tag{21}$

which is to be taken over one atomic polyhedron, may be transformed into a surface integral over the surface of the polyhedron by making use of the Schrödinger equation

$$\frac{\hbar^2}{2m}\nabla^2\psi_{k} + (E_k - V)\psi_{k} = 0,$$

and remembering that $E_k = E_{k'}$. We write

$$\psi_{k'}^{*}\psi_{k}\,\mathrm{grad}\,V = \mathrm{grad}(V\psi_{k'}^{*}\psi_{k}) - V\psi_{k}\,\mathrm{grad}\,\psi_{k'}^{*} - V\psi_{k'}^{*}\,\mathrm{grad}\,\psi_{k}.$$

The right-hand side, by means of the Schrödinger equation, is easily transformed into

$$\mathrm{grad}\left[(V - E_k)\psi_{k'}^{*}\psi_{k} - \frac{\hbar^2}{2m}\psi_{k}\nabla^2\psi_{k'}^{*}\right] + \frac{\hbar^2}{2m}[\psi_{k}\nabla^2\,\mathrm{grad}\,\psi_{k'}^{*} - \nabla^2\psi_{k}\,\mathrm{grad}\,\psi_{k'}^{*}].$$

The first term vanishes. Integrating the second, we obtain for (21)

$$\frac{\hbar^2}{2m} \int \left(\psi_{k'}^* \frac{\partial}{\partial n} \operatorname{grad} \psi_k - \operatorname{grad} \psi_k \frac{\partial}{\partial n} \psi_{k'}^* \right) dS, \tag{22}$$

the integration being over the surface of the polyhedron and $\partial/\partial n$ denoting differentiation normal to the surface. The scattering probability depends, therefore, on the wave functions at the surface of the atomic polyhedron, as one might expect, and not on the field in the interior of the atom.

For monovalent metals, for which we may use the approximation of Wigner and Seitz (p. 80), and write

$$\psi_k = u(r)e^{i(\mathbf{kr})}, \qquad u(r_0) \simeq 1, \qquad u'(r_0) = 0,$$

(22) may easily be evaluated. Since, if u were a constant, (22) would vanish, we obtain for (22)

$$\frac{\hbar^2}{2m} \int e^{-i(\mathbf{k'r})} u''(r) \operatorname{grad} r e^{i(\mathbf{kr})} \, dS.$$

$u''(r)$ may be taken to be constant and set equal to

$$2m[V(r_0) - E_0]u(r_0)/\hbar^2$$

over the surface of the polyhedron; the integral may then be evaluated and gives finally for (22)

$$4\pi r_0^2 \cos\beta [V(r_0) - E_0] \frac{\sin x - x \cos x}{x^2}, \tag{23}$$

where $x = |\mathbf{k} - \mathbf{k}'| r_0$ and β is the angle between the vectors grad V and $\mathbf{k} - \mathbf{k}'$. Note that the transition probability is then a function of $|\mathbf{k} - \mathbf{k}'|$ only. It depends also on the kinetic energy, $V - E_0$, of an electron in the lowest state ($\mathbf{k} = 0$), at the boundary of the atomic polyhedron.

Making the further approximation $E = \hbar^2 k^2/2m$ as for free electrons, so that $kr_0 = (9\pi/4)^{\frac{1}{3}}$, we obtain for the transition probability in terms of the angle θ between $\mathbf{k}$ and $\mathbf{k}'$

$$P \, dS = \frac{3}{\pi} \left(\frac{9\pi}{4} \right)^{\frac{1}{3}} \frac{1}{\hbar} \frac{m}{M} \frac{T}{k\Theta^2} \cos^2\beta [V - E_0]^2 \left(\frac{\sin x - x \cos x}{x^2} \right)^2 d\omega, \tag{24}$$

where

$$x = 2(9\pi/4)^{\frac{1}{3}} \sin\tfrac{1}{2}\theta.$$

5.2. *Validity of the perturbation method.* The perturbation method used here to calculate P must certainly be valid if the displacement X of the atoms is small compared with the radius of the K ring of the atom considered. At ordinary temperatures this is not the case. On the other hand, the fact that $U_{kk'}$ may be transformed into a surface integral makes it probable that the perturbation method remains correct so long as X is small compared with the interatomic distance, which is true up to the melting-point.

5.3. *Calculation taking into account the motion of the atoms.* In the calculation given above it was assumed that, since the vibrating atoms move slowly compared with the electrons, they can be treated as though they were at rest. One may, however, obtain the same result by treating the vibrating atom and electron as a single quantum-mechanical system, as follows:

We assume the atoms to vibrate independently of one another with frequency ν. Consider an atom vibrating parallel to the x-axis; if its displacement from its mean position is X, the change in the potential energy in its neighbourhood is $-X \partial V/\partial x$. If the atom is in the nth vibrational state, its energy is $(n+\tfrac{1}{2})h\nu$, and its wave function a Hermite polynomial which we denote by $\chi_n(X)$. Then the probability that the electron makes a transition from a state **k** to a state **k'**, and that the vibrating atom at the same time jumps to a state n', is given by (14) with

$$-U_{kk'} = \int \int \chi_{n'}^* \psi_{k'}^* X \frac{\partial V}{\partial x} \chi_n \psi_k \, dX...dx...,$$

which may be written

$$-U_{kk'} = \int \chi_{n'}^* X \chi_n \, dX... \int \psi_{k'}^* \frac{\partial V}{\partial x} \psi_k \, dx.... \tag{25}$$

The second integral is the one obtained before; the first, $X_{n'n}$, vanishes unless $n'-n = \pm 1$, and is then equal to

$$\sqrt{\frac{hn}{8\pi^2 M\nu}} \quad \text{or} \quad \sqrt{\frac{hn'}{8\pi^2 M\nu}}, \tag{26}$$

whichever is the larger.

In any scattering process, therefore, the electron gains or loses a quantum $h\nu$ of energy. At high temperatures ($kT > h\nu$), however, this change in the energy can be neglected, and we may assume the electron to be scattered to a state of equal energy.

To obtain the transition probability, we must use formulae (14), (18) as before, but must substitute in these formulae $|X_{n'n}|^2$ instead of X^2, and must average over all initial states. We obtain, adding the squares of the two terms in (26) and averaging over n,

$$\frac{1}{8\pi^2 M\nu^2} \overline{(2n+1)h\nu} = \frac{kT}{4\pi^2 M\nu^2},$$

which is the same expression as that which we obtained for $\overline{X}^2$ in equation (19). Our transition probability is thus the same as before.

We note that none of the collisions between a vibrating atom and an electron are elastic; the atom always gains or loses a quantum of energy.

5.4. *Debye model; scattering probability for low temperatures.* In the preceding section we assumed each atom in the crystal to vibrate independently of all the others. At low temperatures this will not be a good approximation, as emphasized in Chap. I, § 1. We therefore analyse the vibrations of the solid into sound waves, in the manner introduced by Debye.

We consider a sound wave with frequency ν_q and wave number **q**, so that the displacement of an atom at the lattice point $\mathbf{r}_n$ is

$$A_q e^{i(\mathbf{q}\mathbf{r}_n)-2\pi i\nu_q t}+\text{complex conjugate.} \tag{27}$$

The displacement is perpendicular to **q** for transverse waves, parallel for longitudinal waves. A_q represents the amplitude of the wave; at temperature T we have

$$|A_q|^2 = \frac{\hbar^2}{M\nu_q^2} \frac{h\nu_q}{e^{h\nu_q/kT}-1}. \tag{28}$$

Making the same assumptions as in § 5.2, we may take for the change† in the potential due to the lattice wave (27), in the cell of the atom $\mathbf{r}_n$,

$$U = A_q e^{i(\mathbf{q}\mathbf{r}_n)}\text{grad}\, V,$$

and the change in the crystal as a whole is obtained by summing over **q**.

We now treat this change in the potential of the crystal as a whole as our perturbing potential. It will be seen that an integral of the type

$$\int_\Omega \psi_{k'}\, U\psi_k\, d\tau \tag{29}$$

vanishes unless

$$\mathbf{k}-\mathbf{k}'+\mathbf{q} = 2\pi\mathbf{n}/a, \tag{30}$$

where $\mathbf{n} = (n_1, n_2, n_3)$ and n_1, n_2, n_3 are integers or zero and a is the lattice constant.

The transitions for which $\mathbf{n} = 0$ may be called normal transitions; those for which $\mathbf{n} \neq 0$ have been called by Peierls 'Umklappprozesse'.

---

† Bloch and Bethe (loc. cit.) take a different form, viz. $e^{i(\mathbf{q}\mathbf{r})}$ grad V, which is incorrect in the inner shells of the displaced atoms (cf. p. 252) and gives much too large a value of the resistance.

The decision as to which transitions are normal and which 'Umklapp-prozesse' depends, of course, on our definition of $\mathbf{k}$, for, as we have seen in Chap. II, § 4.1, $\mathbf{k}$ is not uniquely defined.

If $\mathbf{n} = 0$, the integral (29) reduces to

$$\frac{\Omega}{\Omega_0} \int_{\Omega_0} \psi_{k'} \operatorname{grad} V \psi_k \, d\tau. \tag{31}$$

If, moreover, we use formula (23) for this integral, we see that it vanishes if the vectors $\mathbf{q}$ and grad V (i.e. the displacement of the atom) are at right angles. Thus only for longitudinal waves does (31) have a non-zero value.

From these remarks and from equation (14) it is clear that the probability that an electron is deflected from a state $\mathbf{k}$ to a state $\mathbf{k}'$ is

$$P \, dS = \frac{1}{4\pi^2 \hbar} \frac{dS}{\Omega_0} |A_q|^2 \left| \int \psi_{k'}^* \operatorname{grad} V \psi_k \, d\tau \right|^2 \Big/ \frac{dE}{dk}. \tag{32}$$

For *low temperatures* $|A_q|^2$ is small unless $h\nu_q \lesssim kT$. Hence, by (28), the transition probability is small for changes in $\mathbf{k}$ except those that satisfy

$$|\mathbf{k} - \mathbf{k}'| \lesssim kT/\hbar c_l,$$

where c_l is the velocity of sound waves in the solid, given by[†]

$$c_l = \frac{k\Theta}{h} \left(\frac{4\pi\Omega_0}{3} \right)^{\frac{1}{3}}.$$

Thus *at low temperatures the electron can only be scattered through small angles.*

If we make the assumption that there is one electron per atom, and that the energy is the same function of $\mathbf{k}$ as for free electrons, so that at the surface of the Fermi distribution

$$|\mathbf{k}| = (\tfrac{3}{2}\pi\Omega_0)^{\frac{1}{3}},$$

then, writing

$$|\mathbf{k} - \mathbf{k}'| = 2|\mathbf{k}| \sin \tfrac{1}{2}\theta,$$

we see that, for P not to be small, we must have

$$\theta \lesssim 2^{\frac{1}{3}} T/\Theta.$$

At high temperatures ($T > \Theta$) we may replace the factor

$$h\nu/(e^{h\nu/kT} - 1)$$

[†] Cf. Chap. I, equation (17); we have put $c_l = c_t$. The formula is thus only valid for an isotropic solid in which the velocities of transverse and longitudinal waves are identical.

by kT. Formula (32) then becomes

$$P \, dS = \frac{dS}{\Omega_0} \left(\frac{3}{4\pi\Omega_0} \right)^{\frac{2}{3}} \frac{1}{|\mathbf{k}-\mathbf{k'}|^2} \frac{\hbar T}{Mk\Theta^2} \left| \int \psi_{k'}^* \operatorname{grad} V \psi_k \, d\tau \right|^2 \bigg/ \frac{dE}{dk}. \quad (33)$$

This formula, however, will be valid only for $|\mathbf{k}-\mathbf{k'}| \leqslant q_{\max}$, where $q_{\max}$ is the wave number of the vibration of maximum frequency, given, according to the Debye theory, by

$$q_{\max} = 2\pi(\tfrac{3}{4}\pi\Omega_0)^{\frac{1}{3}}.$$

For larger values of $|\mathbf{k}-\mathbf{k'}|$ the transitions will be 'Umklappprozesse'.

If we assume as before that the energy is the same function of $\mathbf{k}$ as for free electrons, then we have, if θ is the angle between $\mathbf{k}$ and $\mathbf{k'}$,

$$P \, dS = \frac{dS}{\Omega_0} \frac{\hbar}{4\pi^2 k} \frac{T}{M\Theta^2} \frac{1}{2^{\frac{1}{3}}} \left| \int \psi_{k'}^* \operatorname{grad} V \psi_k \, d\tau \right|^2 \bigg/ (1-\cos\theta), \quad (34)$$

$$\theta < 2\sin^{-1}2^{-\frac{2}{3}} \simeq 79°.$$

Transition probabilities for collisions in which the electron is deflected through an angle θ greater than 79° ('Umklappprozesse') have not yet been calculated; there would be no difficulty in doing so, but it would not at present be possible to use the results obtained to calculate the conductivity, because the transition probability would in general depend on the initial state $\mathbf{k}$, and not merely on $|\mathbf{k}-\mathbf{k'}|$. As we shall see in the next section, a theory of conductivity has not yet been worked out for such a case.

It will be realized that the value 79° for the maximum 'normal' deflexion depends on the Debye model for the vibrations of a solid; other angles would be obtained with the more exact models discussed in Chap. I, § 1.

6. Calculation of the current

In the preceding section we have found an expression for the probability that, owing to any lack of periodicity in the lattice, an electron makes a transition from one state to another. We must now calculate the time of relaxation and hence the conductivity in terms of this transition probability. Unfortunately such a calculation is only possible in certain particularly simple cases. In this section we shall therefore make the two following assumptions:

(1) The energy $E(\mathbf{k})$ of an electron in the state $\mathbf{k}$ is a function of $k = |\mathbf{k}|$ only, so that the surface of the Fermi distribution is spherical.

(2) The transition probability $P(\mathbf{kk'})$ defined in the last section is

a function of $|\mathbf{k}-\mathbf{k}'|$ only; it is thus a function of the angle of scattering θ but not of the initial state; we may thus write

$$P(\mathbf{k}\mathbf{k}')\,dS = P(\theta)k^2\,d\omega. \qquad (35)$$

Let us denote by $f_0(\mathbf{k})$ the Fermi distribution function† in the absence of any external field, which is given by

$$f_0 = \frac{1}{e^{(E-\zeta)/kT}+1}, \qquad (36)$$

so that

$$2\frac{dk_x\,dk_y\,dk_z}{(2\pi)^3}f_0(\mathbf{k}) \qquad (37)$$

is the number of electrons in states having their wave vector in the volume element of k-space $dk_x\,dk_y\,dk_z$. Fig. 95 shows f_0 plotted against k_x. Let us suppose that an external electric field F acts along the x-axis. Then we have seen in Chap. III, § 3 that, under the influence of such a field, the component k_x of the wave vector of any electron increases steadily, the rate of increase being

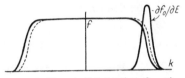

FIG. 95. Fermi distribution function in a metal carrying a current; the full line shows f_0, the dotted line f.

$$\frac{dk_x}{dt} = \frac{eF}{\hbar}.$$

The whole Fermi distribution is thus shifted, and has at the time t the form

$$f(\mathbf{k}) = f_0\left(k_x - \frac{eFt}{\hbar}, k_y, k_z\right), \qquad (38)$$

as shown in Fig. 95 by the dotted line.

If we differentiate this with regard to the time, we obtain, for $t \to 0$,

$$\left(\frac{df}{dt}\right)_{\text{field}} = -\frac{\partial f_0}{\partial k_x}\frac{eF}{\hbar} = -\frac{df_0}{dE}\frac{dE}{dk}\frac{k_x}{k}\frac{eF}{\hbar}. \qquad (39)$$

As shown in Fig. 95, the function df_0/dE is large only near the surface of the Fermi distribution. Thus we see that the density of electrons is increasing only in those states which lie near the surface of the Fermi distribution; the number of electrons having velocities other than the maximum is at first unaltered by the field.‡

In a metal at a finite temperature in the presence of an external field F the whole Fermi distribution will be shifted to the right, as in Fig. 95, until a steady state is reached in which the effect of the field is just balanced by the collision processes considered in the last

† Cf. Chap. VI, § 1.
‡ This would not be the case for Boltzmann statistics.

section. We should expect the Fermi distribution function to be given then by (38), with†

$$t = \tau, \tag{38.1}$$

where τ is the constant which we call the 'time of relaxation'. We shall thus have

$$f(\mathbf{k}) = f_0(\mathbf{k}) + g(\mathbf{k}) \tag{40}$$

with

$$g(\mathbf{k}) = -\frac{df_0}{dk}\frac{k_x}{k}\frac{eF}{\hbar}\tau. \tag{41}$$

Our problem is to calculate τ. The current may then be obtained at once (see below).

We have therefore to find the rate of change of f due to collisions. If we take a volume element of k-space $d\mathbf{k}$, the number of electrons jumping *out* of the corresponding states will be the product of

(1) the number of electrons initially in states $d\mathbf{k}$, i.e.

$$2\frac{dk_x\,dk_y\,dk_z}{(2\pi)^3}f(\mathbf{k});$$

(2) the sum of the transition probabilities $P(\mathbf{kk'})\,dS'$ to all other states $\mathbf{k'}$ with the same energy, multiplied by the probability that the state in question is unoccupied, i.e.

$$1 - f(\mathbf{k'}).$$

We have, therefore, for the number of transitions per unit time

$$2\frac{dk_x\,dk_y\,dk_z}{(2\pi)^3}f(\mathbf{k})\int [1-f(\mathbf{k'})]P(\mathbf{kk'})\,dS',$$

where the integration is over the surface in k-space having the same energy as the initial state $\mathbf{k}$.

The number of electrons jumping *into* our volume element depends on

(1) the number of unoccupied states in the volume element,
(2) the number of *occupied* states having the same energy as $\mathbf{k}$,
(3) the transition probabilities $P(\mathbf{k'k})$. The number per unit time is clearly

$$2\frac{dk_x\,dk_y\,dk_z}{(2\pi)^3}[1-f(\mathbf{k})]\int f(\mathbf{k'})P(\mathbf{k'k})\,dS'.$$

Hence the rate of change of f due to collisions is

$$\left(\frac{df}{dt}\right)_{\text{collisions}} = [1-f(\mathbf{k})]\int f(\mathbf{k'})P(\mathbf{k'k})\,dS' - f(\mathbf{k})\int [1-f(\mathbf{k'})]P(\mathbf{kk'})\,dS'. \tag{42}$$

† Cf. § 1.

Since we are assuming that dE/dk is constant on the surface of the Fermi distribution, we may put (cf. equation (14))

$$P(\mathbf{kk'}) = P(\mathbf{k'k}),$$

so that (42) may be simplified to give

$$\frac{df}{dt} = \int f(\mathbf{k'})P(\mathbf{k'k}) \, dS' - f(\mathbf{k}) \int P(\mathbf{kk'}) \, dS'. \tag{43}$$

This is the formula which we should obtain without taking into account the exclusion principle (i.e. the fact that electrons can only jump into unoccupied states), which therefore has no effect on the conductivity.

If for $f(\mathbf{k}), f(\mathbf{k'})$ we substitute the undisturbed Fermi function $f_0(\mathbf{k})$, it is clear that (43) vanishes, because f_0 depends only on the energy of the state, so that $f_0(\mathbf{k'}) = f_0(\mathbf{k})$. This is, of course, to be expected, because f_0 is the distribution function of electrons in thermodynamical equilibrium in the absence of a field.

We therefore substitute $f = f_0 + g$ as in formula (40) and obtain

$$\frac{df}{dt} = \int [g(\mathbf{k'}) - g(\mathbf{k})]P(\mathbf{kk'}) \, dS'.$$

This must be set equal to *minus* the rate of change due to the field, given by equation (39), whence we obtain

$$\int [g(\mathbf{k'}) - g(\mathbf{k})]P(\mathbf{kk'}) \, dS' = \frac{df_0}{dE} \frac{dE}{dk} \frac{k_x}{k} \frac{eF}{\hbar}. \tag{44}$$

This integral equation for $g(\mathbf{k})$ was first obtained by Bloch (loc. cit.).

Substituting from (41) for $g(\mathbf{k})$, the left-hand side of (44) becomes

$$-\frac{eF}{\hbar} \frac{df_0}{dE} \frac{dE}{dk} \frac{k_x}{k} \tau \int \left(\frac{k'_x}{k_x} - 1\right) P(\mathbf{kk'}) \, dS',$$

giving for the time of relaxation τ

$$-\tau \int \left(\frac{k'_x}{k_x} - 1\right) P(\mathbf{kk'}) \, dS' = 1. \tag{45}$$

We have to show that the integral on the left is independent of the direction of $\mathbf{k}$; otherwise the assumption of a time of relaxation independent of $\mathbf{k}$ would not be justified. We denote by α the angle between $\mathbf{k}$ and the x-axis (direction of the field), by θ, as before, the angle between $\mathbf{k}, \mathbf{k'}$, and by ϕ the angle between the plane $\mathbf{kk'}$ and the plane containing $\mathbf{k}$ and the x-axis, so that

$$k_x = k\cos\alpha, \qquad k'_x = k(\cos\alpha\cos\theta + \sin\alpha\sin\theta\cos\phi),$$
$$dS = k^2\sin\theta \, d\theta d\phi, \qquad P(\mathbf{kk'}) = P(\theta).$$

Carrying out the integration over ϕ, we obtain

$$\frac{1}{\tau} = 2\pi k^2 \int_0^{\pi} (1-\cos\theta)P(\theta)\sin\theta\, d\theta \tag{46}$$

or

$$\frac{1}{\tau} = \int (1-\cos\theta)P(\theta)\, dS. \tag{47}$$

The reciprocal of the time of relaxation is thus equal to the total probability per unit time that an electron is scattered, with the factor $1-\cos\theta$ to give extra weight to large angle collisions.†

Expressed in terms of $U_{kk'}$, formulae (14) and (47) give

$$\frac{1}{\tau} = \frac{1}{2\pi\Omega h}\frac{1}{dE/dk}\int |U_{kk'}|^2(1-\cos\theta)\, dS. \tag{48}$$

The current may be obtained as follows: the current per electron is ev_x; hence the current j per unit volume is, by (37),

$$j = \int\int\int ev_x\frac{2f(\mathbf{k})}{(2\pi)^3}\, dk_x\, dk_y\, dk_z, \tag{49}$$

the integration being over all k-space. For v_x we have, from Chap. III, § 2,

$$v_x = \frac{1}{\hbar}\frac{\partial E}{\partial k_x} = \frac{1}{\hbar}\frac{dE}{dk}\frac{k_x}{k}, \tag{50}$$

and for f, from (40) and (41),

$$f = f_0 - \frac{eF}{\hbar}\frac{df_0}{dk}\frac{k_x}{k}\tau. \tag{51}$$

The term f_0 gives no contribution to the integral (49); we therefore obtain, from (49), (50), and (51),

$$j = -\frac{e^2F}{\hbar^2}\frac{2}{(2\pi)^3}\int\int\int \tau\frac{dE}{dk}\left(\frac{k_x}{k}\right)^2\frac{df_0}{dk}dk_x\, dk_y\, dk_z. \tag{52}$$

We take spherical polar coordinates such that

$$k_x = k\cos\alpha, \qquad dk_x\, dk_y\, dk_z = 2\pi k^2\, dk\sin\alpha\, d\alpha. \tag{53}$$

(52) then gives

$$j = -\frac{e^2F}{3\pi^2\hbar^2}\int_0^{\infty}\frac{dE}{dk}\tau k^2\frac{df_0}{dk}dk. \tag{54}$$

As Fig. 95 shows, df_0/dk is only finite in a small range of k at the surface of the Fermi distribution. In general‡ the other terms in the

---

† This is the same integral that occurs in the theory of gaseous diffusion, cf. Mott and Massey, loc. cit. 230.

‡ For certain exceptions, i.e. transition metals at high temperatures, cf. § 7.

integral may be considered constant in this range, and we obtain for the conductivity, since $-\int (df_0/dk)\, dk = 1$,

$$\sigma = j/F = \frac{e^2}{3\pi^2\hbar^2}\left(\tau k^2 \frac{dE}{dk}\right)_{k=k_{\max}}. \tag{55}$$

The conductivity depends, therefore, only on the value of τ at the surface of the Fermi distribution.

This formula may be expressed in terms of N_{eff}, the effective number of free electrons per unit volume, given in our case, according to Chap. III, § 4, by

$$N_{\text{eff}} = \frac{8\pi}{3}\left(\frac{k}{2\pi}\right)^3 \frac{m}{\hbar^2 k}\frac{dE}{dk}.$$

We thus obtain for the conductivity the 'classical' formula

$$\sigma = \frac{N_{\text{eff}}\, e^2}{m}\tau.$$

It is of interest to express σ in terms of the area $I(\theta)$ introduced in § 5.1. We obtain easily, assuming as before perfectly free electrons,

$$\left.\begin{aligned}
\frac{1}{\sigma} &= \frac{m}{Ne^2}\frac{vA}{\Omega_0}, \\
A &= \int (1-\cos\theta)I(\theta)\, d\omega,
\end{aligned}\right\} \tag{56}$$

where Ω_0 is the volume per atom. A is thus the effective scattering area presented by each atom.

We note that, in the derivation of the above formulae, the transfer of *energy* from the electrons to the lattice vibrations has not been mentioned. The production of Joule heat is, of course, due to this energy transfer. When an electron collides with a vibrating atom, it can, as we have seen, give up energy $\pm h\nu$; when it collides with a foreign atom in solid solution a very small energy transfer can also take place, since the mass of the foreign atom is not infinitely great compared with that of the electrons. It is not, however, necessary to consider this energy transfer explicitly. If it did not take place, the Joule heat produced by the current would be transferred to the electrons alone, whose temperature would therefore rise rapidly, since their heat capacity is small. The rate of transfer of energy between the electrons and the lattice will thus determine only the difference in the temperatures of the electrons and the lattice vibrations.

6.1. *Explicit formula for resistance due to thermal agitation.* If we use Wigner-Seitz wave functions as in § 5.1, we may apply formula (24) for the transition probability P, and thus obtain an explicit formula for τ and hence for σ. We obtain, assuming $E = \alpha \hbar^2 k^2 / 2m$,

$$\frac{1}{\sigma} = \frac{C}{\alpha^2} \frac{m}{Ne^2} \frac{m}{M} \frac{1}{\hbar} \frac{T}{k\Theta^2} [V(r_0) - E_0]^2, \tag{57}$$

where N is the number of electrons per unit volume and C is a numerical factor given, with $f(x) = (\sin x - x \cos x)^2 x^{-4}$, by

$$\frac{3}{\pi} \left(\frac{9\pi}{4} \right)^{\frac{1}{3}} 2\pi \int_0^\pi (1 - \cos \theta) f(3 \cdot 84 \sin \tfrac{1}{2}\theta) \sin \theta \, d\theta$$

$$= 2 \cdot 06 \qquad \text{(Einstein model)}, \tag{58.1}$$

or

$$\frac{3}{\pi} \left(\frac{9\pi}{8} \right)^{\frac{1}{3}} 2\pi \int_0^{79°} f(3 \cdot 84 \sin \tfrac{1}{2}\theta) \sin \theta \, d\theta$$

$$= 0 \cdot 98 \qquad \text{(Debye model)}. \tag{58.2}$$

The result obtained using the Debye model, however, neglects the 'Umklappprozesse' (collisions through more than 79°) and might well be about the same as (58.1) if these were included. We therefore use (58.1) for comparison with experiment.

We note that σ is proportional to α^2, while the free-electron number $N_{\text{eff}} (= N\alpha)$ is proportional to α. If we call m/α the effective mass of an electron, the resistance is proportional to the square of the effective mass.

6.2. *Comparison with experiment.* We give a comparison for the metals silver and sodium, for both of which the free-electron energy formula should be approximately valid. The values of $V(r_0) - E_0$ are taken from the calculations of Wigner and Seitz for sodium and Fuchs for silver (cf. Chap. II, § 4.5). We have put $\alpha = 1$, as for free electrons. For $k\Theta/h$ we must take a mean value of the vibrational frequency of the atoms; we therefore take for Θ three-quarters of the Debye characteristic temperature, as explained on p. 8. The Debye Θ values used are those shown in Table X, and are thus deduced from resistance measurements.

Theoretical and experimental values of the conductivity σ at $273°$ K.

	$\Omega_0 \times 10^{24}$ cm.^{3}	Atomic weight	$\Theta_E = \tfrac{3}{4}\Theta_D$ degrees	$V(r_0)-E_0$ (electron volts)	Resistivity $1/\sigma$ (microhm-cm.)	
					calculated	observed
Na	40·1	23	150	1·3	2·9	4·3
Ag	16·9	108	167	2·3	0·61	1·4

The calculated resistance is too small; we believe this to be due to the neglect of the perturbing potential due to a displaced atom outside its own polyhedron (cf. p. 252). The estimation of Θ is, moreover, uncertain.

6.3. *Divalent and transition metals.* The calculation of the current given in § 5.1 depends on the assumption that the energy E is a function of $|\mathbf{k}|$ only; that is to say, that the surfaces of constant energy in k-space are spheres. It follows that formula (47) is applicable, even approximately, only to the alkalis and noble metals. For all other metals which have been investigated from a theoretical point of view the surface of the Fermi distribution lies actually in two or more Brillouin zones, so that it will have no resemblance to the simple spherical form considered up till now.

We shall consider first the general case when E is an arbitrary function of $\mathbf{k}$. In the steady state, when a current is flowing, the Fermi distribution function will take the form, analogous to (38), (38.1),

$$f(k_x, k_y, k_z) = f_0\left(k_x - \frac{eF}{\hbar}\tau, k_y, k_z\right), \tag{59}$$

where the 'time of relaxation' $\tau = \tau(\mathbf{k})$ is now a function of the wave vector $\mathbf{k}$, instead of, as before, a function of E or $|\mathbf{k}|$ only. The problem, at present unsolved in the general case, is to determine $\tau(\mathbf{k})$. If $\tau(\mathbf{k})$ is known, the current j_x along the x-axis will, as before, be determined by equations (49), (50), which give, analogously to (52),

$$j_x = -\frac{e^2F}{\hbar^2}\frac{2}{(2\pi)^3}\int\int\int\left(\frac{\partial E}{\partial k_x}\right)^2\frac{df_0}{dE}\tau(\mathbf{k})\,dk_x dk_y dk_z. \tag{60}$$

If we write

$$\sigma_x(E) = \frac{2e^2}{(2\pi)^3\hbar^2}\int\left(\frac{\partial E}{\partial k_x}\right)^2\frac{\tau(\mathbf{k})\,dS}{|\mathrm{grad}\,E|}, \tag{61}$$

the integration being over the surface in k-space with energy E, (60) may be written

$$j_x = -F\int\sigma_x(E)\frac{df_0}{dE}\,dE. \tag{62}$$

In general we may take $\sigma(E)$ to be constant in the range of E of breadth kT in which df_0/dE is finite; we obtain, therefore,

$$\sigma_x = j_x/F = \sigma_x(\zeta) \tag{63}$$

for the conductivity in the direction of the x-axis. In certain cases (transition metals at high temperatures, alloys of bismuth) this is not legitimate; (62) then gives (cf. Chap. VI, § 1)

$$\sigma_x = \sigma_x(\zeta) + \frac{\pi^2}{6}(kT)^2 \frac{d^2\sigma_x(\zeta)}{d\zeta^2} + \dots. \qquad (63.1)$$

$\sigma(E)$ represents the value that the conductivity would have if the energy at the surface of the Fermi distribution were E; its variation with E is of importance in the discussion of thermoelectric phenomena (§ 14).

The case where the Fermi distribution lies in two Brillouin zones has been discussed in detail by Mott† and a solution for $\tau(\mathbf{k})$ obtained, subject to the following simplifying conditions:

1. The surface of the Fermi distribution lies in two zones, (a) and (b); zone (a) is nearly full and zone (b) nearly empty, the number of electrons in (b) being equal to the number of holes in (a).

2. The state of an electron in either zone being described by wave vectors $\mathbf{k}_a$, $\mathbf{k}_b$, the energies in the two zones are given by

$$E_a = E_0 - \alpha\hbar^2 k_a^2/2m, \qquad E_b = \beta\hbar^2 k_b^2/2m.$$

3. The transition probabilities $P(\mathbf{k}_a\mathbf{k}_a')$, $P(\mathbf{k}_b\mathbf{k}_b')$, $P(\mathbf{k}_a\mathbf{k}_b)$ are functions only of the angles between the initial and final wave vectors.

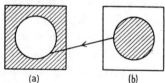

(a)　　　　　　(b)

Fig. 96. k-space of metal with two zones; occupied states are shaded.

The new feature of the problem is that transitions are allowed from one zone to another, as shown in Fig. 96.

With these assumptions it may be shown that τ is independent of $\mathbf{k}$ in either zone; we may thus define two times of relaxation, τ_a, τ_b, one for each zone. Formula (61) for the conductivity then reduces to

$$\sigma = \frac{Ne^2}{m}(\alpha\tau_a + \beta\tau_b), \qquad (64)$$

where N is the number of electrons per unit volume actually in the zone (b), and thus the number of holes in zone (a). The first term represents the current carried by the positive holes, the second by the electrons.

No detailed application of these results has yet been made to the divalent and multivalent metals. For the divalent metals we shall have a certain number of electrons overlapping into the second zone and an equal number of positive holes in the first zone. If this number

† Proc. Roy. Soc. A, 153 (1936), 699.

is small, the conductivity will also be small, as appears to be the case from the discussion of § 3. This then is probably the reason for the small conductivity of the divalent metals.

For the transition metals we have two zones, the d zone and the s zone, to take into consideration. In the d zone the density of states is large; in other words, α is small. From this it follows that, if τ_a and τ_b are comparable, the current is nearly all carried by the electrons in the s zone.

On the other hand, as formula (14) shows, the transition probability from one state to another is proportional to the density of states in the final state. Therefore transitions in which the electron jumps from the s zone to an unoccupied state in the d zone are more probable than the ordinary scattering processes, in which an electron jumps from one s state to another. This appears to be the reason for the low conductivity of the transition metals. If we neglect altogether the ordinary s–s transitions, the time of relaxation of the s electrons is found to be (loc. cit., p. 708)

$$\frac{1}{\tau_s} = N_d(E)\frac{\hbar T}{2Mk\Theta^2}2\pi\int\limits_{0}^{\pi}\left|\int\psi_d^*\frac{\partial V}{\partial x}\psi_s\,d\tau\right|^2\sin\theta\,d\theta, \qquad (65)$$

where $N_d(E)$ is the density of states in the d zone and ψ_s, ψ_d are the wave functions of the states in the s and d bands for which the wave vectors $\mathbf{k}$ make an angle θ with each other.

We thus obtain the following picture of electrical conductivity in the transition metals. The current is carried by s electrons, with effective mass not very different from that of a free electron, as in the noble metals. On the other hand, the resistance is mainly due to scattering processes in which the electron makes a transition from the s to the d band; the probability of such a transition is proportional to $N_d(E)$, the density of states in the d band. Evidence in favour of this hypothesis may be drawn from the resistance of alloys of these metals with Cu, Ag, and Au (§ 13.2), from their thermoelectric properties (§ 15) and from their resistivities at high temperatures (§ 7).

$N_d(E)$ is proportional to the cube root of the number of particles (in this case positive holes) in the d band. Hence the conductivity σ is proportional to

$$\frac{\text{number of electrons in } s \text{ band}}{(\text{number of positive holes in } d \text{ band})^{\frac{1}{3}}}.$$

Finally we give a table of the mean free paths l and times of relaxation τ deduced from the *experimental* conductivities at 0° C. for a number of metals. We have chosen only those metals for which the number of free electrons per atom can be estimated, i.e. the monovalent metals and certain of the transition metals. We use the formulae

$$\sigma = \frac{Ne^2}{m}\tau, \qquad l = 2\tau u, \qquad \tfrac{1}{2}mu^2 = \frac{h^2}{8m}\left(\frac{3N}{\pi}\right)^{\frac{2}{3}}$$

Metal	Electrons per atom, assumed	$u \times 10^{-8}$ cm./sec., calculated	$\sigma \times 10^{-17}$ electro- static units, observed	$\tau \times 10^{15}$ sec.	$l \times 10^8$ cm.
Li	1	1·307	1·06	8·6	225
Na	1	1·067	2·09	31	670
K	1	0·848	1·47	44	745
Rb	1	0·805	0·78	27	440†
Cs	1	0·747	0·49	21	320
Cu	1	1·578	5·76	26·7	842
Ag	1	1·397	6·12	40·9	1,140
Au	1	1·400	4·37	29·0	812
Ni	0·6	1·365	1·36	9·76	266
Co	0·7	1·421	1·45	9·17	260
Fe	0·2	0·912	1·01	24·2	440
Pd	0·55	1·206	0·88	9·2	220
Pt	0·6	1·230	0·92	9·0	220

7. Resistance at high temperatures

At high temperatures the resistance-temperature curves are not quite linear for most metals. For Cu, Ag, Au, W, as is shown in Fig. 97, R/T increases with increasing T. For Cu, Ag, and Au this is probably due to thermal expansion and the consequent decrease in the Debye characteristic temperature Θ, which was discussed in Chap. I, § 4. Assuming that (cf. § 2) $R/T \propto 1/\Theta^2$, we obtain

$$\frac{d}{dT}\left(\log\frac{R}{T}\right) = 2\alpha\gamma,$$

where $\gamma = -d(\log\Theta)/d(\log V)$ and α is the thermal-expansion coefficient. Integrating, we have, for values of T not too large,

$$R/T = \text{const.}(1+2\alpha\gamma T). \tag{66}$$

We may use formula (66) to compare the resistance at any two temperatures T_1 and T_2, as in the following table:

† Recent work by Lovell (*Proc. Roy. Soc.* A, in press) on the resistance of thin films of rubidium on glass supports this value.

Metal	$2\alpha\gamma \times 10^4$	T_1, °K.	T_2, °K.	$\dfrac{R_1}{T_1} \div \dfrac{R_2}{T_2}$ observed	calculated
Cu	1·9	1,273	773	1·11	1·10
Ag	2·7	1,233	773	1·07	1·12
Au	2·6	1,273	773	1·15	1·12
W	0·41	1,273	673	1·17	1·025
Pt	1·3	1,273	773	0·97	1·065
Pd	1·5	773	673	0·97	1·015

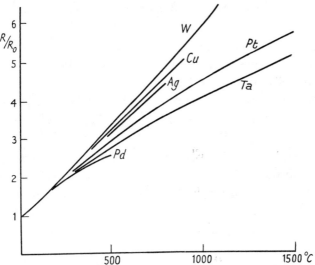

Fig. 97. Resistance of certain metals at high temperatures.†

For the normal metals the agreement is satisfactory, but not for W, Pd, and Pt; for Pt and Pd, and also for Ta, R/T *decreases* with increasing temperature.‡

An explanation of this fact has been given by Mott;‖ for these metals the resistance is determined mainly by scattering processes in which the electron jumps from the s to the d band. The probability of such processes is proportional to the density of states $N(E)$ in the d band in a range of energies kT at the surface of the Fermi distribution. For the transition metals Pd and Pt the d band is nearly full; $N(E)$ therefore decreases very fast with increasing E, and so at high temperatures the electrons with high energy have considerably

† The experimental values are taken from Grüneisen, *Handb. d. Phys.* 13 (1928), 16.
‡ The behaviour of this anomaly for a series of Pd–Au alloys has recently been investigated by Conybeare (*Proc. Phys. Soc.*, in press).
‖ *Proc. Roy. Soc.* A, 153 (1936), 699 and 156 (in press).

smaller probability of being scattered than those of low energy (cf. the discussion of the thermoelectric power, § 15). It will not, therefore, be legitimate to take $\tau(E)$, the time of relaxation, constant over the surface of the range kT in which $\partial f_0/\partial E$ is finite; the resistance must be calculated by formula (63.1). One obtains for the resistance a formula of the type

$$R/T = \text{const.}[1+2\alpha\gamma T][1+AT^2...], \tag{67}$$

where A may in general be positive or negative; if, however, one assumes $N(E) = \text{const.}\sqrt{(E-E_0)}$, one obtains

$$A = -\tfrac{1}{6}\pi^2/T_0^2, \tag{67.1}$$

where T_0 is the degeneracy temperature of the d band (cf. Chap. VI, § 5). Such an assumption will be legitimate for Pd or Pt, where the d band is nearly full, but not for W.

Comparison with experiment for the former metals gives for T_0:

	Pd	Pt
T_0	3,500°	4,900°

The resistance of ferromagnetic metals shows an anomalous behaviour near the Curie temperature; the resistance of nickel measured by Gerlach[†] is shown in Fig. 98. The resistance also decreases in a

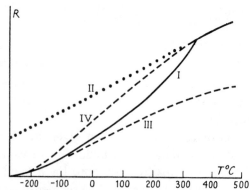

FIG. 98. Resistance of nickel as a function of temperature.

    I. Observed.
    II. Extrapolation of I.
    III. Suggested theoretical curve for ferromagnetic nickel.
    IV. Suggested theoretical curve for paramagnetic nickel.

† For references to a series of papers by Gerlach and his co-workers, cf. Englert, *Ann. d. Physik*, **14** (1932), 589; the change of resistance in a magnetic field has also been measured by Potter, *Proc. Roy. Soc.* A, **132** (1931), 554. Cf. also Krupkowski and de Haas, *Comm. Leiden*, 194 a (1928) and Svensson, *Ann. d. Physik*, **25** (1936), 264, for the corresponding properties of Cu–Ni alloys.

magnetic field. A theoretical discussion has been given by Mott;[†] if, as suggested above, the resistance of a metal such as nickel is due to transitions in which the electron jumps from the s to the d band, it is clear that at low temperatures only *half* the electrons can make such transitions, namely those with spin antiparallel to the spontaneous magnetization, because all places in the d band with spin parallel to the spontaneous magnetization are occupied. At high temperatures electrons with either spin direction can make transitions of this type. If, then, we regard the resistance as a function of two independent variables, temperature and magnetization, the resistance curves for saturated magnetization and for zero magnetization will be somewhat as III and IV in Fig. 98. A quantitative discussion is given in the original papers.

A theoretical discussion has also been given of the resistance of copper-nickel alloys (constantan).[†]

8. Change of resistance under pressure

The resistance of most metals decreases under pressure, but that for certain metals (Li, Ca, Sr, Bi) increases. The decrease for normal metals can be explained qualitatively in the following way. The atoms of a metal under high pressure being closer together are held in position by stronger forces, and therefore at given temperature vibrate with smaller amplitude than at atmospheric pressure. Since the resistance (for $T > \Theta$) is proportional to the mean square amplitude $\overline{X^2}$ of the atomic vibrations, it follows that the resistance is lowered. Quantitatively the effect of the pressure on the mean square amplitude may be calculated as follows: we have seen (§ 2) that $\overline{X^2}$ is proportional to $1/\Theta^2$, where Θ is the Debye characteristic temperature, so that, if V is the volume of the solid,

$$\frac{d(\log \overline{X^2})}{d(\log V)} = -2\frac{d(\log \Theta)}{d(\log V)}. \tag{68}$$

But we saw (Chap. I, § 4.1) that the quantity on the right can be deduced from the thermal-expansion coefficient; or alternatively an estimate can be made from the change of compressibility with pressure.

Table XI shows the values obtained for a series of metals; they are compared with the experimental values of $d(\log \rho)/d(\log V)$, i.e. the

† Mott, loc. cit.

measured† pressure coefficient of resistivity, $d(\log \rho)/dp$, divided by the measured† compressibility, $d(\log V)/dp$. The values given are thus the relative decrease in resistance for unit decrease in the volume.

TABLE XI

Element	$-2\dfrac{d(\log \Theta)}{d(\log V)}$	$\dfrac{d(\log \rho)}{d(\log V)}$	Element	$-2\dfrac{d(\log \Theta)}{d(\log V)}$	$\dfrac{d(\log \rho)}{d(\log V)}$
Li	2·34	−0·85	Mo	3·14	3·7
Na	2·50	2·84	W	3·24	4·3
K	2·68	4·0	Fe	3·20	4·1
Rb	2·96	2·92	Co	3·74	1·76
Cs	2·58	1·8	Ni	3·76	3·35
Cu	3·92	2·7	Pd	4·46	3·75
Ag	4·80	3·6	Pt	5·08	5·6
Au	6·06	5·2	Ca	2·6	−1·95
Al	4·34	2·9	Sr	1·9	−6·7
Pb	5·46	6·0	Ba	2·2	+0·58
Ta	3·50	2·9			

It will be seen that for many metals there is fair agreement between the two columns, showing that for these metals the change in resistance is mainly due to the change in Θ, i.e. to the change in the amplitude of the atomic vibrations. On the other hand, for most metals the agreement is by no means perfect; some of the other factors in the formula for the resistance must therefore be sensitive to pressure. These factors are difficult to estimate.‡ The abnormal behaviour of Ca and Sr has been discussed by Mott,‖ who has shown that such an effect can occur for a divalent metal, but no explanation has been given of why it should occur for these metals only. The abnormal behaviour of lithium has been discussed by N. H. Frank.†† Frank's explanation is that the breadth of the first Brillouin zone contracts as the metal is compressed, as is shown in Fig. 36 on page 83 of this book. The 'effective' number of free electrons would thus decrease.‡‡ According to Frank the anomalous behaviour of Cs, whose resistance first decreases and then increases, may be explained in the same way.

The pressure coefficients of certain metals at low temperatures

---

† Bridgman, *The Physics of High Pressure*, Chaps. VI and IX, London (1931).

‡ Kroll (*Zeits. f. Phys.* 85 (1933), 398) has calculated the change in the scattering power of the ion, but he appears to have taken this proportional to $|V_{kk'}|^2$ instead of $|(\text{grad } V)_{kk'}|^2$. Cf. also Lenssen and Michels (*Physica*, 2 (1935), 1091), who use a formula due to Nordheim.   ‖ *Proc. Phys. Soc.* 46 (1934), 680.

†† *Phys. Rev.* 47 (1935), 282.

‡‡ Alternatively, we may say that the effective mass increases.

down to $-182 \cdot 9°$ C. have been measured by Bridgman;[†] a theoretical discussion appropriate to low temperatures ($T < \Theta$) has been given by Mott.[‡]

9. Resistance at low temperatures ($T < \Theta$)

As we have seen in § 5.1, if T/Θ is small, an electron is unlikely to be scattered through an angle greater than T/Θ. This is because only lattice waves of long wave-length (small wave-number q) are excited, and these can transfer only a small amount of momentum to the electron.

For deflexions through an angle θ less than T/Θ, formula (32) shows that the probability P of scattering is proportional to $T/M\Theta^2$, i.e. to the amplitude of the corresponding lattice wave. We also see from the same formula that the scattering probability is roughly independent of θ for small θ, since both ν_q and, by (23), the integral in (32) are proportional to θ.

We may thus estimate the dependence on temperature of the resistance R from formula (47), namely

$$R \propto \int_0^\pi (1-\cos\theta)P\sin\theta \, d\theta.$$

In our case this becomes

$$R \propto \int_0^{\sim T/\Theta} \frac{T}{M\Theta^2}(1-\cos\theta)\sin\theta \, d\theta$$

or
$$R \propto T^5/M\Theta^6. \tag{69}$$

Thus at low temperatures the resistance is proportional to the fifth power of the temperature.

Unfortunately, however, formula (47) cannot be used to obtain a quantitative expression for the resistance, because it was derived under the assumption that the change in the energy of an electron when it is scattered is small compared with kT. At temperatures comparable with or less than Θ, the change in the energy is comparable with kT.

In the investigation of Bloch,[||] in which this fact was taken into account, the following formula for the resistance at low temperatures

† *Proc. Amer. Acad. Arts Sci.* **67** (1932), 305.
‡ Loc. cit.
|| *Zeits. f. Phys.* **52** (1928), 555. Cf. also *Handb. d. Phys.* **24**/2 (1933), 499–560.

was obtained, subject to the usual assumption, that E is a function of $|\mathbf{k}|$ only (cf. § 6): if T_1 and T_2 are two temperatures such that

$$T_1 \ll \Theta \ll T_2,$$

and R_1 and R_2 are the resistances at these temperatures,

$$\frac{R_1}{R_2} = 497{\cdot}6\left(\frac{T_1}{\Theta}\right)^4 \frac{T_1}{T_2}. \tag{70}$$

The derivation of the numerical factor is not, however, valid unless the energy E is a function of $|\mathbf{k}|$ alone, which is roughly the case only for the noble metals and alkalis. We have, moreover, seen (p. 264) that with the Debye model used by Bloch the theoretical formula for the resistance at high temperatures is too small, because it neglects 'Umklappprozesse' (collisions through more than 79°). Such collisions cannot occur for low temperatures. The numerical factor in (70) is therefore probably too large by a factor 2 or more.

For intermediate values of the temperature Grüneisen† has suggested the formula

$$R/T = \text{const. } G(\Theta/T),$$

where

$$G(x) = 4x^{-4}\left[5\int_0^x \frac{z^4\,dz}{e^z-1} - \frac{x^5}{e^x-1}\right]. \tag{71}$$

The formula may be derived from quantum mechanics only by making certain further approximations.‡ This formula includes formula (70), and the criticisms made above apply to it also. The function $G(x)$ has been tabulated by Grüneisen and has the following values:

$x = \Theta/T$	$G(x)$	$x = \Theta/T$	$G(x)$
0	1·0000	10	0·0465
1	0·9465	12	0·0235
2	0·8074	14	0·01289
3	0·6309	16	0·00758
4	0·4607	20	0·00311
5	0·3208	30	0·0$_3$6145
6	0·2196	40	0·0$_3$1944
7	0·1471	50	0·0$_4$7964
8	0·0991	60	0·0$_4$384

It is of interest that $G(\Theta/T)$ plotted against T/Θ is very similar to the Debye curve, $C_v/(C_v)_{\text{classical}}$, as Fig. 99 shows.

Formula (71) is in good agreement with experiment for most metals

† *Ann. d. Physik,* **16** (1933), 530; *Leipziger Vortr.* (1930), 46.
‡ Cf. Sommerfeld and Bethe, loc. cit. 570.

over a considerable temperature range, if Θ is suitably chosen.† For instance, for copper, according to measurements quoted by Meissner,‡ the following are the values:

$T°$ abs.	$(R/R_0)_{\mathrm{red}}$ $\zeta = 0.003$	$(R/R_0)_{\mathrm{calc}}$ $\Theta = 333$
1·32	0·00000	0·00000
4·20	0·00000	0·00000
20·42	0·00051	0·000553
81·2	0·141	0·1413
90·2	0·1804	0·1804
273·2	1·0	1·0

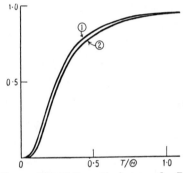

Fig. 99. Comparison of theoretical curves for R/T and C_v.
(1) $C_v/(C_v)_{\mathrm{cl}}$.   (2) $G(\Theta/T)$.

Here $(R/R_0)_{\mathrm{red}}$ denotes the 'reduced' values, obtained by subtracting from the observed resistance the estimated residual resistance ζ. The very good agreement between theory and experiment is surprising in view of the criticism made above. It must, however, be remembered that Θ, a high power of which occurs in the formula, is in effect an arbitrary parameter.

Deviations for low temperatures (below $20°$ K.) occur for lithium and sodium, probably owing to the drop in Θ at low temperatures (Fig. 4). Also the formula does not agree with experiment for many of the transition metals; this will be discussed below.

A further criticism of formula (71) has been made by Peierls.‖ If the electrons in a metal behave like free electrons (i.e. if the energy is proportional to $|\mathbf{k}|^2$), then, when a current is flowing, the momentum is in the opposite direction to the conventional direction of the current. If, on the other hand, the current is carried by positive holes in a nearly full zone, the momentum and current will be in the same direction. In a metal, therefore, which contains only electrons or only positive holes, the momentum transferred to the lattice by the electrons when a current flows will be all in the same direction. At low temperatures, according to Peierls, the lattice requires a comparatively long time to get rid of this momentum and to return to equilibrium, so that, after the current has been flowing for a short time, there will be a large excess of momentum in the direction in which the electrons or positive holes are moving. Now lattice waves moving in this direction cannot decrease the current except by 'Umklappprozesse',† which involve a large change of $\mathbf{q}$ and are therefore very improbable at low temperatures. The current should therefore rise to much greater values than formula (71) for the conductivity suggests. Peierls shows that the resistance should follow the law

$$R \propto e^{-E/kT}, \tag{72}$$

where E is the energy interval from the surface of the Fermi distribution to the first plane of energy discontinuity in k-space.

These considerations are not, of course, applicable to metals where the current is carried by electrons and by positive holes (e.g. the divalent metals). For the monovalent metals, where the electrons lie in one Brillouin zone, Peierls has shown that it is not applicable if the surface of the Fermi distribution touches the planes of energy-discontinuity, as illustrated in Fig. 26 (a). This represents an intermediate case between 'free electrons' and 'positive holes'.

If the electrons behaved as though they were free, E and E/k would have the following values for the alkalis and noble metals:

	Na	K	Rb	Cu	Au
E (volts) . . .	0·67	0·45	0·39	1·52	1·19
E/k (degrees K.) . .	7,800	5,300	4,550	17,700	14,000

If, therefore, formula (72) were valid, the resistance would become exceedingly small in the region $T < \Theta_D$ where these considerations are applicable. The observed resistances, both of the alkali and noble metals,‡ are, however, given as regards their orders of magnitude by Grüneisen's formula (71) We must conclude that the surface of the Fermi distribution touches the first plane of energy discontinuity. In view of the other evidence that the electrons

† Cf. § 5.4.   ‡ Cf. the summary by Meissner quoted above.

behave as if they were free in these metals,[†] this conclusion is surprising.

We see, therefore, that the T^5 law is likely to be valid only if the Fermi surface is far from spherical. In this case, however, it is necessary to consider collisions of one electron with another. If the electrons behave as though they were free, it is clear that collisions between them cannot affect the current, since the total momentum is conserved. When, however, as in the divalent metals, positive holes and electrons are present, it is clear that collisions between them will change the current. The effect on the current is, however, small compared with the scattering by lattice waves at ordinary temperatures, the reason being that both electrons must be in states lying within an energy range kT at the surface of the Fermi distribution, both before and after the collision. Detailed analysis,[‡] however, shows that the term in the resistance due to this cause is proportional to T^2; it will therefore be the dominant term at low temperatures. Moreover, it may be shown that the term will be the greater the smaller is the 'degeneracy temperature' T_0 of the Fermi distribution.

Actually the resistance of the metals Ni, Pd, and Pt can be represented at low temperatures by a formula of the type[||]

$$R = \alpha T^2 + R_G,$$

where R_G is a 'normal' resistance curve given by the Grüneisen formula (71). For platinum the term αT^2 is the dominating one below $10°\,K$. The occurrence of such a term for these metals only is in agreement with the fact, deduced in other ways,[††] that T_0 for the positive holes in these metals is exceptionally small.

10. Resistance of liquid metals

Table XII gives the *change* of resistance of certain metals on melting.[‡‡]

For the monovalent metals it is unlikely that the number (or effective number) of free electrons in the metal changes on melting. The increase in the resistance on melting must therefore be due to the increased disorder of the liquid phase. A measure of the increase in the disorder is given by the increase in the entropy on melting.

---

† Cf. Appendix I.

‡ Baber, *Proc. Roy. Soc.* A, (in press).

|| Cf. de Haas and de Boer, *Physica*, **1** (1934), 609.

†† Cf. Appendix I.

‡‡ The experimental values are taken from the article by Grüneisen, *Handb. d. Phys.* **13** (1928), 28; cf. also Perlitz, *Phil. Mag.* **2** (1926), 1148.

TABLE XII

Metal	Melting-point (degrees K.)	Resistance of liquid / Resistance of solid
Li	459	1·68
Na	370·5	1·45
K	335·3	1·55
Rb	311·5	1·61
Cs	299	1·66
Cu	1356	2·1
Ag	1233·5	1·9
Au	1337	2·3
Al	933	1·64
Zn	692·4	2·1
Cd	593·9	2·0
Sn	504·8	2·1
Pb	600·5	2·1
Tl	580·5	2·0
Hg	234	3·2–4·9†
Ga	302·7	0·58
Sb	903·5	0·67
Bi	544	0·43

Attempts to relate the change in the entropy to the change in the resistance have been made by Simon,‡ Schubin,‖ and Mott.†† We note here that the change ΔS of the entropy per gm. atom on melting, like the change of resistance shown in Table XII, is less for the alkalis than for the close-packed elements, as the following table shows:

	Li	Na	K	Rb	Cs	Cu	Ag	Au	Pb	Zn
Latent heat L (cal. gm. atom) . .	850	630	570	520	500	2,750	2,650	3,200	1,200	1,700
Melting-point T (degrees K.) .	459	370	335	311	299	1,356	1,233	1,336	600	692
Change of entropy $\Delta S = L/T$ . .	1·85	1·7	1·7	1·7	1·7	2·0	2·1	2·4	2·0	2·5

The fact that, for the close-packed metals with more than one valence electron per atom (Al, Cd, Pb, Zn), the increase in the resistance is about the same as for Cu, Ag, and Au suggests that the effective number of free electrons (N_{eff}) in these metals is about the same in the liquid as in the solid phase, and therefore (cf. § 6.3)

---

† The resistance of solid mercury is strongly anisotropic, cf. Grüneisen and Sckell, *Ann. d. Physik*, **19** (1934), 387.

‡ *Zeits. f. Phys.* **27** (1924), 157.   ‖ *Phys. Zeits. d. Sowjetunion*, **5** (1934), 83.

‡‡ *Proc. Roy. Soc.* A, **146** (1934), 465.

considerably less than unity. This fact is of interest in connexion with the optical properties of liquid metals (Chap. III, § 8.5). On the other hand, the fact that the resistance of Zn *decreases* as the temperature is raised above the melting-point suggests that further distortion of the lattice may increase the effective number of free electrons.† For the typically monovalent metals Li, Na, K, where N_{eff} must remain equal to unity, Bridgman‡ finds that

$$\frac{1}{\rho}\frac{d\rho}{dT} \sim \frac{1}{T} \tag{73}$$

for the liquid as for the solid phase.

The change in the resistance of mercury is exceptionally large. This fact is probably connected with the change in the crystal structure, the mean interatomic distance in liquid mercury corresponding to a close-packed structure.‖ We may assume that for close-packed mercury the number of electrons overlapping the first Brillouin zone is very small, but that for the less symmetrical zone of the actual rhombohedral structure (Chap. V, § 2.4) the overlap is greater.

Liquid mercury is also anomalous in that the solution of most other metals in it decreases the resistance.†† This may perhaps be explained in the same way; any distortion of the first Brillouin zone will tend to increase the number of electrons which overlap it.

Ga, Sb, and Bi have a *higher* conductivity in the liquid than the solid phase. We know (Chap. VI, § 6) that the high diamagnetism of Sb and Bi disappears on melting, and it is highly probable that these two metals are close-packed in the liquid phase (cf. p. 314). It follows that the particular Brillouin zone characteristic of the bismuth structure,‡‡ containing just five electronic states per atom, disappears in the liquid phase. We should therefore expect liquid bismuth to have a 'normal' resistance comparable with that of lead. This is in fact observed, as the following values show:‖‖

	Solid	Liquid
Resistance of lead at 325° C. . . .	50	100 microhm-cm.
Resistance of bismuth at 250° C. . . .	220	120 ,,

† Cf. Mott, loc. cit.
‡ *Proc. Amer. Acad. Arts Sci.* **56** (1921), 59; **60** (1925), 395.
‖ Cf., for instance, the discussion by Kratky, *Phys. Zeits.* **34** (1933), 482.
†† Cf. *Handb. d. Metallphysik*, **1** (1935), 344.
‡‡ Chap. V, § 2.4.   ‖‖ Pietenpol and Miley, *Phys. Rev.* **34** (1929), 1588.

Finally we give the mean free path l for some liquid metals at the melting-point, calculated from the formula

$$\frac{1}{\rho} = \sigma = \frac{Ne^2}{m}\frac{l}{2v}, \qquad \tfrac{1}{2}mv^2 = E_{\max}, \qquad (74)$$

with one electron per atom except in the case of Ni. The metals chosen are those for which $N_{\text{eff}} \simeq N$, which is not the case for Zn, Hg, Pb, etc.

	K	Rb	Cu	Ag	Au	Ni†
Resistivity‡ (microhm-cm.) . .	12	25	20	18	30	110
l (Å. U.)	300	200	70	90	54	10

11. Effect of a magnetic field

The two most important effects which we have to consider are the Hall effect and the change of resistance in a magnetic field. To find the magnitude of both these effects, we must investigate the form of the Fermi function when both an electric and a magnetic field act on the electron. We take a magnetic field H parallel to the z-axis and an electric field with components F_x, F_y, F_z. Then the rate of change of the wave number $\mathbf{k}$ of an electron will be given initially by the equations, analogous to (14.1) of Chapter III,

$$\hbar \dot{k}_x = Hev_y/c + eF_x,$$
$$\hbar \dot{k}_y = -Hev_x/c + eF_y,$$
$$\hbar \dot{k}_z = eF_z.$$

For free electrons these reduce, of course, to the classical equations of motion.

If τ is the time of relaxation at any point of the surface of the Fermi distribution, then we may write

$$\frac{f - f_0}{\tau} + \frac{\partial f}{\partial k_x}\dot{k}_x + \frac{\partial f}{\partial k_y}\dot{k}_y + \frac{\partial f}{\partial k_z}\dot{k}_z = 0, \qquad (75)$$

where f is the Fermi distribution function and f_0 the function in the absence of the field. We shall write

$$f - f_0 = \frac{\partial f_0}{\partial E}\chi.$$

† 0·6 electrons per atom.
‡ *Handb. d. Metallphysik*, **1** (1935), 329.

Substituting for $\dot{\mathbf{k}}$, and introducing the operator Ω defined by

$$\hbar\Omega = v_y\frac{\partial}{\partial k_x} - v_x\frac{\partial}{\partial k_y},$$

(75) then becomes $\dfrac{\chi}{\tau} + e(\mathbf{vF}) + \dfrac{eH}{c}\Omega\chi = 0,$ (76)

since the term in Hf_0 vanishes by symmetry. We have neglected here the product χF, which is legitimate since all effects considered are linear in F. The equation is otherwise exact. Equation (76) was first obtained by Peierls.[†]

To make further progress we must make some assumption about the way in which τ and E depend on the wave number $\mathbf{k}$. If one assumes that τ is a function of E only and that E is the same function of $\mathbf{k}$ as for free electrons, then for the Hall coefficient A_H Sommerfeld has found

$$A_H = c/eN,$$

where e is the electronic charge (in e.s.u.) and hence a negative quantity, and N the number of electrons per unit volume. This formula is valid only if kT/ζ is small; at high temperatures it must tend to the classical value[‡]

$$(A_H)_{\text{class}} = 3\pi c/8Ne.$$

It may further be shown that, with this model, the change of resistance $\Delta\rho$ vanishes as $T \to 0$. At finite temperatures there is a very small effect given by

$$\Delta\rho/\rho = BH^2/(1+CH^2),$$

where $B = \tfrac{1}{3}\pi^2(emlkT)^2/(\hbar|\mathbf{k}|)^6,$

$$C = (el/\hbar|\mathbf{k}|)^2,$$

and where l is the mean free path. This formula, however, gives a change in resistance smaller by a factor 10^6 than that observed for normal metals, and also gives a wrong dependence on temperature. As Peierls was the first to point out, the vanishing of $\Delta\rho/\rho$ as $T \to 0$ depends on the assumption that E is the same function of $\mathbf{k}$ as for free electrons, and when this assumption is abandoned one obtains values for $\Delta\rho/\rho$ of a higher order of magnitude.

A general solution of equation (76) can be obtained[||] for *any* form

† *Ann. d. Physik*, **10** (1931), 97; *Ergebn. d. exakt. Naturw.* **11** (1932), 264. Cf. also Jones and Zener, *Proc. Roy. Soc.* A, **145** (1934), 269, equation (4).

‡ Cf. Sommerfeld and Frank, *Rev. Mod. Phys.* **3** (1931), 1.

|| Blochinzev and Nordheim, *Zeits. f. Phys.* **84** (1933), 168; Jones and Zener, loc. cit.

of the function E_k subject to the assumption that τ is constant over the surface of the Fermi distribution and that H is small. The results will thus be valid for the region where A_H is independent of H and $\Delta\rho/\rho$ proportional to H^2. The solution is

$$\chi = -e\tau\Big\{(\mathbf{vF}) - \frac{eH\tau}{c}\Omega(\mathbf{vF}) + \Big(\frac{eH\tau}{c}\Big)^2\Omega^2(\mathbf{vF})...\Big\}.$$

To obtain the Hall coefficient we set $F_z = 0$, and equate the current in the y direction to zero; this gives us†

$$A_H = \frac{2}{eN_x N_y}\int\Big\{\Big(\frac{\partial E}{\partial k_y}\Big)^2\frac{\partial^2 E}{\partial k_x^2} - \frac{\partial E}{\partial k_y}\frac{\partial E}{\partial k_x}\frac{\partial^2 E}{\partial k_x\partial k_y}\Big\}\frac{dS}{|\text{grad }E|}, \quad (77.1)$$

where the integration is over the surface of the Fermi distribution and

$$N_x = 2\int\Big(\frac{\partial E}{\partial k_x}\Big)^2\frac{dS}{|\text{grad }E|},$$

with a similar expression for N_y.

For the change of resistance we get a complicated expression, namely
$$\Delta\rho/\rho = BH^2,$$

$$B = \Big(\frac{2}{ec\rho}\Big)^2\frac{1}{N_x^4}\Big[\int\frac{\partial f_0}{\partial E}\frac{\partial E}{\partial k_y}\Omega\frac{\partial E}{\partial k_x}\,d\mathbf{k}\int\frac{\partial f_0}{\partial E}\frac{\partial E}{\partial k_x}\Omega\frac{\partial E}{\partial k_y}\,d\mathbf{k} -$$
$$-\int\frac{\partial f_0}{\partial E}\Big(\frac{\partial E}{\partial k_y}\Big)^2 d\mathbf{k}\int\frac{\partial f_0}{\partial E}\frac{\partial E}{\partial k_x}\Omega^2\frac{\partial E}{\partial k_x}\,d\mathbf{k}\Big]. \quad (77.2)$$

Both these formulae are correct only to the first order in kT/ζ; they are derived on the assumption that τ is constant and H small. They apply, moreover, only to temperatures above the Debye characteristic temperature.

The expression for the Hall coefficient may be simplified in certain cases. We shall consider four special cases:

(a) The energy is given by

$$E = \frac{\hbar^2}{2m}\alpha k^2.$$

The Hall coefficient is then given by

$$A_H = c/Ne, \quad (78)$$

which is the same expression as for free electrons. In other words, the constant α does not affect the Hall coefficient.

† Cf. Jones and Zener, loc. cit.

(b) The surfaces of constant energy form a family of similar ellipsoids

$$E = \frac{\hbar^2}{2m}(\alpha_1 k_x^2 + \alpha_2 k_y^2 + \alpha_3 k_z^2).$$

Again (77.1) reduces to Sommerfeld's formula c/Ne, so that deviations from spherical symmetry of the energy surfaces in k-space do not necessarily imply that this formula is inapplicable.

(c) A zone is nearly full; the surface of the Fermi distribution consists of a sphere or ellipsoid enclosing positive holes. We obtain

$$A_H = -c/Ne,$$

where N is the number of *positive holes* per unit volume. The 'anomalous' sign of the Hall coefficient of certain metals (Zn and Cd) is thereby explained qualitatively.

(d) The surface of the Fermi distribution consists of two spherical surfaces, one enclosing electrons, the other positive holes. Such a case arises in connexion with the transition elements. Over one spherical surface let the energy be given by

$$E = \frac{\hbar^2}{2m}\alpha_1 k^2$$

and over the other by

$$E = \text{const.} - \frac{\hbar^2}{2m}\alpha_2 |\mathbf{k} - \mathbf{k}_0|^2.$$

Let the number of electrons per atom be n_1 and the number of holes n_2. The Hall coefficient according to (77.1) is then

$$A_H = \frac{c}{N_a e}\frac{n_1 \alpha_1^2 - n_2 \alpha_2^2}{(n_1 \alpha_1 + n_2 \alpha_2)^2}, \tag{79}$$

where N_a is the number of atoms per cm.[3]

For the transition metals we have two zones to consider, the s zone and the d zone; if α_2 refers to the latter, $\alpha_2 \ll \alpha_1$ and hence A_H reduces approximately to $c/N_a e n_1$. For such metals the Hall coefficient is thus given almost entirely by electrons in the s zone, the holes in the d band making no essential contribution. This is in rough agreement with experiment; for example, at room temperature the Hall coefficient for Ag is -8.0×10^{-4} e.m.u. and for Pd -6.8×10^{-4} e.m.u., corresponding to 1·3 and 1·35 electrons per atom.

The alkali metals agree very well with Sommerfeld's formula (78),

assuming one electron per atom, as the following table shows.† Both
the experimental and calculated values are given in e.m. units.

	Li	Na	K	Rb	Cs
$-10^4 A_H$ exp. . .	17·5	25·5	42·1	..	78·1
$-10^4 A_H$ calc. . .	13·7	24·5	47·6	58	73

We give below the measured Hall coefficients for some further
metals, together with the value of n_0 (number of electrons per atom)
deduced from (78). It must be remembered that, in metals where
electrons and positive holes are responsible for the conductivity, the
value of n_0 deduced will be considerably greater than the actual
number of electrons or holes, because, as formula (79) shows, the
effects of the positive holes and electrons tend to cancel out.

	Cu	Ag	Au	Pd	Pt	Sb	Bi
$A_H \times 10^4$	−4·95	−8·01	−7·27	−6·75	−2·00	+12,000	−64,400
n_0	1·50	1·33	1·46	1·35	4·68	$1·5 \times 10^{-3}$‡	$3·4 \times 10^{-4}$

The very small number for polycrystalline bismuth will be noted.

For the change of resistance in a magnetic field few detailed com-
parisons with experiment have been made. For free electrons (energy
surfaces spherical) formula (77.2) gives zero, as already stated, and to
obtain a finite change of resistance one must evaluate the formulae
to the next power to kT/ζ_0. Blochinzev and Nordheim‖ have considered a divalent metal with the form of energy surface shown in
Fig. 39 (b). They assume that the energy of an electron is given as a
function of **k** by the approximation of nearly free electrons (Chap. II,
§ 4.2). With this assumption the number of overlapping electrons
and the curvature of the energy surfaces are functions of a single
parameter ΔE, giving the energy-gap separating the first and second
zones. Blochinzev and Nordheim find that to obtain agreement with
experiment the energy-gap must be taken to be about 0·03 e.v.,
while the true value must be considerably larger. On the other hand,
it is possible that the assumption of nearly free electrons is at fault.
Jones and Zener†† have considered the deviations from the spherical
form of the energy surfaces for the metal lithium, and have shown
that with plausible assumptions the observed change of resistance
can be obtained.

† Zener, *Phys. Rev.* **47** (1935), 636.
‡ Since A_H is positive, n_0 is here the number of positive holes.
‖ Loc. cit.   †† Loc. cit.

Strong fields. Kapitza[†] has found that for strong fields $\Delta\rho/\rho$ is not proportional to H^2, but that for most metals it is approximately proportional to H. When the change of resistance deviates from the square law, the Hall coefficient is in general no longer independent of H.

It is not possible in general to obtain a solution of equation (76) in a closed form valid for all fields, but, if the energy depends upon **k** in a simple manner corresponding to some particular model of the metal, it is possible in certain cases to obtain solutions applicable to all field strengths. Blochinzev and Nordheim (loc. cit.) have obtained such a solution for a model intended to apply to divalent metals. In this model the energy surfaces for the electrons are three sets of ellipses (cf. Fig. 39) and for the positive holes a set of spheres. These authors show that for high fields the $\Delta\rho/\rho, H$ curve is of roughly the same form as the experimental curves of Kapitza.

Jones[‡] has found an exact expression for the change of resistance and Hall coefficient of bismuth, assuming that the electrons have energy surfaces bounded by a series of *similar* ellipses in k-space, and that the energy surface for the positive holes is also bounded by an ellipse (cf. Chap. VI, § 6.3). He also assumes that the times of relaxation τ_+ and τ_- are independent of **k** on either surface. For ideally pure bismuth the results are as follows: For the Hall coefficient with H parallel to the principal axis it is found that

$$(A_H)_\| = \frac{c}{eN_0}(2x-1),$$

where x denotes the ratio of the conductivity perpendicular to the axis due to electrons alone to the total conductivity due to both electrons and positive holes. N_0 is the number of electrons or positive holes per cm.[3] For the change of resistance when the current is perpendicular and H parallel to the principal axis, the coefficient B (cf. equation (77.2)) is given by

$$B = x(1-x)\left(\frac{\sigma_\perp c}{eN_0}\right)^2.$$

$\sigma_\perp$ is the conductivity perpendicular to the axis.

Both these formulae are valid for all field strengths. They contain only two unknowns, x and N_0. Using for $(A_H)_\|$ the value $+7$ given

† *Proc. Roy. Soc.* A, **123** (1929), 292 and 342.
‡ Ibid. **155** (1936), 653.

by Heaps† and for B Kapitza's value $7 \cdot 6 \times 10^{-8}$, we find for N_0 about $10^{-4} \times$ the number of atoms per cm.³, a result already obtained in Chapter VI from the diamagnetic properties.

The formulae obtained above for B and A_H are valid only if the number of electrons and positive holes are *equal*; they will, however, be unequal if minute traces of, say, lead, are present. Jones has shown that in this case A_H is no longer independent of H and that $\Delta\rho/\rho$ is not proportional to H^2. Kapitza has in fact shown that, for bismuth, the field strength at which $\Delta\rho/\rho$ becomes linear in H depends on the purity of the specimen.

12. Resistance of alloys; dilute solutions

12.1. *Matthiessen's rule.*

The resistance of a metal which contains foreign atoms in solid solution is nearly always greater than that of the pure metal, the increase being in many cases considerable; if one atomic per cent. of tin is added to copper (specific resistance at room temperature $1 \cdot 55$ microhm-cm.), the increase in its resistance is $2 \cdot 6$ microhm-cm.

As was first shown by Matthiessen,‡ the increase in the resistance of a metal due to a *small concentration* of another metal in solid solution is in general independent of the temperature. This is shown in Fig. 100, where the resistance of copper both in the pure state and with various small concentrations of other metals in solid solution is plotted against temperature. Since the lines are parallel, the increase ρ_0 is independent of temperature.∥

An alternative statement of Matthiessen's rule is that, if ρ is the resistivity of the alloy, $d\rho/dT$ is independent of concentration. Fig. 101 shows $d\rho/dT$ plotted against the concentration for various metals in solid solution in gold. It will be seen that Co and Cr form exceptions to the rule.

The high resistance of alloys and Matthiessen's rule find a particularly simple explanation in the quantum theory of conductivity. As explained above in § 2, an electron can move quite freely through a perfect lattice, which would consequently have no resistance; the resistance of a pure metal is due to the thermal agitation of the atoms, which destroys the periodicity of the lattice. When, moreover, a

---

† *Phys. Rev.* **30** (1928), 401.
‡ Matthiessen and Vogt, *Ann. d. Phys. u. Chem.* **122** (1864), 19.
∥ Mn is an exception.

foreign atom is present in solid solution, the periodicity of the field within the lattice is broken at that point. Thus electrons may be

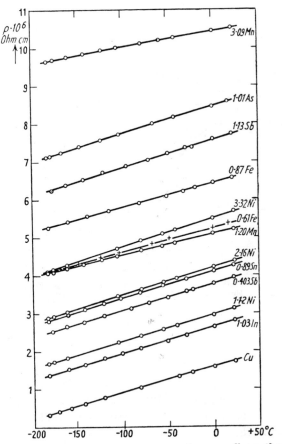

FIG. 100. Resistance-temperature curves of copper alloys; the figures show the atomic percentages of the metal named. (From Linde, *Ann. d. Physik*, **15** (1932), 219.)

deflected by that atom and a resistance will arise, even in the absence of any temperature agitation.

The resistivity of a metal may be written (cf. equation (1))

$$\rho = \frac{m}{Ne^2}\frac{1}{\tau},\tag{80}$$

where N is the (effective) number of electrons per unit volume, and $1/\tau$ the number of times per second that an electron is deflected. Let $1/\tau_0$ be the number of times per second that an electron is deflected by a foreign atom, and $1/\tau_T$ the number of times that it is deflected owing to the thermal oscillation of the atoms;† then

$$\frac{1}{\tau} = \frac{1}{\tau_0} + \frac{1}{\tau_T}.$$

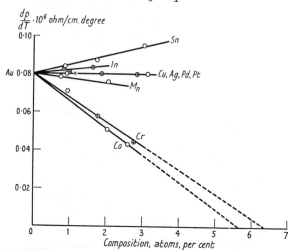

FIG. 101. Change in $d\rho/dT$ with composition for some gold alloys. (From Borelius, *Metallwirtschaft*, **12** (1933), 173.)

Thus we may write for the resistivity

$$\rho = \rho_0 + \rho_T,$$

where

$$\rho_0 = \frac{m}{Ne^2}\frac{1}{\tau_0}, \qquad \rho_T = \frac{m}{Ne^2}\frac{1}{\tau_T}.$$

ρ_0 is equal to the resistance of the alloy at the absolute zero of temperature, and is independent of temperature; ρ_T will be approximately proportional to the temperature, as for a pure metal.

Matthiessen's rule is satisfied if

(1) The effective number of free electrons is unaltered by the addition of foreign atoms.

(2) The thermal vibrations of the foreign atoms give the same scattering as those of the atoms of the solvent metal. Neither of these conditions will be fulfilled exactly, but we should expect the change of ρ_T due to the admixture of, say, 1 per cent. of the foreign metal to be of the order 1 per cent., while the change in ρ_0 may be very much greater.

† The assumption that the effects of foreign atoms and of thermal agitation are additive is justified in § 12.4.

As we have seen, the resistance ρ_0 of an alloy at the absolute zero is due to the break-down in the periodicity of the field in the lattice. This may be due to two causes: the lattice may be distorted by the foreign atom, the neighbouring atoms being pushed out of position; or, even if this does not occur, the field within the foreign atom will be different from what it would be at that lattice point in the pure metal. The latter effect is in general the more important and is the only one which will be considered here.† It is then important to remember that the extra resistance is due to the *difference* between the fields in the two atoms.

12.2. *Theoretical calculation of the increase ρ_0 in the resistance of a noble metal due to foreign atoms in solid solution.* In formula (56) we gave an expression for the resistivity of a metal or alloy in terms of the probability that an electron makes a transition from a state **k** to a state **k′**. Let us consider a metal at the absolute zero of temperature containing a small proportion x of foreign atoms in solid solution. The whole resistance ρ_0 is then due to scattering by the foreign atoms, and formula (56) may conveniently be written

$$\rho_0 = \frac{m}{Ne^2} \frac{vx}{\Omega_0} A, \qquad (81)$$

where A is the effective area presented by each foreign atom. For this we obtained, subject to certain assumptions,

$$A = \int_0^\pi (1-\cos\theta) \left| \frac{2\pi m}{h^2} \int \psi_{k'}^* U \psi_k \, d\tau \right|^2 2\pi \sin\theta \, d\theta, \qquad (82)$$

where U is the difference between the potential in the dissolved and solvent atoms. In the perturbation method used in obtaining this formula, U was assumed to be small. The method is the same as Born's approximation in ordinary collision theory, which is known to give incorrect results for slow electrons and heavy atoms;‡ for such problems the exact method, due to Faxén and Holtsmark,‖ must be used. In our problem, therefore, the perturbation method should give fair results for pairs of atoms, such as copper and zinc, with nearly the same atomic numbers, but not for pairs such as copper (29) and gold (79).

---

† An estimate of the increase in the resistance due to distortion has not at present been made.

‡ Mott and Massey, *The Theory of Atomic Collisions*, p. 126.

‖ Ibid., Chap. II; *Zeits. f. Phys.* **45** (1927), 307.

A method analogous to that of Faxén and Holtsmark, and applic-able when both dissolved and solvent metals are monovalent, has been given by Mott.† Let $u(r)$ be the wave function for an electron in the lowest state in the solvent metal (i.e. the wave function satisfy-ing the boundary condition $\partial u/\partial n = 0$ at the boundary of the atomic polyhedron (cf. Chap. II, § 4.5)). Let E_0 be the corresponding energy. As is shown in Chap. II, § 4.5, the wave function of any other state $\mathbf{k}$ may be written, to a fair approximation, in the form

$$\psi_k = u(r)e^{i(\mathbf{kr})}, \tag{83}$$

with energy
$$E_k = \hbar^2 k^2/2m + E_0.$$

Now consider one of the dissolved atoms: an atomic polyhedron may be drawn surrounding this atom, and a solution $u'(r)$ of Schrödinger's equation obtained such that $\partial u'/\partial n$ vanishes over the surface of the polyhedron. Let E_0' be the corresponding energy. The wave function within the atomic polyhedron of the dissolved atom for energy E_k is then, to the same approximation,

$$u'(r)e^{i(\mathbf{k'r})}, \tag{84}$$

where
$$k'^2 = \frac{2m}{\hbar^2}(E_k - E_0') = k^2 + \frac{2m}{\hbar^2}(E_0 - E_0').$$

We consider now the wave function of an electron of energy E_k moving in the field of the lattice as a whole. The wave function must consist of terms of the type (83) outside and of the type (84) inside the atomic polyhedron of the dissolved atom. In other words, the wave-length $2\pi/k$ will be different inside and outside the dissolved atom.

Both the wave function itself and its gradient must be continuous over the boundary of the polyhedron. These boundary conditions are already satisfied by the functions $u(r)$, $u'(r)$. We have therefore only to consider the exponential terms in the wave functions, $e^{i(\mathbf{kr})}$ and $e^{i(\mathbf{k'r})}$.

We wish to know the probability that an electron is deflected by the dissolved atom. We therefore require a wave function, which, at large distances from the dissolved atom, takes the form of an incident wave and a scattered wave. The problem is therefore the same as that which arises in investigating the scattering of a beam

† *Proc. Camb. Phil. Soc.* **32** (1936), 281.

of free electrons by a field with potential

$$U = E_0 - E_0' \quad \text{within} \atop = 0 \quad\quad \text{outside} \Big\} \text{ the atomic polyhedron.}$$

The solution of this problem is well known; in the cases considered here $|E_0 - E_0'| \ll \frac{1}{2}mv^2$, so that the problem may be treated by Born's approximation. We obtain† for the effective scattering area presented by each foreign atom to the beam of oncoming electrons

$$A = 0 \cdot 86 \pi r_0^2 \left(\frac{E_0 - E_0'}{E_{\max}} \right)^2. \tag{85}$$

Here r_0 is the 'atomic radius' (p. 77) and $E_{\max} = \frac{1}{2}mv^2$, the energy of the electrons with the maximum energy in the Fermi distribution. The increase in the resistance of a metal due to the admixture of $100x$ per cent. of another metal in solid solution is thus given by (81) and (85).

The following table gives the observed‡ increase ρ_0 in the resistance of copper, silver, and gold due to the admixture of 1 per cent. of these metals. We show also the atomic volumes. Ag and Au have the same atomic volume, and hence the same values of $E_{\max}$ and r_0. Formula (85) therefore predicts that ρ_0 *will be the same for 1 per cent. of silver in gold as for 1 per cent. of gold in silver.* The table shows that this is in agreement with experiment.

Increase of resistance in microhm-cm.

Due to	In Cu	In Ag	In Au
1 per cent. of Cu	..	0·068	0·485
„ of Ag	0·14	..	0·38
„ of Au	0·55	0·38	..
Atomic volume, cm.³ per gm. atom .	7·1	10·3	10·3

We show below the areas A and the values of $|E_0 - E_0'|$ for the pairs of metals considered, deduced from the experimental values of ρ_0. The latter must be compared with the calculated values of E_0 obtained by the method of Wigner and Seitz (Chap. II, § 4.5). Fig. 58 on p. 145 shows E_0 for Cu and Ag plotted against r_0. The calculated values are shown in the last column of the following table. For gold no theoretical curve has been obtained; but since the ionization potential of gold is 1·8 volts greater than that of silver and the

† Mott, loc. cit.   ‡ Linde, *Ann. d. Physik*, **15** (1932), 239.

binding energy of the metal 0·8 volts greater, we should expect that for this pair of metals $|E_0 - E_0'|$ would be $1·8 + 0·8 = 2·6$ volts. For copper dissolved in silver it should be $2·6 - 0·25 \simeq 2·4$. These values are shown bracketed. The agreement is fair.

	$A \times 10^{16}\ cm.^2$	$|E - E_0'|$ (e.v.)	
		from (85), using observed resistance	calculated in Chap. IV; cf. Fig. 58
1 per cent. of Cu in Ag	0·14	0·8	0·25
,, of Ag in Cu	0·25	1·6	1·3
,, of Au in Ag ⎱ ,, of Ag in Au ⎰	0·76	1·9	(2·6)
,, of Cu in Au	0·76	2·1	(2·4)
,, of Au in Cu	0·98	3·0	..

12.3. *Dependence of resistance on the properties of the solvent and dissolved metals* (*dilute solutions*). Norbury,† in a paper published in 1921, has summarized the experimental material available at that date. Of the work carried out since then, we shall frequently refer to that of Linde‡ on the resistance of dilute solid solutions in Cu, Ag, and Au.

Norbury‖ was the first to point out the connexion between the increase in resistance ρ_0 and the *valencies*, or horizontal position in the periodic table, of the solvent and dissolved metals. If the metals have the same valency, the resistance ρ_0 due to one atomic per cent. of one metal dissolved in the other will be small; if the valencies are different, it will be large. Figures illustrating this are given in Norbury's paper. Linde (loc. cit.) has found a numerical relationship between the increase of resistance of a noble metal due to the admixture of 1 per cent. of a foreign metal and the valency of the latter. This is illustrated in Fig. 102. If $z+1$ is the number of electrons outside a closed d shell in the dissolved atom, and one the number in the solvent, Linde finds that ρ_0 is proportional to z^2.

The results of Linde have a simple theoretical explanation.†† The core of a solvent atom carries a positive charge $+e$. The core of the dissolved atom carries a positive charge $+(z+1)e$, which is greater by ze than for an atom of the solvent metal. It is the field of this extra charge that must be regarded as causing the scattering of the

† *Trans. Faraday Soc.* **16** (1921), 570.
‡ *Ann. d. Physik*, **10** (1931), 52; **14** (1932), 353; **15** (1932), 219.
‖ Loc. cit.; cf. also Hume-Rothery, *The Metallic State*, Chap. II, Oxford (1931).
†† Mott, *Proc. Camb. Phil. Soc.* **32** (1936), 281.

electrons, and hence the resistance ρ_0. But, by the Rutherford scattering law, the intensity of the scattering is proportional to the square of the scattering charge, and hence to z^2. It follows that ρ_0 is proportional to z^2.

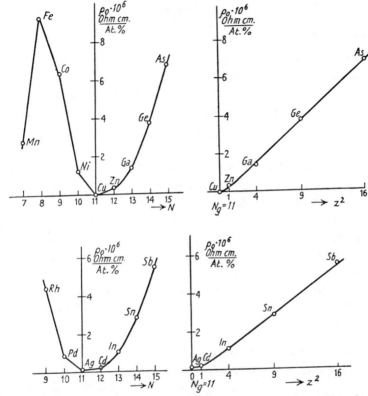

Fig. 102. Increase in the resistance of copper due to one atomic per cent. of various metals in solid solution. N denotes the number of electrons outside an inert gas shell; z denotes $(N-11)$. (From Linde, *Ann. d. Physik*, **15** (1932), 219.)

To obtain a numerical estimate of the resistance we must know the field round the charge. It will not be correct to take the field of a bare positive charge ze, because, as explained in Chap. II, § 5, the charge is screened by the surrounding electrons. An unscreened charge would actually give an infinite resistance.† If we take for the potential

$$V(r) = \frac{ze^2}{r}e^{-qr},$$

† This may be seen by putting $q = 0$ in formula (86) below.

and assume that the electrons behave as if they were free, then the scattering probability may be calculated using Born's approximation, i.e. formula (82), and we obtain for the resistance due to 1 per cent. of the foreign metal in solid solution

$$\rho_0 = \frac{1}{100} \frac{2\pi z^2 e^2}{mv^3} \left\{ \log\left(1 + \frac{1}{y}\right) - \frac{1}{1+y} \right\}, \qquad (86)$$

where

$$y = q^2 \hbar^2 / 4m^2 v^2.$$

v is here the velocity of the electrons. Agreement with experiment may be obtained by calculating v from the Sommerfeld formula (Chap. II, § 3), and taking $1/q \sim 0.3$ Å. U. This is somewhat less than the value obtained by the Thomas-Fermi method in Chap. II, § 5, which may perhaps be due to the errors inherent in that method.

Increase of resistance of pairs of metals. With two metals A and B which are mutually soluble in one another, it is interesting to compare the increase ρ_A in the resistance of A, due to one atomic per cent. of B in solid solution, with the increase ρ_B, due to one atomic per cent. of A dissolved in B. If the two metals A and B have the same atomic volume, the same crystal structure, and the same effective number of free electrons, we should expect ρ_A and ρ_B to be equal, as already stated in the discussion of Cu, Ag, and Au. The following table shows the extent to which this prediction of the theory is fulfilled:

Change in specific resistance ρ_0 (microhm-cm.) due to one atomic per cent. of the first-named metal dissolved in the second

Ratio of atomic volume of first-named metal to second		ρ_0		ρ_0
0·54	Na in K†	1·3	K in Na†	0·95–1·1
0·70	Cu in Au‡	0·485	Au in Cu‡	0·55
0·70	Cu in Ag‡	0·068	Ag in Cu‡	0·14
1·01	Ag in Au‡	0·38	Au in Ag‡	0·38
1·17	Mg in Cd‖	0·40–0·45	Cd in Mg‖	0·5–0·6
0·97	Pd in Pt‖	0·55–0·60	Pt in Pd‖	0·7

The case of Mg and Cd is interesting, the metals having quite

† Molten alloys at 100° C.; cf. the summary by Norbury, loc. cit.
‡ Linde, *Ann. d. Physik*, **15** (1932), 239.
‖ Norbury, loc. cit.

different corrected conductivities $\sigma/M\Theta^2$. The data[†] for the two elements are as follows:

	Mg	Cd
$\sigma/M\Theta^2$	1·25	0·47
Atomic volume cm.$^3 \times 10^{24}$.	23·2	19·8

The fact that ρ_0 is the same for both seems to us good evidence that the higher resistance of pure cadmium is not due to the effective number of free electrons, but to some other cause, e.g. the size and scattering power of the ions.

If we compare a monovalent with a divalent metal, we find, using the scanty data available, that the increase is greater in the divalent metal, as the following table shows:

	Microhm-cm.		Microhm-cm.
Mg in Ag[†]	0·8–1·3	Ag in Mg[†]	3·0–3·5
Cd in Au[‡]	0·64	Au in Cd[†]	1·7–1·9

It has been suggested[||] that this is due to the same cause as the drop in the conductivity in passing from a monovalent to a divalent metal,[††] namely the smaller 'effective number of free electrons' in a divalent metal.

Finally we consider alloys of the transition metals with copper or silver or gold. The following are some of the values:

Pd in Cu[‡]	0·89	Cu in Pd[‡‡]	1·27
Pd in Ag[†]	0·436	Ag in Pd[‡‡]	1·4
Pd in Au[‡]	0·407	Au in Pd[‡‡]	1·0
Ni in Cu[‡]	1·25	Cu in Ni[‡‡]	1·0
Pt in Ag[‡]	1·59	Ag in Pt[||||]	(2·3)
Pt in Au[‡]	1·02	Au in Pt[†††]	1·55

In considering these values we must remember that, for *dilute* solid solutions of Cu, Ag, or Au in Pd or Ni, transitions from the s to the d band will not be possible for the following reason: the transition probability (cf. § 6.3) is proportional to

$$\left| \int \psi_d^* U \psi_s \, d\tau \right|^2,$$ (87)

† Norbury, loc. cit.   ‡ Linde, loc. cit.
|| Mott, *Proc. Phys. Soc.* 46 (1934), 680.   †† Cf. §§ 3, 6.3.
‡‡ Svensson, *Ann. d. Physik*, 14 (1932), 699.
|||| Johansson and Linde, ibid. 6 (1930), 458.   ††† Ibid. 5 (1930), 762.

where U is the perturbing potential. Now this perturbing potential is only finite within a foreign atom, i.e. one of Cu, Ag, or Au. But certainly in a silver atom, and almost certainly in one of Cu and Au, the d shell is full; the wave function of any *empty d* state will have very small amplitude within any of the dissolved atoms. Thus the quantity (87) will be small.

It follows that, so far as the resistance of dilute solid solutions is concerned, only the s electrons need be taken into account in calculating the resistance, and we should not expect any striking difference between the values given in the two halves of the table above.

One might perhaps explain the rather larger values of ρ_0 for a noble metal in Pd than for Pd in a noble metal by the small number (~ 0.55) of s electrons per atom in Pd compared with one in the noble metal. The similarity of the orders of magnitude of the numbers in the two columns of the above table shows, in any case, that the low conductivities of these transition metals are not due to an abnormally small effective number of free electrons.†

13. Resistance of alloys; concentration of both components comparable

In this section we shall assume that the alloy under consideration consists of a single phase, i.e. that all the small crystals of which it is built up have the same composition and crystal structure.

This is not the case for any range of composition for Sn–Pb, Sn–Zn, Sn–Cd, Cd–Zn, which are insoluble in each other. Also in the brasses Cu–Zn, and in many other alloys, there are ranges of composition for which the alloy consists of a mixture of two phases. For the resistance in these cases cf. *Handb. d. Metallphysik*, **1** (1935), 333.

As explained in § 12, we may, from the theoretical point of view, divide up the resistance of an alloy into two parts

$$\rho = \rho_0 + \rho_T. \tag{88}$$

ρ_T is due to the thermal vibration of the atoms, as for a pure metal; ρ_0 is due to the fact that, in an alloy, the lattice field is not periodic.‡

In alloys such as Ag–Au, which do not form any superlattice, ρ_0 is

---

† Iron may be an exception, since we saw (Chap. VI, § 5.1) that there appear to be only 0.2 conduction or s electrons per atom. We should therefore expect the increase in the resistance of iron due to foreign atoms in solid solution to be abnormally high.

‡ A theoretical justification of this separation is given below (p. 300).

independent of temperature.† We may therefore equate ρ_0 to the resistance at the absolute zero of temperature. In alloys such as Cu–Au, however, the degree of order in the lattice depends on temperature, and hence ρ_0 depends on temperature. This is discussed in Chap. I, § 7.

A theory of the resistance of a completely *disordered* alloy is given in the next section. For a completely *ordered* alloy ρ_0 should, in theory, vanish. Actually, the residual resistance found by extrapolation to the absolute zero is found to be much smaller than for the disordered alloy.‡

For a *partially* ordered alloy no theory has yet been given; the assumption made by Bragg and Williams‖ is purely for simplicity.

The term ρ_T in the resistance is in general linear in T. Hence we may write, assuming that no superlattice is in process of formation,

$$\rho_T \simeq T \frac{d\rho}{dT}.$$

13.1. *Resistance of a totally disordered alloy.*†† We shall suppose that the structure remains unchanged through the whole range of composition, and that no superlattice is formed (examples Ag–Au, Pd–Pt). Let the alloy consist of two kinds of atom, A and B, present in the ratio $x : (1-x)$, and suppose that in the crystal the potential energy of an electron in the neighbourhood of an atom of A is $V_A(r)$ and in the neighbourhood of B is $V_B(r)$. As we have seen, an electron can move quite freely through a periodic field, and will only be scattered when deviations from the periodicity occur; the greater the deviations from periodicity the greater will be the probability of scattering. For *dilute* solid solutions (x small), we supposed each A atom to represent a break in the periodicity of the perfect lattice of B atoms; but in the present case, where x and $1-x$ are comparable, it is clearly necessary to take for our perfect lattice the periodic field which resembles most closely the actual field, and to regard deviations from periodicity as arising both in A and B atoms.

We therefore take for our periodic field the field of which the

---

† Certain alloys, such as CuNi (constantan), are exceptions; cf. Mott, *Proc. Roy. Soc.* A, **153** (1936), 699.

‡ Borelius, Johansson, and Linde, *Ann. d. Physik*, **86** (1928), 291.

‖ *Proc. Roy. Soc.* A, **145** (1934), 699; cf. Chap. I, § 7.

†† Nordheim, *Ann. d. Physik*, **9** (1931), 641.

potential in each atom is

$$V = xV_A + (1-x)V_B. \tag{89}$$

In each A atom the divergence from this potential is

$$V - V_A = (1-x)(V_B - V_A), \tag{90.1}$$

and in each B atom

$$V - V_B = x(V_A - V_B). \tag{90.2}$$

The probability per unit time that an atom will be scattered by any one A atom will therefore† be proportional to

$$(1-x)^2|U|^2, \qquad U = \int \psi_k^*(V_B - V_A)\psi_k \, d\tau,$$

and by any B atom

$$x^2|U|^2.$$

Since there are x of the A atoms and $(1-x)$ of the B atoms, the total probability of scattering is proportional to

$$[x(1-x)^2 + (1-x)x^2]|U|^2 = x(1-x)|U|^2.$$

It follows that, if the atomic volume, crystal structure, and number of free electrons remain constant throughout the range, the resistance ρ_0 of the alloys at the absolute zero depends on the composition according to the formula

$$\rho_0 = \text{const.}\, x(1-x). \tag{91}$$

Fig. 103 (a) shows measurements of the resistance of Ag–Au and of Pd–Pt. In both cases the curves lie very close to the theoretical form (91). Fig. 103 (b) shows a similar curve obtained when the resistances of the two pure metals (In and Pb) differ considerably.

The form of the curve of ρ_0 (resistance extrapolated to the absolute zero) plotted against atomic composition for alloys of the ferromagnetic and strongly paramagnetic metals‡ Ni, Pd, Pt with Cu, Ag, and Au is strikingly different from that of the curves illustrated in

---

† Cf. equation (14). The perturbation method is used, but the same result would follow from the analysis of § 12.2.

‡ Au–Pt has been investigated by Johansson and Linde, *Ann. d. Physik*, **6** (1930), 458, and Ag–Pt by the same authors and by Kurnakow and Nemilow, *Zeits. f. an. Chem.* **168** (1928), 339. In the latter there is a 'solubility gap', and in the former case unless the alloy is quenched from above 1,160° C. Cu–Pd has been investigated by the same authors, *Ann. d. Physik*, **82** (1927), 449, and by Svensson (ref. below), with more points on the curve near 50 per cent.; the alloy may be obtained in the disordered, and, for certain compositions, in the ordered state. The resistance and temperature coefficients of Ag–Pd and Au–Pd have been determined by Geibel, *Zeits. f. an. Chem.* **69** (1911), 38; **70** (1911), 246, and by Svensson, *Ann. d. Physik*, **14** (1932), 699. Cu–Ni has been investigated by Chevenard, *Comptes rendus*, **182** (1925), 1388, Krupkowski and de Haas, *Comm. Leiden*, **194** a (1930), and Svensson, *Handb. d. Metallphysik*, **1** (1935), 341.

Figs. 103 (*a*) and (*b*). A typical example is shown in Fig. 103 (*c*). The type of curve, reminiscent of the silhouette of the Matterhorn, may be explained as follows:†

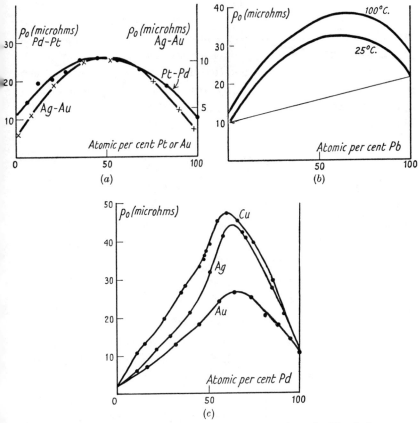

FIG. 103. Resistance of alloys where a continuous range of solid solutions is formed.

(*a*) × Ag–Au, at 0° C. (Beckman, *Thesis*, Upsala (1911). Cf. *International Critical Tables*).
    • Pt–Pd, at 0° C. (Geibel, *Zeits. f. an. Chem.* **70** (1911), 240).

(*b*) In–Pb at 25° C. and 100° C. (Kurnakow and Žemčužny, *Zeits. f. an. Chem.* **64** (1909), 149).

(*c*) Alloys of Cu, Ag, and Au in Pd, at 0° C.
    Cu–Pd, Ag–Pd (Svensson, *Ann. d. Physik*, **14** (1932), 699).
    Au–Pd (Geibel, loc. cit.).

In these alloys the *d* band in the transition metals is incomplete for all percentages of the noble metal up to about 60 per cent. We

† Mott, *Proc. Phys. Soc.* **47** (1935), 571.

have already explained that the resistance of a transition metal is mainly due to transitions in which the electron jumps from the s to the d band, but that isolated atoms of Cu, Ag, or Au dissolved in the metal cannot cause such transitions, because within the dissolved atom the d states are full. But for nearly equal compositions *both* kinds of atom act as scattering centres. Taking, for instance, the case of Ag–Pd, and denoting by $1-x$ the atomic proportion of Pd in the alloy, we see from (90) that the perturbation in the *mean* periodic field (89) represented by any Pd atom is proportional to x, and thus the scattering probability to x^2. The transition probability from an s state to a d state due to the $1-x$ Pd atoms will therefore be

$$(1-x)x^2 N(E_{\max})\left|\int \psi_d(V_A-V_B)\psi_s \, d\tau\right|^2,$$

where $N(E)$ is the density of states in the d band.

We may take the variation of $N(E)$ to be the same as the variation of the paramagnetic susceptibility with concentration x. This is given approximately by†

$$\text{const.}(p-x)^2, \qquad p \simeq 0\cdot6.$$

Hence, finally, the probability that an $s \to d$ transition takes place, considered as a function of x, is proportional to

$$(p-x)^2(1-x)x^2.$$

The probability of an s–s transition will be proportional to $x(1-x)$

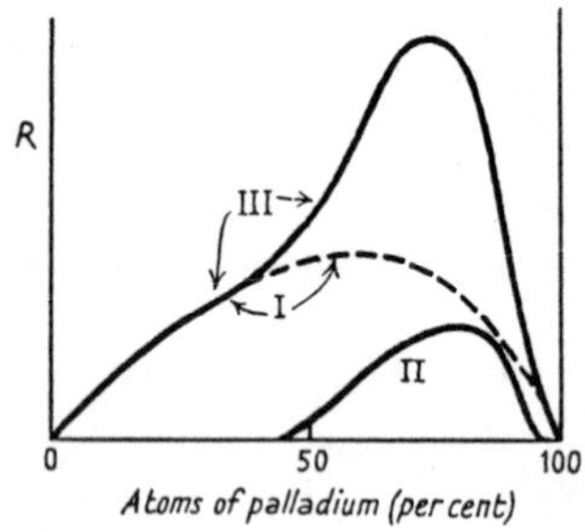

FIG. 104. Theoretical curve for the resistance, at $0°$ K., of Ag–Pd alloys. I, due to s–s transitions; II, due to s–d transitions; III total, viz. I plus an arbitrary multiple of II.

as before, so that the total resistance will be proportional to

$$A(p-x)^2(1-x)x^2+B(1-x)x, \quad (92)$$

where A and B are constants.

Fig. 104 shows the form of this curve with arbitrary values of A and B, which should be compared with the experimental curves of Fig. 103 (c).

We give finally a justification of the assumption made throughout this work, that the resistances due to disorder and to thermal agitation are additive. If an atom A in a disordered alloy is displaced a distance X from its mean position, the probability that an

† Cf. Chap. VI, § 5.4 (c).

electron will be caused by that atom to make a transition from a
state $\mathbf{k}$ to a state $\mathbf{k}'$ will be proportional to a term of the type

$$\left| \int \psi_{k'}^* \left(U - X \frac{\partial V_A}{\partial x} \right) \psi_k \, d\tau \right|^2 ;$$

we have assumed that this may be replaced by

$$\left| \int \psi_{k'}^* U \psi_k \, d\tau \right|^2 + X^2 \left| \int \psi_{k'}^* \frac{\partial V_A}{\partial x} \psi_k \, d\tau \right|^2 .$$

Since U is approximately spherically symmetrical and $\partial V_A / \partial x$ is of
the form $f(r)\cos\theta$, this is justified if we may assume for $\psi_k(\mathbf{r})$ the form
$e^{i(\mathbf{kr})}u(r)$ with $u(r)$ spherically symmetrical; for then it may easily
be shown that the matrix element of U is real and that of $\partial V_A / \partial x$
imaginary.

We shall therefore assume the two terms in the resistance to be
additive in our subsequent discussion; but it must be remembered
that the theorem is not *exact*.

13.2. *Temperature coefficient of resistance of alloys.* As we have
stated, the quantity

$$\rho_T = T \frac{d\rho}{dT}$$

is approximately equal to that part of the resistance of an alloy which
is due to the thermal agitation of the atoms, i.e. to the same cause as
the resistance of a pure metal.

For pairs of metals with similar electronic structure, such as Ag–Au,
Pd–Pt, we should expect ρ_T plotted against atomic composition to
give approximately a straight line. Fig. 105 shows this to be the case
for Ag–Au. The deviations for Pd–Pt may be compared with those for
the paramagnetic susceptibility of the same series of alloys, illustrated
in Fig. 84. According to the theories of Chap. VI, §§ 4 and 5, and
Chap. VII, § 6.3, both ρ_T and the susceptibility are proportional to
the density of states $N(E)$ in the d band.

Fig. 105 (*b*) shows also ρ_T for the Au–Pd series of alloys; it will be
seen that it falls sharply to a value comparable with that of pure
gold at an atomic composition of about 50 per cent. Au. Again the
ρ_T curve should be compared with the curve in Fig. 84 showing the
paramagnetic susceptibility of these alloys. As we have seen, at a
composition of 55 per cent. Au the incomplete d shells in the Pd
atoms become full. The form of the curve is thus direct evidence

that the high resistance of a transition metal is due to the presence of incomplete shells, and hence lends support to the hypothesis of

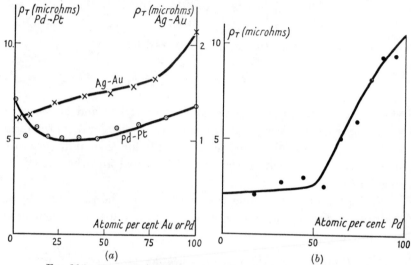

Fig. 105. ρ_T ($= T d\rho/dT$) for various alloys in the range 0°–100° C.
(a) Ag–Au (Beckman, loc. cit.).
    Pd–Pt (Geibel, *Zeits. f. an. Chem.* **70** (1911), 240).
(b) Au–Pd (Geibel, loc. cit.).

§ 6.3 that the resistance is due to electrons making transitions from the s band to the d band.†

14. Resistance of bismuth and its alloys

As we have seen (Chap. V, § 2.4), in bismuth there exists a Brillouin zone which can just contain five electrons per atom. The five valence electrons, however, overlap slightly into the next zone; so in the zone considered there are a certain number of 'positive holes', and an equal number of overlapping electrons in the next zone. We have estimated this number n_0' to be $\sim 10^{-4}$ per atom.

The negative Hall coefficient of polycrystalline bismuth (§ 11), and also the negative thermoelectric power (§ 15) both parallel and perpendicular to the principal axis, suggest that the 'electrons' rather than the positive holes are mainly responsible for the electrical conductivity. We shall therefore work out the *effective* number of free electrons per atom (n_{eff}) for these overlapping electrons.

† Rosenhall (*Ann. d. Physik*, **24** (1935), 297) has found that the addition of hydrogen to Pd–Ag alloys, which we know (p. 200) fills up the positive holes in the d band, also *decreases* the total resistance.

In Chap. VI, § 6.3 we found that the energy surfaces were of the form

$$E = \frac{\hbar^2}{2m}(\alpha_1 k_x^2 + \alpha_2 k_y^2 + \alpha_3 k_z^2),$$

where $\alpha_3 \sim 1$ refers to the direction of the principal axis, and $\alpha_1 \sim \alpha_2 \sim 10^2$. The effective number of free electrons, $\alpha n_0'$, will thus be of the order 10^{-2} perpendicular and 10^{-4} parallel to the principal axis.

The resistance of bismuth at room temperature is actually 0·02 of the resistance of gold; so the time of relaxation, τ, needed to obtain the observed electrical resistance is not abnormally large, if one takes for n_{eff} the value 10^{-2} deduced for motion perpendicular to the principal axis.

For motion parallel to the principal axis, however, one must assume *either* that τ for this direction is a hundred times larger than for gold *or* that the current is carried by positive holes. As we have seen, the negative thermoelectric power seems to rule out the latter hypothesis; but it is difficult to understand why τ should depend so strongly on the direction of the current.

The velocities of the overlapping electrons parallel and perpendicular to the principal axes have already been estimated, and are

$$v_\parallel \sim 10^{-\frac{2}{3}} v_f, \qquad v_\perp \sim v_f.$$

Here v_f is the velocity of 'free' electrons, given by

$$v_f = (3/\pi\Omega_0)^{\frac{1}{3}} h/2m.$$

Since the mean free path is given by $l = 2v\tau$, we do not in any case need to assume an abnormally large mean free path for bismuth. We believe, therefore, that experiments such as those of Eucken and Förster[†] on the anomalous resistance of bismuth wires of about 10^{-4} cm. cross-section must have another explanation.

Resistance of alloys of bismuth. Fig. 106 shows the resistance of alloys of bismuth containing small percentages of other metals in solid solution. The measurements are due to N. Thompson;[‡] earlier measurements on Bi–Sn and Bi–Pb have been made by Ufford[||] and by Thomas and Evans.[††] It will be noticed that Sn and Pb, which have fewer electrons in the outermost shell, increase the resistance

---

[†] *Göttinger Nachrichten*, Math.-Phys. Klasse, Fachgruppe 2, **1** (1934), 43.
[‡] *Proc. Roy. Soc.* A, **155** (1936), 111.
[||] *Proc. Amer. Acad. Arts Sci.* **63** (1928), 309.
[††] *Phil. Mag.* **16** (1933), 329.

and produce an anomalous temperature coefficient; tellurium, on the other hand, decreases the resistance, except at low temperatures. As we saw in Chap. VI, § 6.3, Sn and Pb on the one hand, and Te on the other, have opposite effects on the magnetic anisotropy.

We saw in Chapter VI that ζ_0, measured from the bottom of the second zone for pure bismuth, was about 0·3 e.v., corresponding to a degeneracy temperature of about 4,000°. In the alloys with Sn

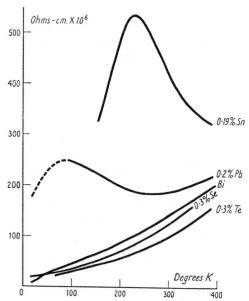

FIG. 106. Resistivity of bismuth alloys parallel to the principal axis, according to N. Thompson.

and Pb, on the other hand, the number of overlapping electrons will be less, so that ζ_0 may be considerably smaller. For a certain concentration the number of overlapping electrons will vanish at the absolute zero of temperature; but with increasing temperature it will increase, as in a semi-conductor. If, then, one assumes that the electric current is carried mainly by the 'electrons', and not by the positive holes, it is possible that in a certain temperature range the resistance will decrease with temperature. The addition of Te would increase the number of overlapping electrons, and would not thus produce any anomaly, but would decrease the resistance, as is observed. A quantitative explanation, however, does not seem possible at present.

15. Thermal conductivity and thermoelectricity

Up to this point we have considered the conduction in metals at uniform temperatures. We now consider a metal in which there is a temperature gradient dT/dx as well as an electric field F. If $f(\mathbf{k})$ is the Fermi distribution function when a steady state is reached, then the electric current j and the rate of flow of heat Q are given by

$$\left.\begin{aligned} j &= \frac{2e}{\hbar} \frac{1}{8\pi^3} \int \frac{\partial E}{\partial k_x} f(\mathbf{k}) \, d\mathbf{k}, \\ Q &= \frac{2}{\hbar} \frac{1}{8\pi^3} \int E \frac{\partial E}{\partial k_x} f(\mathbf{k}) \, d\mathbf{k}, \end{aligned}\right\} \tag{93}$$

where E is the energy of an electron in the state $\mathbf{k}$.

Now the rate of change of the distribution function f due to the field and to the temperature gradient is

$$-\frac{df}{dt} = \frac{\partial f_0}{\partial k_x} \frac{eF}{\hbar} + \frac{\partial f_0}{\partial x} \frac{dx}{dt}, \tag{94}$$

where $dx/dt = \hbar^{-1}\partial E/\partial k_x$ is the velocity along the x-axis of an electron in the state $\mathbf{k}$. The second term arises from the temperature gradient. We obtain easily from the definition (36) of f_0

$$\frac{\partial f_0}{\partial x} = -\frac{\partial f_0}{\partial E}\left(\frac{\partial \zeta}{\partial T} + \frac{E-\zeta}{T}\right)\frac{\partial T}{\partial x}.$$

Hence, from (94), the rate of change of f is

$$-\frac{1}{\hbar}\frac{\partial f_0}{\partial E}\frac{\partial E}{\partial k_x}\left[eF - \left(\frac{\partial \zeta}{\partial T} + \frac{E-\zeta}{T}\right)\frac{\partial T}{\partial x}\right]. \tag{95}$$

This must now be equated to the rate of change of f due to collisions, and hence the time of relaxation $\tau(\mathbf{k})$ determined, giving the displacement of the Fermi-distribution function. The problem is *exactly* the same as that treated in § 6.3, where the current in the absence of a temperature gradient was determined. The only difference is that, instead of eF, the constant in the square brackets in (95) occurs. This constant is independent of $\mathbf{k}$, and depends only on the energy. Therefore, we may write, analogously to (59), for the displaced Fermi-distribution function

$$f = f_0 - \frac{\tau(\mathbf{k})}{\hbar}\frac{\partial E}{\partial k_x}\frac{\partial f_0}{\partial E}\left[eF - \left(\frac{\partial \zeta}{\partial T} + \frac{E-\zeta}{T}\right)\frac{\partial T}{\partial x}\right], \tag{96}$$

where $\tau(\mathbf{k})$ is just the same function of $\mathbf{k}$ as it would be in the absence of a temperature gradient.

We proceed to calculate the current j and flow of heat Q from equation (93), using (96) for f. The term in f_0 will not make any contribution; therefore we have

$$\frac{j}{e} = \left[eF + \left(\frac{\zeta}{T} - \frac{\partial \zeta}{\partial T} \right) \frac{\partial T}{\partial x} \right] K_0 - \frac{1}{T} \frac{\partial T}{\partial x} K_1, \tag{97}$$

$$Q = \left[eF + \left(\frac{\zeta}{T} - \frac{\partial \zeta}{\partial T} \right) \frac{\partial T}{\partial x} \right] K_1 - \frac{1}{T} \frac{\partial T}{\partial x} K_2, \tag{98}$$

where

$$K_n = - \frac{2}{\hbar^2} \frac{1}{8\pi^3} \int E^n \left(\frac{\partial E}{\partial k_x} \right)^2 \frac{\partial f_0}{\partial E} \tau(\mathbf{k}) \, d\mathbf{k}. \tag{99}$$

K_n may be evaluated as follows: let us denote by $\phi(E')$ the integral

$$\phi(E') = \frac{2}{\hbar^2} \frac{1}{8\pi^3} \int \int \left(\frac{\partial E}{\partial k_x} \right)^2 \tau(\mathbf{k}) \frac{dS}{|\text{grad } E|},$$

the integration being over the surface $E(\mathbf{k}) = E'$ in k-space. Then

$$-K_n = \int_0^\infty \phi(E) E^n \frac{\partial f_0}{\partial E} \, dE. \tag{100}$$

Since $\partial f_0 / \partial E$ vanishes except in a small range about the point $E = \zeta$, (100) may be expanded in ascending powers of T; we obtain (cf. Chap. VI, § 1)

$$K_n = \zeta^n \phi(\zeta) + \frac{\pi^2}{6} (kT)^2 \frac{d^2}{d\zeta^2} \{ \zeta^n \phi(\zeta) \} \dots . \tag{101}$$

15.1. *Thermal conductivity.* When no electrical current is flowing, we must put $j = 0$ in equations (97) and (98). Hence we have, multiplying (97) by K_1 / K_0 and subtracting the product from (98),

$$Q = \left(\frac{K_1^2}{K_0} - K_2 \right) \frac{1}{T} \frac{\partial T}{\partial x}.$$

Substituting from (101) for K_0, K_1, and K_2, we obtain, after a short calculation, to the first order in T, for the thermal conductivity κ,

$$\kappa = -Q \Big/ \frac{\partial T}{\partial x} = \frac{\pi^2}{3} k^2 T \phi(\zeta). \tag{102}$$

For the electrical conductivity, σ, we must put $\partial T / \partial x = 0$ in (97), and obtain, as in § 6.3,

$$\sigma = e^2 K_0 = e^2 \phi(\zeta), \tag{103}$$

again to the first order in T. Dividing (102) by (103) we obtain the Wiedemann-Franz law,

$$\frac{\kappa}{\sigma} = \frac{\pi^2}{3} \frac{k^2}{e^2} T, \tag{104}$$

and hence for the 'Lorenz number'

$$L = \frac{\kappa}{\sigma T} = \frac{\pi^2}{3}\left(\frac{k}{e}\right)^2 = 2\cdot45\times10^{-8}\left(\frac{\text{volts}}{\text{degree}}\right)^2.$$

The derivation of (104) given here does not depend on any assumption about the form of the energy surfaces, and is therefore valid for all metals and not merely the monovalent metals. It is valid whether the resistance is mainly due to impurities, or to disorder in alloys, or to the thermal agitation of the atoms. In the latter case, however, it is only valid if $T > \Theta_D$. It is, moreover, only correct to the first order in kT/ζ, and therefore, for metals for which ζ is small,† deviations may be expected at high temperatures. It neglects, further, the contribution made by the lattice vibrations to the thermal conductivity, and will therefore give in general too low a value for the thermal conductivity, especially for poor conductors (e.g. bismuth or alloys with high resistance).‡

The following are the Lorenz numbers for some metals:‖

Metal	Lorenz number $L\times10^8$ volt degree		Metal	Lorenz number $L\times10^8$ volt degree	
	0°	100°		0°	100°
Mg	..	2·31	Pt	2·51	2·60
Al	..	2·23	Cu	2·23	2·33
Mo	2·61	2·79	Ag	2·31	2·37
W	3·04	3·20	Au	2·35	2·40
Fe	2·47	..	Zn	2·31	2·33
Rh	2·57	2·54	Cd	2·42	2·43
Ir	2·49	2·49	Sn	2·52	2·49
Ni	..	2·28	Pb	2·47	2·56
Pd	2·59	2·74	Bi	3·31	2·89

Theoretical value for electronic conduction only, $2\cdot45\times10^{-8}$.

Low temperatures, $T < \Theta$. We shall give only the results of the theory;†† the Wiedemann-Franz law is not satisfied, the dependence of κ on temperature being given by

$$1/\kappa = \text{const. } T^2$$

† Transition metals, or bismuth.

‡ C. S. Smith and E. W. Palmer, *American Inst. of Mining and Metallurgical Engineers, Technical Publication,* **648,** New York (1935), have examined the electrical and thermal conductivities of a number of copper alloys, and have found a relation of the form

$$\kappa = L\sigma T + \kappa_0$$

to exist between them, where κ_0 is roughly the same for all the alloys; for the alloys of high resistance, κ_0 may be as big as $L\sigma T$.

‖ From *Handb. d. Metallphysik,* **1** (1935), 379.

†† Cf. the report by Sommerfeld and Bethe, *Handb. d. Phys.* **24/2** (1933), 535.

for pure metals. This law is in agreement with the results of Grüneisen and Goens,† who measured κ at $21\cdot2°$K. and $83\cdot2°$K. for certain metals. According to the theory the conductivities should be in the ratio $(83\cdot2/21\cdot2)^2 = 15\cdot4$; the experimental results were

					Ratio	Θ
Cu	.	.	.	.	15·8	320
W	.	.	.	.	14·8	310
Au	.	.	.	.	4·7	170

Probably $83\cdot2$ is not sufficiently low in comparison with the characteristic temperature Θ of gold for the T^{-2} law to be valid.

For that part of the resistance which is due to impurities the Wiedemann-Franz law is, however, valid at all temperatures. Thus, if ρ_0 is the 'Restwiderstand', we shall have for the 'residual thermal resistance'

$$\frac{1}{\kappa_0} = \frac{3}{\pi^2}\left(\frac{e}{k}\right)^2 \frac{\rho_0}{T}.$$

A metal with small ρ_0 will have, therefore, for temperatures below Θ, a heat conductivity given by

$$\frac{1}{\kappa} = \frac{3}{\pi^2}\left(\frac{e}{k}\right)^2 \frac{\rho_0}{T} + \text{const. } T^2. \tag{105}$$

15.2. *Thermoelectricity.* When a current flows in a metal, the production of Joule heat is, of course, irreversible. If, however, a current flows through a metal in which a temperature gradient exists, it is observed that a certain amount of heat is developed in a reversible manner. For instance, if for a certain metal this additional heat is given up when the current flows from points of high to points of low temperature, then, if the direction of the current is reversed, the same additional heat will be absorbed.

We consider a rod of unit cross-section, and take the x-axis parallel to the rod. Let there be a field F, a temperature gradient $\partial T/\partial x$, and a current j along the x-axis. The reversible heat which is given up by the current per unit volume in unit time is proportional to j and to $\partial T/\partial x$; we shall write it

$$-\mu j \frac{\partial T}{\partial x}.$$

The coefficient μ defined in this way is known as the Thomson coefficient.

† *Zeits. f. Phys.* **44** (1927), 615; **46** (1927), 151.

We shall now consider a circuit composed of two metals a and b, as shown in Fig. 107, the temperatures of the joints being T_0 and T; under these conditions there exists an electromotive force ϕ acting round the circuit. This is due partly to the contact potential difference which always exists at a junction of dissimilar metals, and which varies with the temperature. We shall denote this contact potential difference by Π. If we take unit charge round the circuit and equate, according to the first law of thermodynamics, the work done to the heat developed, we obtain the equation

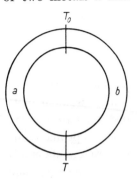

FIG. 107.

$$\phi = \int_{T_0}^{T} \left\{ \frac{d\Pi}{dT} - (\mu_a - \mu_b) \right\} dT. \tag{106}$$

The rate of change of ϕ with T is known as the thermoelectric power, and will be written
$$\epsilon = d\phi/dT. \tag{107}$$

We obtain therefore from (106)

$$\epsilon = \frac{d\Pi}{dT} - (\mu_a - \mu_b). \tag{108}$$

ϵ and Π are both quantities which are associated with a *pair* of metals.

Since the process of carrying unit charge round the circuit is reversible, we may equate the total change in the entropy to zero as follows:
$$\int_{T_0}^{T} \left\{ \frac{d}{dT}\left(\frac{\Pi}{T}\right) - \frac{\mu_a - \mu_b}{T} \right\} dT = 0, \tag{109}$$

and, since this equation holds for all T and T_0, we obtain with the help of (108)
$$\Pi = \epsilon T. \tag{110}$$

Finally, combining (108) and (110), we have the result

$$\epsilon = \int_{0}^{T} \frac{\mu_a}{T} dT - \int_{0}^{T} \frac{\mu_b}{T} dT. \tag{111}$$

It is convenient to define a quantity known as the absolute thermoelectric power, which we shall denote by S, by the equation

$$S = \int_0^T \frac{\mu}{T}\, dT. \tag{112}$$

S is a quantity defined for a single metal. If S is known for any standard metal, its value for any other metal may be obtained by a measurement of the thermoelectric power ϵ of a circuit composed of these two metals. From the theoretical standpoint it is clear that S is the primary quantity which we must consider, since all thermo-electric effects can be derived from it. We shall obtain it by finding μ from the expression for the flow of energy when both a heat current and an electric current pass through a conductor.

We consider, as in the preceding section, a rod of unit cross-section in which there is an electric current j, a heat current Q, a temperature gradient $\partial T/\partial x$, and an electric field F. The energy developed per unit time per unit volume, which we denote by U, is

$$U = Fj - \partial Q/\partial x.$$

We evaluate U by using formulae (97) and (98); eliminating F from the expression for Q, we have

$$Q = \frac{K_1 j}{K_0 e} - \frac{K_0 K_2 - K_1^2}{K_0} \frac{1}{T} \frac{\partial T}{\partial x},$$

and hence, neglecting $\partial^2 T/\partial x^2$ and $(\partial T/\partial x)^2$,

$$\frac{\partial Q}{\partial x} = \frac{j}{e} \frac{\partial}{\partial x}\left(\frac{K_1}{K_0}\right).$$

Hence we obtain easily

$$U = \frac{j^2}{K_0 e^2} - \frac{j}{e} T \frac{\partial T}{\partial x} \frac{\partial}{\partial T}\left\{\frac{1}{T}\left(\frac{K_1}{K_0} - \zeta\right)\right\}. \tag{113}$$

The term in j^2 gives the heat emitted irreversibly; we thus have for μ

$$\mu = \frac{T}{e} \frac{\partial}{\partial T}\left\{\frac{1}{T}\left(\frac{K_1}{K_0} - \zeta\right)\right\} \tag{114}$$

and for S

$$S = \frac{1}{e}\left\{\frac{1}{T}\left(\frac{K_1}{K_0} - \zeta\right)\right\}. \tag{115}$$

Substituting for K_1, K_0 from (101) we have, to the first order in kT/ζ,

$$S = \frac{\pi^2}{3} \frac{k^2 T}{e}\left\{\frac{\partial(\log \sigma(E))}{\partial E}\right\}_{E=\zeta}, \tag{116}$$

where
$$\sigma(E) = \frac{2e^2}{\hbar^2} \int \left(\frac{\partial E}{\partial k_x}\right)^2 \tau(\mathbf{k}) \frac{dS}{|\text{grad } E|}. \tag{117}$$

If we write in the neighbourhood of $E = \zeta$, $\sigma(E) = \text{const.} E^x$, this becomes

$$\left.\begin{aligned} S &= \frac{\pi^2}{3} \frac{k^2 T}{e\zeta} x \\ &= -\frac{2\cdot45 \times 10^{-2} Tx}{\zeta \text{ (e.v.)}} \text{ microvolts per degree.} \end{aligned}\right\} \tag{118}$$

Formula (116) for the thermoelectric power S is valid whatever the relation between E and $\mathbf{k}$, and may thus be applied to all metals and not merely to the monovalent metals. It is valid both for pure metals and for alloys at temperatures above the characteristic temperature Θ, and at lower temperatures provided the 'Restwiderstand' is large compared with the resistance due to the thermal motion of the atoms. It is not, however, valid for *pure* metals if $T < \Theta$. The formula is correct only to the first order in kT/ζ, and thus deviations may be expected at very high temperatures ($\sim 1,000°$) for the transition metals, for which ζ is small.

The quantity $\sigma(E)$ has already been defined;† it represents the electrical conductivity of the metal when the maximum energy ζ of the Fermi distribution is equal to E. $\sigma(\zeta)$ is the actual conductivity of the metal. The thermoelectric power depends, therefore, on the way in which the mean free path and effective mass of the electrons, which determine $\sigma(E)$, vary with the energy. If the electrons with higher energy have greater mean free path, smaller effective mass, etc., than those with less energy, the thermoelectric power will be negative; if the converse is the case, S will be positive. We shall discuss a number of special cases.

(a) The energy surfaces in k-space are spheres

$$E = \frac{\hbar^2}{2m} \alpha |\mathbf{k}|^2.$$

This is the case discussed in § 6.1. We found (equation (55))

$$\sigma(E) \propto |\mathbf{k}|^3 \tau, \tag{119}$$

but the variation of τ with E is not easy to obtain. We should expect that faster electrons (i.e. those with larger $\mathbf{k}$) would be less easily scattered than slower ones, so we shall put $\tau \propto |\mathbf{k}|^s$, with s positive. Since $E \propto |\mathbf{k}|^2$, formula (118) applies with $2x = 3+s$.

† Cf. § 6.3.

We give below a table of the thermoelectric powers observed for two alkali metals, and the values of s and x that must be assumed to give agreement with experiment:†

	ζ_0 (theoretical) in volts	S (observed) microvolts per degree	s	x
Na . . .	3·16	− 5·0	1·8	2·2
K . . .	2·06	−12·5	4·7	3·8

(b) An energy band is almost completely filled with electrons, so that the energy near the surface of the Fermi distribution is given by

$$E = E_0 - \frac{\hbar^2}{2m}\beta|\mathbf{k}|^2,$$

where E_0 represents the highest energy level of the band. In such cases we may speak of conduction by 'positive holes'. $\sigma(E)$ is given by formula (119), and we obtain

$$S = -\frac{\pi^2 k^2 T}{3e(E_0-\zeta)}\frac{3+s}{2}.$$

We should expect s to be positive as before, so that S will be positive.

(c) Noble metals, copper, silver, and gold. These have positive thermoelectric power, and must represent an intermediate case between (a) and (b). Expressing $\sigma(E)$ in the form const. E^x the following are the values of x required to give agreement with experiment:

	Cu	Ag	Au
$S_{100}-S_0$ (microvolts obs.).	+0·5	+0·7	+0·5
ζ_0 (volts calculated) . .	7·1	5·52	5·56
x	−1·4	−1·6	−1·1

(d) Bismuth group. In bismuth, as we have seen,‡ a Brillouin zone is nearly full, there being about 10^{-4} electrons per atom overlapping into the second zone. Since the thermoelectric power is negative, we must assume that the 'overlapping' electrons in the second zone are more important for the conductivity than the 'positive holes' (this agrees with the fact that the Hall coefficient of polycrystalline bismuth is negative, cf. § 11). With the form of the energy surfaces given in § 14, we find that the energy interval

† $x = 3$ is sometimes quoted as the 'theoretical' value for free electrons, but this can only be derived by making certain special assumptions about the scattering field.
‡ Chap. V, § 2.4.

between the top of the Fermi distribution and the lowest energy in the second zone is 0·3 e.v.; equating this to ζ in formula (118), and putting $s = 3$, we obtain $S = -70$ microvolts per degree at 0° C., which compares favourably with the experimental values -110 and -54 microvolts per degree parallel and perpendicular respectively to the principal axis.

Antimony has a large positive thermoelectric power as well as a positive Hall coefficient. The positive holes appear, therefore, to carry most of the electric current in this metal.

(e) *Transition metals, Ni, Pd, Pt.*† These metals have large negative thermoelectric powers, at 100° C., $-21·6$, $-9·5$, $-7·3$ microvolts per degree, as compared with silver ($+2·0$). This is easily to be explained in terms of the theory of conduction in these metals given in § 6.3. The scattering processes mainly responsible for the resistance are those in which an electron makes a transition from the s to the d band. The probability $1/\tau$ of such transitions is proportional to the density of states $N_d(E)$ in the d band. Now $N_d(E)$ decreases rapidly with increasing energy as Fig. 80 shows; therefore τ increases rapidly, so that, by (116), S is large and negative.

If we assume that the variation with energy of all factors in $\sigma(E)$ other than $N_d(E)$ is negligible, we have, setting

$$N_d(E) = \text{const. } \sqrt{(E_0 - E)} \quad \text{and} \quad 1/\tau \propto N_d(E),$$

$$S = \frac{\pi^2}{6} \frac{k^2 T}{e(E_0 - \zeta)} = -\frac{1·22 \times 10^{-2} T}{(E_0 - \zeta) \text{ e.v.}} \text{ microvolts per degree.}$$

The following values of $E_0 - \zeta$, in electron volts, have to be assumed to give agreement with experiment:

	Ni	Pd	Pt
$S_{100} - S_0$ (microvolts obs.)	$-3·8$	$-2·8$	$-2·9$
$E_0 - \zeta$ (deduced)	0·32	0·44	0·42
$E_0 - \zeta$ (from sp. heat)	0·2	0·17	0·27
$E_0 - \zeta$ (from resistance at high temperatures) .	..	0·30	0·42

We show also the values deduced from specific heat measurements at low temperatures (Chap. VI, § 5.2) and from resistance measurements at high temperatures (§ 7).

If copper, silver, or gold‡ be added to a transition metal, the positive holes in the d band are partially filled up, so that $E_0 - \zeta$ will decrease. We should therefore expect S to increase with increasing composition

† Cf. note on p. 314.   ‡ Or hydrogen, cf. Heimburg, *Phys. Zeits.* **24** (1923), 149.

of the noble metal, until about the 55 per cent. composition is reached, when the positive holes are all full. We should then expect S to drop rapidly to a value comparable with that for silver or gold. Fig. 108 shows the results obtained by Geibel[†] for Pd–Ag and Pd–Au. The drop when the concentration of Ag or Au passes the critical value is not as sharp as the theory leads one to expect.

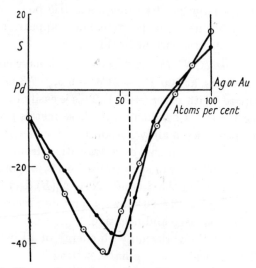

FIG. 108. Thermoelectric power S, in millivolts, of Pd–Ag and Pd–Au alloys, at 900° C., against Pt.

⊙ Pd–Ag.   ● Pd–Au.

(f) *Low temperatures*, $T \ll \Theta_D$. According to Bethe,[‡] in this region

$$S \propto T^3 dT/dx.$$

The factor giving the sign and numerical value has not, however, been evaluated.

[†] *Zeits. f. an. Chem.* **69** (1910), 38; **70** (1911), 240. See also Sedström, *Diss.* Lund (1924).

[‡] Sommerfeld and Bethe, loc. cit. 579.

Notes of recent developments

p. 279. Mr. J. T. Randall has informed us by letter that he has obtained X-ray diffraction bands from liquid bismuth, and that they are similar to those from lead, which is close-packed in the solid state.

p. 313. The thermoelectric behaviour of these metals and also of nickel near the Curie temperature is discussed in greater detail by Mott (*Proc. Roy. Soc.* **A, 156** (1936), 368).

THE EFFECTIVE NUMBER OF FREE ELECTRONS IN
CERTAIN METALS

Alkali metals. There is a good deal of evidence that the electrons in these metals may be treated as free, in the sense that the energy is given in terms of the wave number by the formula $E = \alpha \hbar^2 k^2 / 2m$ with $\alpha \sim 1$. The evidence is deduced from the optical constants (Chap. III, §§ 7, 8), the Hall coefficient (Chap. VII, § 11), soft X-ray emission (Chap. III, § 9.1), the magnetic susceptibility (Chap. VI, § 4), and from theoretical calculations of Wigner and Seitz (Chap. II, § 4.5). The evidence from the optical properties is not entirely conclusive, because optical measurements always give N_{eff} equal to N for frequency large compared with that for which photoelectric absorption takes place, which is probably in the infra-red for the alkalis; the Hall coefficient, moreover, does not depend on α. However, the calculations for sodium of Wigner and Seitz suggest that α is about unity, while those of Seitz for lithium give a rather smaller value. The X-ray emission shows that the breadth of the Fermi distribution for Li and Na is in accordance with the assumption $\alpha \sim 1$. On the other hand, according to Peierls (Chap. VII, § 9), the electrical conductivity at low temperatures suggests that the surface of the Fermi distribution touches the first planes of energy discontinuity.

Noble metals, copper, silver, and gold. It is certain that these are monovalent in the metallic state, because otherwise the d shells would be ionized and the metals would be ferromagnetic or strongly paramagnetic (Chap. VI, § 5), and have the other properties, such as low conductivity (Chap. VII, § 6.3), characteristic of transition metals. Further, in alloy structures determined by the Hume-Rothery rule (Chap. V, § 3.1) these metals always contribute just one electron.

Optical measurements give for n_{eff} (the effective number of free electrons per atom) 0·37, 0·89, 0·73 for Cu, Ag, and Au respectively. Evidence from the resistance of Ag–Au alloys (Chap. VII, § 12.3) shows, however, that n_{eff} must be almost exactly the same for silver as for gold. Comparison of the resistances of copper and silver containing small amounts of foreign atoms in solid solution suggests also that n_{eff} is not much smaller for copper than for silver (Chap. VII,

§ 12.3) and the value obtained by optical measurements is probably too low, possibly owing to oxidization. For copper, theoretical calculations by the method of Wigner and Seitz give $n_{\text{eff}} \sim 1$. The electronic specific heat of silver at low temperatures also agrees with the assumption that the electrons behave as if free (Chap. VI, § 2).

The absorption bands, and hence the colours of these metals, may be due to ejection of a d electron or to absorption by the conduction electrons (Chap. III, § 8.2).

According to the considerations of Peierls (Chap. VII, § 9) the behaviour at low temperatures of the resistance of these metals shows that the surface of the Fermi distribution must touch the planes in k-space across which the energy is discontinuous. If this is the case, the energy-gap ΔE must give the low-frequency limit of internal photoelectric absorption, which is 4·0 e.v. for Ag, unless the absorption is due to ejection of electrons from the d band, in which case ΔE will be greater.

Transition metals. In these metals the band of levels which corresponds to an inner d state of the free atom is not fully occupied. To the 'positive holes' in the d band are due:

(1) The ferromagnetism or high paramagnetism shown by these metals (Chap. VI, § 5).

(2) The low electrical conductivity and high thermoelectric power and anomalous behaviour of the resistance both at high and low temperatures (Chap. VII).

(3) The low reflection coefficient for long wave-lengths (Chap. III, § 8).

(4) The high electronic specific heat (Chap. VI, § 5.2).

For direct evidence of the existence of these 'holes' cf. the discussion on p. 131 of the X-ray absorption spectrum.

The triad Ni, Pd, Pt are discussed most fully in this book. These metals have from 0·55–0·6 positive holes per atom, and an equal number of electrons in the s band, which are responsible for the electrical conductivity and the cohesion.

The evidence for the number of positive holes is drawn from:

(1) The saturation magnetic moment of nickel and of alloys of nickel with copper and other elements (Chap. VI, § 5.1).

(2) The paramagnetic susceptibility of alloys of palladium and platinum with noble metals and with hydrogen (Chap. VI, § 5.4).

(3) The electrical conductivities and thermoelectric powers of these alloys (Chap. VII, §§ 6.3 and 15.2).

The behaviour of the specific heat (p. 198), optical constants (p. 125), and thermal expansion (p. 200) of these alloys is also relevant.

Evidence that the effective numbers of s electrons (N_{eff}) are not very different for these metals and for the noble metals is deduced in Chap. VII, § 13.2 from the resistance of alloys, and from the Hall effect in § 12.

The 'degeneracy temperature' T_0 for the positive holes may be deduced from:

(1) The specific heat in the liquid helium range and also at high temperatures (Chap. VI, § 5.2).

(2) The electrical resistance at high temperatures (Chap. VII, § 7).

(3) The thermoelectric power (Chap. VII, § 15.2).

The results obtained lie between 2,000° and 4,000°, compared with a theoretical value of about 50,000° for the conduction electrons.

Bismuth. In bismuth the electrons almost fill a Brillouin zone, there being a small number of electrons overlapping into the next zone and an equal number of positive holes. This number we estimate to be about 10^{-4} per atom; this is deduced from:

(1) The large effect on the magnetic susceptibility (Chap. VI, § 6.3) and on the electrical conductivity (Chap. VII, § 14) obtained by adding small amounts of Sn, Pb, Te, etc.

(2) The de Haas-van Alphen effect in pure bismuth, considered in relation to the absolute magnitude of the diamagnetic susceptibility (Chap. VI, § 6.4).

(3) The high values of the Hall coefficient and the change of resistance in a magnetic field (Chap. VII, § 11).

In spite of the small number of 'overlapping electrons' n_0', the effective number of free electrons n_{eff} is much larger, and is about 0·03–0·04 (Chap. VII, § 14). We do not believe the mean free path in bismuth to be abnormally great.

The degeneracy temperature T_0 of the 'overlapping electrons' is about 2,700° K., corresponding to an energy of 0·23 e.v. (cf. p. 213).

APPENDIX II

SOME CONSTANTS OF METALS

Metal	Atomic number	Crystal structure	Atomic volume V, cm.³ per gm. atom	Atomic radius r_0, cm.×10⁸	Compressibility at 20° C., χ, dynes/cm.² ×10¹²	Thermal-expansion coefficient, α, degree⁻¹×10⁶	$\gamma = \dfrac{\alpha V_0}{\chi_0 C_v}$
Li	3	b.c. C.	13·0	1·72	8·9	180	1·17
Be	4	Hex.	4·90	1·24	..	..	
Na	11	b.c. C.	23·7	2·10	15·8	216	1·25
Mg	12	Hex.	14·0	1·76	3·0	75	1·51
Al	13	f.c. C.	10·03	1·58	1·37	67·8	2·17
K	19	b.c. C.	45·5	2·61	33	250	1·34
Ca	20	{ f.c. C. / Hex.	25·85	2·16	5·7	..	..
Ti	22	Hex.	10·7	1·62	..	..	
V	23	b.c. C.	8·92	1·52	..	..	
Cr	24	b.c. C.	7·32	1·42	0·8	..	..
Mn	25	{ Cubic / Tetrag.	7·52	1·43	0·84	63	2·42
Fe	26	{ b.c. C. / f.c. C.	7·10	1·40	0·60	33·6	1·60
Co	27	{ f.c. C. / Hex.	6·70	1·38	0·55	37·2	1·87
Ni	28	f.c. C.	6·67	1·38	0·54	38·1	1·88
Cu	29	f.c. C.	7·10	1·41	0·75	49·2	1·96
Zn	30	Hex.	9·16	1·53	1·72	90	2·01
Ga	31	..	11·82	1·67	2·1	..	..
Ge	32	diamond	13·44	1·74	1·4	..	..
As	33	Trig.	13·11	1·73	4·5	16	0·19
Rb	37	b.c. C.	56·2	2·81	40	270	1·48
Sr	38	f.c. C.	33·7	2·36	8·2	..	..
Y	39	..	19·45	1·97	..	..	..
Zr	40	{ Hex. / b.c. C.	13·97	1·71	..	..	..
Nb	41	b.c. C.	11·0	1·63	..	..	
Mo	42	b.c. C.	9·42	1·55	0·36	15·0	1·57
Ru	44	Hex.	8·28	1·48	..	..	
Rh	45	f.c. C.	8·37	1·49	0·37	..	..
Pd	46	f.c. C.	9·28	1·54	0·54	34·5	2·23
Ag	47	f.c. C.	10·27	1·59	1·01	57	2·40
Cd	48	Hex.	13·01	1·73	2·25	93	2·19
In	49	Tetrag.	15·83	1·84	2·7	..	..
Sn	50	{ Tetrag. / diamond	16·30	1·86	1·91	64	2·14
Sb	51	Trig.	18·20	1·93	2·7	33	0·92
Cs	55	b.c. C.	71·0	3·04	61	290	1·29
Ba	56	b.c. C.	38·2	2·47	10·3	..	..
La	57	..	22·59	2·07	3·5	..	..
Hf	72	Hex.	13·98	1·76	0·9	..	..
Ta	73	b.c. C.	11·2	1·64	0·49	19·2	1·75
W	74	b.c. C.	9·63	1·56	0·30	13·0	1·62
Re	75	Hex.	8·78	1·51	..	..	..
Os	76	Hex.	8·49	1·49	..	..	..

SOME CONSTANTS OF METALS (contd.)

Metal	Atomic number	Crystal structure	Atomic volume V, cm.^{3} per gm. atom	Atomic radius r_0, cm.$\times 10^8$	Compressibility at $20°$ C., χ, dynes/cm.^{2} $\times 10^{12}$	Thermal-expansion coefficient, α, degree$^{-1}\times 10^6$	$\gamma = \dfrac{\alpha V_0}{\chi_0 C_v}$
Ir	77	f.c. C.	8·62	1·50	0·27	..	..
Pt	78	f.c. C.	9·12	1·53	0·38	26·7	2·54
Au	79	f.c. C.	10·22	1·59	0·54	43·2	3·03
Hg	80	Trig.	14·26	1·76	3·8	..	..
Tl	81	Tetrag.	17·25	1·89	2·3	90	2·73
Pb	82	f.c. C.	18·27	1·93	2·30	86·4	2·73
Bi	83	Trig.	21·33	2·03	2·97	40	1·14

PHYSICAL CONSTANTS AND CONVERSION FACTORS

Velocity of light	c	$2·99796 \times 10^{10}$ cm. sec.$^{-1}$
Electronic charge	e	$4·770 \times 10^{-10}$ e.s.u.
Planck's constant	h	$6·547 \times 10^{-27}$ erg. sec.
	$\hbar = h/2\pi$	$1·0420 \times 10^{-27}$ erg. sec.
Electronic mass	m	$9·035 \times 10^{-28}$ gm.
Mass of hydrogen atom	M	$1·6617 \times 10^{-24}$ gm.
Atomic unit of length	$\hbar^2/me^2$	$0·5284 \times 10^{-8}$ cm.
Atomic unit of energy (ionization potential of hydrogen)	$me^4/2\hbar^2$	$13·53$ electron volts.
Bohr magneton	$e\hbar/2mc$	$9·174 \times 10^{-21}$ erg. gauss^{-1}.
		$5·766 \times 10^{-9}$ electron volt gauss^{-1}.
Loschmidt's number (molecules per gram-molecule)	L	$6·064 \times 10^{23}$.
Boltzmann's constant	k	$1·3708 \times 10^{-16}$ erg. degree^{-1}.
		$0·8615 \times 10^{-4}$ electron volt degree^{-1}.
Mechanical equivalent of heat	J	$4·1852 \times 10^7$ erg. cal.$^{-1}$
Gas constant (Lk/J)	R	$1·9864$ cal. degree^{-1} mol.$^{-1}$

One electron volt	=	$1·5911 \times 10^{-12}$ erg.
One erg	=	$0·6285 \times 10^{12}$ electron volt.
One electron volt per atom	=	$23·05$ kilo-cal. per gm. atom.
One absolute electrostatic unit of resistance	=	$8·98776 \times 10^{11}$ ohm.
Wave-length of light with quantum energy one electron volt	=	$1·2336 \times 10^{-4}$ cm.
Velocity of electron with energy one electron volt	=	$0·5935 \times 10^8$ cm. sec.$^{-1}$
Wave number k ($= mv/\hbar$) of electron with energy one electron volt	=	$0·5146 \times 10^8$ cm.$^{-1}$

SOME BOOKS AND ARTICLES DEALING WITH THE PROPERTIES OF METALS AND ALLOYS

F. BLOCH. 'Elektronentheorie der Metalle', *Handbuch d. Radiologie*, VI/1, 2te Aufl., pp. 226–75. Leipzig, 1933.

—— 'Moleculartheorie des Magnetismus', ibid. VI/2, pp. 354–484. Leipzig, 1934.

G. BORELIUS. 'Physikalische Eigenschaften der Metalle', *Handbuch d. Metallphysik*, i, pp. 181–520. Berlin, 1934.

W. H. and W. L. BRAGG. *The Crystalline State*. London, 1933.

L. BRILLOUIN. *Die Quantenstatistik*. Berlin, 1933.

U. DEHLINGER. 'Gitteraufbau metallischer Systeme', *Handbuch d. Metallphysik*, i, pp. 1–180. Berlin, 1934.

A. EUKEN. 'Energie- und Wärme-Inhalt', *Handbuch d. exp. Physik*, **13**, 1928.

R. H. FOWLER. *Statistical Mechanics*, 2nd ed. Cambridge, 1936.

G. GRÜNEISEN. 'Zustand des festen Körpers', *Handbuch d. Physik*, **10**, pp. 1–59. Berlin, 1926.

—— 'Metallische Leitfähigkeit', ibid. **10**, pp. 1–75. Berlin, 1928.

W. HUME-ROTHERY. *The Metallic State*. Oxford, 1931.

—— *The Structure of Metals and Alloys*. London, 1936.

W. MEISSNER. 'Elektronenleitung', *Handbuch d. exp. Physik*, **11**/2, 1935.

L. W. NORDHEIM. 'Kinetische Theorie des metallischen Zustandes' in Müller-Pouillet's *Lehrbuch d. Physik*, i, pp. 243–876. Braunschweig, 1934.

R. PEIERLS. 'Elektronentheorie der Metalle', *Ergebnisse d. exakt. Naturwiss.* **11**, pp. 264–322, 1932.

E. SCHMID and W. BOAS. *Kristallplastizität*. Berlin, 1935.

J. C. SLATER. 'Electronic Structure of Metals', *Rev. Mod. Phys.* **6**, pp. 208–80, 1934.

A. SOMMERFELD and H. BETHE. 'Elektronentheorie der Metalle', *Handbuch d. Physik*, **24**/2, pp. 333–622, 1933.

E. C. STONER. *Magnetism and Matter*. London, 1934.

J. H. VAN VLECK. *Electric and Magnetic Susceptibilities*. Oxford, 1933.

E. VOGT. 'Magnetismus der metallischen Elemente', *Ergebnisse d. exakt. Naturwiss.* **11**, pp. 323–51, 1932.

INDEX OF NAMES

INDEX OF SUBJECTS

CATALOGUE OF DOVER BOOKS

PHYSICS

General physics

FOUNDATIONS OF PHYSICS, R. B. Lindsay & H. Margenau. Excellent bridge between semi-popular works & technical treatises. A discussion of methods of physical description, construction of theory; valuable for physicist with elementary calculus who is interested in ideas that give meaning to data, tools of modern physics. Contents include symbolism, mathematical equations; space & time foundations of mechanics; probability; physics & continua; electron theory; special & general relativity; quantum mechanics; causality. "Thorough and yet not overdetailed. Unreservedly recommended," NATURE (London). Unabridged, corrected edition. List of recommended readings. 35 illustrations. xi + 537pp. 5⅜ x 8.
S377 Paperbound **$3.00**

FUNDAMENTAL FORMULAS OF PHYSICS, ed. by D. H. Menzel. Highly useful, fully inexpensive reference and study text, ranging from simple to highly sophisticated operations. Mathematics integrated into text—each chapter stands as short textbook of field represented. Vol. 1: Statistics, Physical Constants, Special Theory of Relativity, Hydrodynamics, Aerodynamics, Boundary Value Problems in Math. Physics; Viscosity, Electromagnetic Theory, etc. Vol. 2: Sound, Acoustics, Geometrical Optics, Electron Optics, High-Energy Phenomena, Magnetism, Biophysics, much more. Index. Total of 800pp. 5⅜ x 8.
Vol. 1 S595 Paperbound **$2.25**
Vol. 2 S596 Paperbound **$2.25**

MATHEMATICAL PHYSICS, D. H. Menzel. Thorough one-volume treatment of the mathematical techniques vital for classic mechanics, electromagnetic theory, quantum theory, and relativity. Written by the Harvard Professor of Astrophysics for junior, senior, and graduate courses, it gives clear explanations of all those aspects of function theory, vectors, matrices, dyadics, tensors, partial differential equations, etc., necessary for the understanding of the various physical theories. Electron theory, relativity, and other topics seldom presented appear here in considerable detail. Scores of definitions, conversion factors, dimensional constants, etc. "More detailed than normal for an advanced text . . . excellent set of sections on Dyadics, Matrices, and Tensors," JOURNAL OF THE FRANKLIN INSTITUTE. Index. 193 problems, with answers. x + 412pp. 5⅜ x 8.
S56 Paperbound **$2.00**

THE SCIENTIFIC PAPERS OF J. WILLARD GIBBS. All the published papers of America's outstanding theoretical scientist (except for "Statistical Mechanics" and "Vector Analysis") Vol I (thermodynamics) contains one of the most brilliant of all 19th-century scientific papers—the 300-page "On the Equilibrium of Heterogeneous Substances," which founded the science of physical chemistry, and clearly stated a number of highly important natural laws for the first time; 8 other papers complete the first volume. Vol II includes 2 papers on dynamics, 8 on vector analysis and multiple algebra, 5 on the electromagnetic theory of light, and 6 miscellaneous papers. Biographical sketch by H. A. Bumstead. Total of xxxvi + 718pp. 5⅝ x 8⅜.
S721 Vol I Paperbound **$2.50**
S722 Vol II Paperbound **$2.00**
The set **$4.50**

BASIC THEORIES OF PHYSICS, Peter Gabriel Bergmann. Two-volume set which presents a critical examination of important topics in the major subdivisions of classical and modern physics. The first volume is concerned with classical mechanics and electrodynamics: mechanics of mass points, analytical mechanics, matter in bulk, electrostatics and magnetostatics, electromagnetic interaction, the field waves, special relativity, and waves. The second volume (Heat and Quanta) contains discussions of the kinetic hypothesis, physics and statistics, stationary ensembles, laws of thermodynamics, early quantum theories, atomic spectra, probability waves, quantization in wave mechanics, approximation methods, and abstract quantum theory. A valuable supplement to any thorough course or text.
Heat and Quanta: Index. 8 figures. x + 300pp. 5⅜ x 8½. S968 Paperbound **$2.00**
Mechanics and Electrodynamics: Index. 14 figures. vii + 280pp. 5⅜ x 8½.
S969 Paperbound **$1.85**

THEORETICAL PHYSICS, A. S. Kompaneyets. One of the very few thorough studies of the subject in this price range. Provides advanced students with a comprehensive theoretical background. Especially strong on recent experimentation and developments in quantum theory. Contents: Mechanics (Generalized Coordinates, Lagrange's Equation, Collision of Particles, etc.), Electrodynamics (Vector Analysis, Maxwell's equations, Transmission of Signals, Theory of Relativity, etc.), Quantum Mechanics (the Inadequacy of Classical Mechanics, the Wave Equation, Motion in a Central Field, Quantum Theory of Radiation, Quantum Theories of Dispersion and Scattering, etc.), and Statistical Physics (Equilibrium Distribution of Molecules in an Ideal Gas, Boltzmann statistics, Bose and Fermi Distribution, Thermodynamic Quantities, etc.). Revised to 1961. Translated by George Yankovsky, authorized by Kompaneyets. 137 exercises. 56 figures. 529pp. 5⅜ x 8½. S972 Paperbound **$2.50**

ANALYTICAL AND CANONICAL FORMALISM IN PHYSICS, André Mercier. A survey, in one volume, of the variational principles (the key principles—in mathematical form—from which the basic laws of any one branch of physics can be derived) of the several branches of physical theory, together with an examination of the relationships among them. Contents: the Lagrangian Formalism, Lagrangian Densities, Canonical Formalism, Canonical Form of Electrodynamics, Hamiltonian Densities, Transformations, and Canonical Form with Vanishing Jacobian Determinant. Numerous examples and exercises. For advanced students, teachers, etc. 6 figures. Index. viii + 222pp. 5⅜ x 8½. S1077 Paperbound **$1.75**

Acoustics, optics, electricity and magnetism, electromagnetics, magnetohydrodynamics

THE THEORY OF SOUND, Lord Rayleigh. Most vibrating systems likely to be encountered in practice can be tackled successfully by the methods set forth by the great Nobel laureate, Lord Rayleigh. Complete coverage of experimental, mathematical aspects of sound theory. Partial contents: Harmonic motions, vibrating systems in general, lateral vibrations of bars, curved plates or shells, applications of Laplace's functions to acoustical problems, fluid friction, plane vortex-sheet, vibrations of solid bodies, etc. This ᴉs the first inexpensive edition of this great reference and study work. Bibliography. Historical introduction by R. B. Lindsay. Total of 1040pp. 97 figures. 5⅜ x 8.
S292, S293, Two ˌvolume set, paperbound, **$4.70**

THE DYNAMICAL THEORY OF SOUND, H. Lamb. Comprehensive mathematical treatment of the physical aspects of sound, covering the theory of vibrations, the general theory of sound, and the equations of motion of strings, bars, membranes, pipes, and resonators. Includes chapters on plane, spherical, and simple harmonic waves, and the Helmholtz Theory of Audition. Complete and self-contained development for student and specialist; all fundamental differential equations solved completely. Specific mathematical details for such important phenomena as harmonics, normal modes, forced vibrations of strings, theory of reed pipes, etc. Index. Bibliography. 86 diagrams. viii + 307pp. 5⅜ x 8.
S655 Paperbound **$2.00**

WAVE PROPAGATION IN PERIODIC STRUCTURES, L. Brillouin. A general method and application to different problems: pure physics, such as scattering of X-rays of crystals, thermal vibration in crystal lattices, electronic motion in metals; and also problems of electrical engineering. Partial contents: elastic waves in 1-dimensional lattices of point masses. Propagation of waves along 1-dimensional lattices. Energy flow. 2 dimensional, 3 dimensional lattices. Mathieu's equation. Matrices and propagation of waves along an electric line. Continuous electric lines. 131 illustrations. Bibliography. Index. xii + 253pp. 5⅜ x 8.
S34 Paperbound **$2.00**

THEORY OF VIBRATIONS, N. W. McLachlan. Based on an exceptionally successful graduate course given at Brown University, this discusses linear systems having 1 degree of freedom, forced vibrations of simple linear systems, vibration of flexible strings, transverse vibrations of bars and tubes, transverse vibration of circular plate, sound waves of finite amplitude, etc. Index. 99 diagrams. 160pp. 5⅜ x 8.
S190 Paperbound **$1.50**

LIGHT: PRINCIPLES AND EXPERIMENTS, George S. Monk. Covers theory, experimentation, and research. Intended for students with some background in general physics and elementary calculus. Three main divisions: 1) Eight chapters on geometrical optics—fundamental concepts (the ray and its optical length, Fermat's principle, etc.), laws of image formation, apertures in optical systems, photometry, optical instruments etc.; 2) 9 chapters on physical optics—interference, diffraction, polarization, spectra, the Rayleigh refractometer, the wave theory of light, etc.; 3) 23 instructive experiments based directly on the theoretical text. "Probably the best intermediate textbook on light in the English language. Certainly, it is the best book which includes both geometrical and physical optics," J. Rud Nielson, PHYSICS FORUM. Revised edition. 102 problems and answers. 12 appendices. 6 tables. Index. 270 illustrations. xi +489pp. 5⅜ x 8½.
S341 Paperbound **$2.50**

PHOTOMETRY, John W. T. Walsh. The best treatment of both "bench" and "illumination" photometry in English by one of Britain's foremost experts in the field (President of the International Commission on Illumination). Limited to those matters, theoretical and practical, which affect the measurement of light flux, candlepower, illumination, etc., and excludes treatment of the use to which such measurements may be put after they have been made. Chapters on Radiation, The Eye and Vision, Photo-Electric Cells, The Principles of Photometry, The Measurement of Luminous Intensity, Colorimetry, Spectrophotometry, Stellar Photometry, The Photometric Laboratory, etc. Third revised (1958) edition. 281 illustrations. 10 appendices. xxiv + 544pp. 5½ x 9¼.
S319 Paperbound **$3.00**

EXPERIMENTAL SPECTROSCOPY, R. A. Sawyer. Clear discussion of prism and grating spectrographs and the techniques of their use in research, with emphasis on those principles and techniques that are fundamental to practically all uses of spectroscopic equipment. Beginning with a brief history of spectroscopy, the author covers such topics as light sources, spectroscopic apparatus, prism spectroscopes and graphs, diffraction grating, the photographic process, determination of wave length, spectral intensity, infrared spectroscopy, spectrochemical analysis, etc. This revised edition contains new material on the production of replica gratings, solar spectroscopy from rockets, new standard of wave length, etc. Index. Bibliography. 111 illustrations. x + 358pp. 5⅜ x 8½.
S1045 Paperbound **$2.25**

FUNDAMENTALS OF ELECTRICITY AND MAGNETISM, L. B. Loeb. For students of physics, chemistry, or engineering who want an introduction to electricity and magnetism on a higher level and in more detail than general elementary physics texts provide. Only elementary differential and integral calculus is assumed. Physical laws developed logically, from magnetism to electric currents, Ohm's law, electrolysis, and on to static electricity, induction, etc. Covers an unusual amount of material; one third of book on modern material: solution of wave equation, photoelectric and thermionic effects, etc. Complete statement of the various electrical systems of units and interrelations. 2 Indexes. 75 pages of problems with answers stated. Over 300 figures and diagrams. xix +669pp. 5⅜ x 8.
S745 Paperbound **$3.50**

SUPERFLUIDS: MACROSCOPIC THEORY OF SUPERCONDUCTIVITY, Vol. I, Fritz London. The major work by one of the founders and great theoreticians of modern quantum physics. Consolidates the researches that led to the present understanding of the nature of super-conductivity. Prof. London here reveals that quantum mechanics is operative on the macro-scopic plane as well as the submolecular level. Contents: Properties of Superconductors and Their Thermodynamical Correlation; Electrodynamics of the Pure Superconducting State; Relation between Current and Field; Measurements of the Penetration Depth; Non-Viscous Flow vs. Superconductivity; Micro-waves in Superconductors; Reality of the Domain Structure; and many other related topics. A new epilogue by M. J. Buckingham discusses developments in the field up to 1960. Corrected and expanded edition. An appreciation of the author's life and work by L. W. Nordheim. Biography by Edith London. Bibliography of his publications. 45 figures. 2 Indices. xviii + 173pp. 5⅝ x 8⅜.   S44 Paperbound **$1.75**

SELECTED PAPERS ON PHYSICAL PROCESSES IN IONIZED PLASMAS, Edited by Donald H. Menzel, Director, Harvard College Observatory. 30 important papers relating to the study of highly ionized gases or plasmas selected by a foremost contributor in the field, with the assistance of Dr. L. H. Aller. The essays include 18 on the physical processes in gaseous nebulae, covering problems of radiation and radiative transfer, the Balmer decrement, electron temperatures, spectrophotometry, etc. 10 papers deal with the interpretation of nebular spectra, by Bohm, Van Vleck, Aller, Minkowski, etc. There is also a discussion of the intensities of "forbidden" spectral lines by George Shortley and a paper concern-ing the theory of hydrogenic spectra by Menzel and Pekeris. Other contributors: Goldberg, Hebb, Baker, Bowen, Ufford, Liller, etc. viii + 374pp. 6⅛ x 9¼.   S60 Paperbound **$2.95**

THE ELECTROMAGNETIC FIELD, Max Mason & Warren Weaver. Used constantly by graduate engineers. Vector methods exclusively: detailed treatment of electrostatics, expansion meth-ods, with tables converting any quantity into absolute electromagnetic, absolute electrostatic, practical units. Discrete charges, ponderable bodies, Maxwell field equations, etc. Introduc-tion. Indexes. 416pp. 5⅜ x 8.   S185 Paperbound **$2.25**

THEORY OF ELECTRONS AND ITS APPLICATION TO THE PHENOMENA OF LIGHT AND RADIANT HEAT, H. Lorentz. Lectures delivered at Columbia University by Nobel laureate Lorentz. Unabridged, they form a historical coverage of the theory of free electrons, motion, absorption of heat, Zeeman effect, propagation of light in molecular bodies, inverse Zeeman effect, optical phenomena in moving bodies, etc. 109 pages of notes explain the more advanced sections. Index. 9 figures. 352pp. 5⅜ x 8.   S173 Paperbound **$2.00**

FUNDAMENTAL ELECTROMAGNETIC THEORY, Ronold P. King, Professor Applied Physics, Harvard University. Original and valuable introduction to electromagnetic theory and to circuit theory from the standpoint of electromagnetic theory. Contents: Mathematical Description of Matter—stationary and nonstationary states; Mathematical Description of Space and of Simple Media—Field Equations, Integral Forms of Field Equations, Electromagnetic Force, etc.; Transformation of Field and Force Equations; Electromagnetic Waves in Unbounded Regions; Skin Effect and Internal Impedance—in a solid cylindrical conductor, etc.; and Electrical Circuits—Analytical Foundations, Near-zone and quasi-near zone circuits, Balanced two-wire and four-wire transmission lines. Revised and enlarged version. New preface by the author. 5 appendices (Differential operators: Vector Formulas and Identities, etc.). Problems. Indexes. Bibliography. xvi + 580pp. 5⅜ x 8½.   S1023 Paperbound **$3.00**

Hydrodynamics

A TREATISE ON HYDRODYNAMICS, A. B. Basset. Favorite text on hydrodynamics for 2 genera-tions of physicists, hydrodynamical engineers, oceanographers, ship designers, etc. Clear enough for the beginning student, and thorough source for graduate students and engineers on the work of d'Alembert, Euler, Laplace, Lagrange, Poisson, Green, Clebsch, Stokes, Cauchy, Helmholtz, J. J. Thomson, Love, Hicks, Greenhill, Besant, Lamb, etc. Great amount of docu-mentation on entire theory of classical hydrodynamics. Vol I: theory of motion of frictionless liquids, vortex, and cyclic irrotational motion, etc. 132 exercises. Bibliography. 3 Appendixes. xii + 264pp. Vol II: motion in viscous liquids, harmonic analysis, theory of tides, etc. 112 exercises, Bibliography. 4 Appendixes. xv + 328pp. Two volume set. 5⅜ x 8.
S724 Vol I Paperbound **$1.75**
S725 Vol II Paperbound **$1.75**
The set **$3.50**

HYDRODYNAMICS, Horace Lamb. Internationally famous complete coverage of standard refer-ence work on dynamics of liquids & gases. Fundamental theorems, equations, methods, solutions, background, for classical hydrodynamics. Chapters include Equations of Motion, Integration of Equations in Special Gases, Irrotational Motion, Motion of Liquid in 2 Dimen-sions, Motion of Solids through Liquid-Dynamical Theory, Vortex Motion, Tidal Waves, Surface Waves, Waves of Expansion, Viscosity, Rotating Masses of liquids. Excellently planned, ar-ranged; clear, lucid presentation. 6th enlarged, revised edition. Index. Over 900 footnotes, mostly bibliographical. 119 figures. xv + 738pp. 6⅛ x 9¼.   S256 Paperbound **$3.75**

HYDRODYNAMICS, H. Dryden, F. Murnaghan, Harry Bateman. Published by the National Research Council in 1932 this enormous volume offers a complete coverage of classical hydrodynamics. Encyclopedic in quality. Partial contents: physics of fluids, motion, turbulent flow, compressible fluids, motion in 1, 2, 3 dimensions; viscous fluids rotating, laminar motion, resistance of motion through viscous fluid, eddy viscosity, hydraulic flow in channels of various shapes, discharge of gases, flow past obstacles, etc. Bibliography of over 2,900 items. Indexes. 23 figures. 634pp. 5⅜ x 8. **S303 Paperbound $2.75**

Mechanics, dynamics, thermodynamics, elasticity

MECHANICS, J. P. Den Hartog. Already a classic among introductory texts, the M.I.T. professor's lively and discursive presentation is equally valuable as a beginner's text, an engineering student's refresher, or a practicing engineer's reference. Emphasis in this highly readable text is on illuminating fundamental principles and showing how they are embodied in a great number of real engineering and design problems: trusses, loaded cables, beams, jacks, hoists, etc. Provides advanced material on relative motion and gyroscopes not usual in introductory texts. "Very thoroughly recommended to all those anxious to improve their real understanding of the principles of mechanics." MECHANICAL WORLD. Index. List of equations. 334 problems, all with answers. Over 550 diagrams and drawings. ix + 462pp. 5⅜ x 8.
S754 Paperbound $2.00

THEORETICAL MECHANICS: AN INTRODUCTION TO MATHEMATICAL PHYSICS, J. S. Ames, F. D. Murnaghan. A mathematically rigorous development of theoretical mechanics for the advanced student, with constant practical applications. Used in hundreds of advanced courses. An unusually thorough coverage of gyroscopic and baryscopic material, detailed analyses of the Coriolis acceleration, applications of Lagrange's equations, motion of the double pendulum, Hamilton-Jacobi partial differential equations, group velocity and dispersion, etc. Special relativity is also included. 159 problems. 44 figures. ix + 462pp. 5⅜ x 8.
S461 Paperbound $2.25

THEORETICAL MECHANICS: STATICS AND THE DYNAMICS OF A PARTICLE, W. D. MacMillan. Used for over 3 decades as a self-contained and extremely comprehensive advanced undergraduate text in mathematical physics, physics, astronomy, and deeper foundations of engineering. Early sections require only a knowledge of geometry; later, a working knowledge of calculus. Hundreds of basic problems, including projectiles to the moon, escape velocity, harmonic motion, ballistics, falling bodies, transmission of power, stress and strain, elasticity, astronomical problems. 340 practice problems plus many fully worked out examples make it possible to test and extend principles developed in the text. 200 figures. xvii + 430pp. 5⅜ x 8. **S467 Paperbound $2.25**

THEORETICAL MECHANICS: THE THEORY OF THE POTENTIAL, W. D. MacMillan. A comprehensive, well balanced presentation of potential theory, serving both as an introduction and a reference work with regard to specific problems, for physicists and mathematicians. No prior knowledge of integral relations is assumed, and all mathematical material is developed as it becomes necessary. Includes: Attraction of Finite Bodies; Newtonian Potential Function; Vector Fields, Green and Gauss Theorems; Attractions of Surfaces and Lines; Surface Distribution of Matter; Two-Layer Surfaces; Spherical Harmonics; Ellipsoidal Harmonics; etc. "The great number of particular cases . . . should make the book valuable to geophysicists and others actively engaged in practical applications of the potential theory," Review of Scientific Instruments. Index. Bibliography. xiii + 469pp. 5⅜ x 8. **S486 Paperbound $2.50**

THEORETICAL MECHANICS: DYNAMICS OF RIGID BODIES, W. D. MacMillan. Theory of dynamics of a rigid body is developed, using both the geometrical and analytical methods of instruction. Begins with exposition of algebra of vectors, it goes through momentum principles, motion in space, use of differential equations and infinite series to solve more sophisticated dynamics problems. Partial contents: moments of inertia, systems of free particles, motion parallel to a fixed plane, rolling motion, method of periodic solutions, much more. 82 figs. 199 problems. Bibliography. Indexes. xii + 476pp. 5⅜ x 8. **S641 Paperbound $2.50**

MATHEMATICAL FOUNDATIONS OF STATISTICAL MECHANICS, A. I. Khinchin. Offering a precise and rigorous formulation of problems, this book supplies a thorough and up-to-date exposition. It provides analytical tools needed to replace cumbersome concepts, and furnishes for the first time a logical step-by-step introduction to the subject. Partial contents: geometry & kinematics of the phase space, ergodic problem, reduction to theory of probability, application of central limit problem, ideal monatomic gas, foundation of thermo-dynamics, dispersion and distribution of sum functions. Key to notations. Index. viii + 179pp. 5⅜ x 8. **S147 Paperbound $1.50**

ELEMENTARY PRINCIPLES IN STATISTICAL MECHANICS, J. W. Gibbs. Last work of the great Yale mathematical physicist, still one of the most fundamental treatments available for advanced students and workers in the field. Covers the basic principle of conservation of probability of phase, theory of errors in the calculated phases of a system, the contributions of Clausius, Maxwell, Boltzmann, and Gibbs himself, and much more. Includes valuable comparison of statistical mechanics with thermodynamics: Carnot's cycle, mechanical definitions of entropy, etc. xvi + 208pp. 5⅜ x 8. **S707 Paperbound $1.45**

FOUNDATIONS OF POTENTIAL THEORY, O. D. Kellogg. Based on courses given at Harvard this is suitable for both advanced and beginning mathematicians. Proofs are rigorous, and much material not generally avaliable elsewhere is included. Partial contents: forces of gravity, fields of force, divergence theorem, properties of Newtonian potentials at points of free space, potentials as solutions of Laplace's equations, harmonic functions, electrostatics, electric images, logarithmic potential, etc. One of Grundlehren Series. ix + 384pp. 5⅜ x 8.
S144 Paperbound **$2.00**

THERMODYNAMICS, Enrico Fermi. Unabridged reproduction of 1937 edition. Elementary in treatment; remarkable for clarity, organization. Requires no knowledge of advanced math beyond calculus, only familiarity with fundamentals of thermometry, calorimetry. Partial Contents: Thermodynamic systems; First & Second laws of thermodynamics; Entropy; Thermodynamic potentials: phase rule, reversible electric cell; Gaseous reactions: van't Hoff reaction box, principle of LeChatelier; Thermodynamics of dilute solutions: osmotic & vapor pressures, boiling & freezing points; Entropy constant. Index. 25 problems. 24 illustrations. x + 160pp. 5⅜ x 8.
S361 Paperbound **$1.75**

THE THERMODYNAMICS OF ELECTRICAL PHENOMENA IN METALS and A CONDENSED COLLECTION OF THERMODYNAMIC FORMULAS, P. W. Bridgman. Major work by the Nobel Prizewinner: stimulating conceptual introduction to aspects of the electron theory of metals, giving an intuitive understanding of fundamental relationships concealed by the formal systems of Onsager and others. Elementary mathematical formulations show clearly the fundamental thermodynamical relationships of the electric field, and a complete phenomenological theory of metals is created. This is the work in which Bridgman announced his famous "thermomotive force" and his distinction between "driving" and "working" electromotive force. We have added in this Dover edition the author's long unavailable tables of thermodynamic formulas, extremely valuable for the speed of reference they allow. Two works bound as one. Index. 33 figures. Bibliography. xviii + 256pp. 5⅜ x 8. S723 Paperbound **$1.75**

TREATISE ON THERMODYNAMICS, Max Planck. Based on Planck's original papers this offers a uniform point of view for the entire field and has been used as an introduction for students who have studied elementary chemistry, physics, and calculus. Rejecting the earlier approaches of Helmholtz and Maxwell, the author makes no assumptions regarding the nature of heat, but begins with a few empirical facts, and from these deduces new physical and chemical laws. 3rd English edition of this standard text by a Nobel laureate. xvi + 237pp. 5⅜ x 8.
S219 Paperbound **$1.85**

THE MATHEMATICAL THEORY OF ELASTICITY, A. E. H. Love. A wealth of practical illustration combined with thorough discussion of fundamentals—theory, application, special problems and solutions. Partial Contents: Analysis of Strain & Stress, Elasticity of Solid Bodies, Elasticity of Crystals, Vibration of Spheres, Cylinders, Propagation of Waves in Elastic Solid Media, Torsion, Theory of Continuous Beams, Plates. Rigorous treatment of Volterra's theory of dislocations, 2-dimensional elastic systems, other topics of modern interest. "For years the standard treatise on elasticity," AMERICAN MATHEMATICAL MONTHLY. 4th revised edition. Index. 76 figures. xviii + 643pp. 6⅛ x 9¼.
S174 Paperbound **$3.25**

STRESS WAVES IN SOLIDS, H. Kolsky, Professor of Applied Physics, Brown University. The most readable survey of the theoretical core of current knowledge about the propagation of waves in solids, fully correlated with experimental research. Contents: Part I—Elastic Waves: propagation in an extended plastic medium, propagation in bounded elastic media, experimental investigations with elastic materials. Part II—Stress Waves in Imperfectly Elastic Media: internal friction, experimental investigations of dynamic elastic properties, plastic waves and shock waves, fractures produced by stress waves. List of symbols. Appendix. Supplemented bibliography. 3 full-page plates. 46 figures. x + 213pp. 5⅜ x 8½.
S1098 Paperbound **$1.75**

Relativity, quantum theory, atomic and nuclear physics

SPACE TIME MATTER, Hermann Weyl. "The standard treatise on the general theory of relativity" (Nature), written by a world-renowned scientist, provides a deep clear discussion of the logical coherence of the general theory, with introduction to all the mathematical tools needed: Maxwell, analytical geometry, non-Euclidean geometry, tensor calculus, etc. Basis is classical space-time, before absorption of relativity. Partial contents: Euclidean space, mathematical form, metrical continuum, relativity of time and space, general theory. 15 diagrams. Bibliography. New preface for this edition. xviii + 330pp. 5⅜ x 8.
S267 Paperbound **$2.25**

ATOMIC SPECTRA AND ATOMIC STRUCTURE, G. Herzberg. Excellent general survey for chemists, physicists specializing in other fields. Partial contents: simplest line spectra and elements of atomic theory, building-up principle and periodic system of elements, hyperfine structure of spectral lines, some experiments and applications. Bibliography. 80 figures. Index. xii + 257pp. 5⅜ x 8.
S115 Paperbound **$2.00**

SELECTED PAPERS ON QUANTUM ELECTRODYNAMICS, edited by **J. Schwinger.** Facsimiles of papers which established quantum electrodynamics, from initial successes through today's position as part of the larger theory of elementary particles. First book publication in any language of these collected papers of Bethe, Bloch, Dirac, Dyson, Fermi, Feynman, Heisenberg, Kusch, Lamb, Oppenheimer, Pauli, Schwinger, Tomonoga, Weisskopf, Wigner, etc. 34 papers in all, 29 in English, 1 in French, 3 in German, 1 in Italian. Preface and historical commentary by the editor, xvii + 423pp. 6⅛ x 9¼. S444 Paperbound **$2.75**

THE FUNDAMENTAL PRINCIPLES OF QUANTUM MECHANICS, WITH ELEMENTARY APPLICATIONS, E. C. Kemble. An inductive presentation, for the graduate student or specialist in some other branch of physics. Assumes some acquaintance with advanced math; apparatus necessary beyond differential equations and advanced calculus is developed as needed. Although a general exposition of principles, hundreds of individual problems are fully treated, with applications of theory being interwoven with development of the mathematical structure. The author is the Professor of Physics at Harvard Univ. "This excellent book would be of great value to every student . . . a rigorous and detailed mathematical discussion of all of the principal quantum-mechanical methods . . . has succeeded in keeping his presentations clear and understandable," Dr. Linus Pauling, J. of the American Chemical Society. Appendices: calculus of variations, math. notes, etc. Indexes. 611pp. 5⅜ x 8. S472 Paperbound **$3.00**

QUANTUM MECHANICS, H. A. Kramers. A superb, up-to-date exposition, covering the most important concepts of quantum theory in exceptionally lucid fashion. 1st half of book shows how the classical mechanics of point particles can be generalized into a consistent quantum mechanics. These 5 chapters constitute a thorough introduction to the foundations of quantum theory. Part II deals with those extensions needed for the application of the theory to problems of atomic and molecular structure. Covers electron spin, the Exclusion Principle, electromagnetic radiation, etc. "This is a book that all who study quantum theory will want to read," J. Polkinghorne, PHYSICS TODAY. Translated by D. ter Haar. Prefaces, introduction. Glossary of symbols. 14 figures. Index. xvi + 496pp. 5⅜ x 8⅜. S1150 Paperbound **$2.75**

THE THEORY AND THE PROPERTIES OF METALS AND ALLOYS, N. F. Mott, H. Jones. Quantum methods used to develop mathematical models which show interrelationship of basic chemical phenomena with crystal structure, magnetic susceptibility, electrical, optical properties. Examines thermal properties of crystal lattice, electron motion in applied field, cohesion, electrical resistance, noble metals, para-, dia-, and ferromagnetism, etc. "Exposition . . . clear . . . mathematical treatment . . . simple," Nature. 138 figures. Bibliography. Index. xiii + 320pp. 5⅜ x 8. S456 Paperbound **$2.00**

FOUNDATIONS OF NUCLEAR PHYSICS, edited by **R. T. Beyer.** 13 of the most important papers on nuclear physics reproduced in facsimile in the original languages of their authors: the papers most often cited in footnotes, bibliographies. Anderson, Curie, Joliot, Chadwick, Fermi, Lawrence, Cockcroft, Hahn, Yukawa. UNPARALLELED BIBLIOGRAPHY. 122 double-columned pages, over 4,000 articles, books, classified. 57 figures. 288pp. 6⅛ x 9¼. S19 Paperbound **$2.00**

MESON PHYSICS, R. E. Marshak. Traces the basic theory, and explicitly presents results of experiments with particular emphasis on theoretical significance. Phenomena involving mesons as virtual transitions are avoided, eliminating some of the least satisfactory predictions of meson theory. Includes production and study of π mesons at nonrelativistic nucleon energies, contrasts between π and μ mesons, phenomena associated with nuclear interaction of π mesons, etc. Presents early evidence for new classes of particles and indicates theoretical difficulties created by discovery of heavy mesons and hyperons. Name and subject indices. Unabridged reprint. viii + 378pp. 5⅜ x 8. S500 Paperbound **$1.95**

Prices subject to change without notice.

Dover publishes books on art, music, philosophy, literature, languages, history, social sciences, psychology, handcrafts, orientalia, puzzles and entertainments, chess, pets and gardens, books explaining science, intermediate and higher mathematics, mathematical physics, engineering, biological sciences, earth sciences, classics of science, etc. Write to:

Dept. catrr.
Dover Publications, Inc.
180 Varick Street, N.Y. 14, N.Y.